AF847043

Series in Contemporary Mathematics

Volume 6

Editor-in-Chief

Tatsien Li, School of Mathematical Sciences, Fudan University, Shanghai, China

Series Editors

Philippe G. Ciarlet, City University of Hong Kong, Hong Kong, Germany

Jean-Michel Coron, Laboratoire Jacques-Louis Lions, Université Pierre et Marie Curie, Paris, France

Weinan E, Department of Mathematics, Princeton University, PRINCETON, USA

Jianshu Li, Department of Mathematics, The Hong Kong University of Science and Technology, Hong Kong, Germany

Jun Li, Department of Mathematics, Stanford University, STANFORD, USA

Tatsien Li, School of Mathematical Sciences, Fudan University, Shanghai, China

Fanghua Lin, Courant Inst. Mathematical Science, New York University, NEW YORK, USA

Zhi-ming Ma, Academy of Mathematics and Systems Science, Beijing, Germany

Andrew J. Majda, Department of Mathematics, New York University, New York, USA

Cédric Villani, Institut Henri Poincaré, Paris, France

Ya-xiang Yuan, Institute of Computational Mathematics and Science/Engineering Computing, Academy of Mathematics and Systems Science, Beijing, China

Weiping Zhang, Chern Institute of Mathematics, Nankai University, Tianjin, China

Series in Contemporary Mathematics (SCM), featuring high-quality mathematical monographs, is to presents original and systematic findings from the fields of pure mathematics, applied mathematics and math-related interdisciplinary subjects. It has a history of over fifty years since the first title was published by Shanghai Scientific & Technical Publishers in 1963. Professor HUA Luogeng (Lo-Keng Hua) served as Editor-in-Chief of the first editorial board, while Professor SU Buqing acted as Honorary Editor-in-Chief and Professor GU Chaohao as Editor-in-Chief of the second editorial board since 1992. Now the third editorial board is established and Professor LI Tatsien assumes the position of Editor-in-Chief. The series has already published twenty-six monographs in Chinese, and among the authors are many distinguished Chinese mathematicians, including the following members of the Chinese Academy of Sciences: SU Buqing, GU Chaohao, LU Qikeng, ZHANG Gongqing, CHEN Hanfu, CHEN Xiru, YUAN Yaxiang, CHEN Shuxing etc. The monographs have systematically introduced a number of important research findings which not only play a vital role in China, but also exert huge influence all over the world. Eight of them have been translated into English and published abroad. The new editorial board will inherit and carry forward the former traditions and strengths of the series, and plan to further reform and innovation in terms of internalization so as to improve and ensure the quality of the series, extend its global influence, and strive to forge it into an internationally significant series of mathematical monographs.

Editor-in-Chief
Tatsien Li
Series Editors
Philippe G. Ciarlet, Jean-Michel Coron, Weinan E,
Jianshu Li, Jun Li, Tatsien Li, Fanghua Lin,
Zhi-ming Ma, Andrew J. Majda, Cédric Villani,
Ya-xiang Yuan, Weiping Zhang

Yanheng Ding • Tian Xu

Localization Approaches in Strongly Indefinite Problems

Springer

Yanheng Ding
School of Mathematics
Jilin University
Changchun, China

Tian Xu
(Corresponding Author)
Department of Mathematics
Zhejiang Normal University
Jinhua, Zhejiang, China

ISSN 2364-009X ISSN 2364-0103 (electronic)
Series in Contemporary Mathematics
ISBN 978-981-97-9538-3 ISBN 978-981-97-9539-0 (eBook)
https://doi.org/10.1007/978-981-97-9539-0

Mathematics Subject Classification: 35A15, 35Q40, 35R01, 49J35, 53C27

© The Editor(s) (if applicable) and The Author(s), under exclusive license to Springer Nature Singapore Pte Ltd. 2025

This work is subject to copyright. All rights are solely and exclusively licensed by the Publisher, whether the whole or part of the material is concerned, specifically the rights of translation, reprinting, reuse of illustrations, recitation, broadcasting, reproduction on microfilms or in any other physical way, and transmission or information storage and retrieval, electronic adaptation, computer software, or by similar or dissimilar methodology now known or hereafter developed.
The use of general descriptive names, registered names, trademarks, service marks, etc. in this publication does not imply, even in the absence of a specific statement, that such names are exempt from the relevant protective laws and regulations and therefore free for general use.
The publisher, the authors and the editors are safe to assume that the advice and information in this book are believed to be true and accurate at the date of publication. Neither the publisher nor the authors or the editors give a warranty, expressed or implied, with respect to the material contained herein or for any errors or omissions that may have been made. The publisher remains neutral with regard to jurisdictional claims in published maps and institutional affiliations.

This Springer imprint is published by the registered company Springer Nature Singapore Pte Ltd.
The registered company address is: 152 Beach Road, #21-01/04 Gateway East, Singapore 189721, Singapore

If disposing of this product, please recycle the paper.

Foreword

Several important problems arising in Physics, Differential Geometry, and other topics have prompted the consideration of semilinear variational equations of strongly indefinite type, leading to a significant amount of research in this area. From a mathematical perspective, the main interest lies in the fact that the tools of Nonlinear Functional Analysis, which rely on compactness arguments and non-degenerate structure, cannot be straightforwardly applied. Therefore, new techniques need to be developed.

On the other hand, there are several strongly indefinite problems that exhibit a perturbative nature. In some cases, there exists a natural perturbation parameter, such as in the study of semiclassical standing waves for nonlinear Dirac equations, singularly perturbed Hamiltonian Partial Differential Equations, or the bifurcation from the spectrum. In some other circumstances, perturbations are studied either as a preliminary step toward obtaining global results or because they often provide localized studies on PDE problems.

For these perturbation problems, a specific approach that takes advantage of the perturbative setting appears to be most appropriate. The abstract tools are provided by localization methods in critical point theory. Remarkably, such a framework can be used to address a wide variety of equations, which were typically considered distinct in nature. The objective of this monograph is to discuss these abstract methods alongside their applications to several localization problems. These problems share a common feature, namely, involving semilinear partial differential equations with a strongly indefinite structure.

The results presented here are based on papers of the authors carried out in the last years. Many of them are works in collaboration with esteemed professors such as T. Bartsch, J.C. Wei, B. Ruf, and C. Lee. The second author would like to express gratitude to Professors B. Ammann and I.A. Taimanov for introducing him to the field of Spin Geometry and for their continuous encouragement and assistance. We extend our warmest thanks to all of them.

Acknowledgements

The first author is supported by the National Science Foundation of China (NSFC12031015, 12271508).

The second author is supported by the National Science Foundation of China (NSFC12371117).

Contents

Notation

- $\mathbb{R}^n$ is the Euclidean n-dimensional space with points $x = (x_1, \dots, x_n)$, $x_k \in \mathbb{R}$.
- $\mathbb{C}^n$ is the complex n-dimensional space with points $x = (x_1, \dots, x_n)$, $x_k \in \mathbb{C}$.
- $x \cdot y$ denotes the scalar product of x, y in $\mathbb{R}^n$ or $\mathbb{C}^n$; we also set $|x|^2 = x \cdot \bar{x}$, where $\bar{x} = (\bar{x}_1, \dots, \bar{x}_n)$ stands for the complex conjugate of x.
- S^n denotes the unit n-dimensional sphere: $S^n = \{x \in \mathbb{R}^{n+1} : |x| = 1\}$.
- If Ω is an open subset of $\mathbb{R}^n$ and $u : \Omega \to \mathbb{R}$ is smooth, we denote $\nabla u : \Omega \to \mathbb{R}^n$ the gradient of u and $\partial_i u$ the partial derivative with respect to x_i.
- Id denotes the identity map in $\mathbb{R}^n$ or a general topological space.
- $L^p(\Omega, \mathbb{R}^m)$, $L^p(\Omega, \mathbb{C}^m)$ $L^p_{loc}(\mathbb{R}^n, \mathbb{R}^m)$, $L^p_{loc}(\mathbb{R}^n, \mathbb{C}^m)$, etc. denote the usual Lebesgue spaces.
- $C_c^\infty(\Omega, \mathbb{R})$, $C_c^\infty(\mathbb{R}^n, \mathbb{R}^m)$, and $C_c^\infty(\mathbb{R}^n, \mathbb{C}^m)$ denote the space of smooth functions with compact support.
- $W^{m,p}(\mathbb{R}^n, \mathbb{R}^m)$, $W^{m,p}(\mathbb{R}^n, \mathbb{C}^m)$, $W^{m,p}(\Omega, \mathbb{R}^m)$, $W^{m,p}(\Omega, \mathbb{C}^m)$ etc. denote the usual Sobolev spaces. If M is a smooth manifold, $H^m(M, \mathbb{R}^m)$ and $H^m(M, \mathbb{C}^m)$ denote the Sobolev space $W^{m,2}(M, \mathbb{R}^m)$ and $W^{m,2}(M, \mathbb{C}^m)$, respectively.
- If X and Y are Banach spaces, $\mathscr{L}(X, Y)$ denotes the space of bounded linear maps from X to Y.
- If $f \in C^k(X, Y)$, $k \geq 1$, $\nabla f(u)$, $\nabla^2 f(u)$, denote the Fréchet derivatives of f at $u \in X$. They are, respectively, a linear bounded map from X to Y, and a bilinear continuous map from $X \times X$ to Y.
- If $\Phi \in C^k(E, \mathbb{R})$, $k \geq 1$, is a functional on a (real) Hilbert space E, $\nabla\Phi(u)$ is identified with the gradient of Φ at $u \in E$, defined by means of the Riesz representation Theorem. Similarly, $\nabla^2\Phi(u)$ is a bilinear operator from $E \times E \to \mathbb{R}$. Sometimes, the notation Φ' and Φ'' will also be used to shorten $\nabla\Phi$ and $\nabla^2\Phi$, respectively.
- $u = o(\varepsilon)$ means that $u\varepsilon^{-1}$ tends to zero as $\varepsilon \to 0$.
- $u = O(\varepsilon)$ means that $|u\varepsilon^{-1}| \leq C$ as $\varepsilon \to 0$, where C is a constant.
- $o_n(1)$ or $o_\varepsilon(1)$ signifies a quantity that diminishes toward 0 as $n \to \infty$ or $\varepsilon \to 0$.
- i the imaginary unit.
- The notation "$\sim$" denotes quantities which, in the limit are of the same order.

Chapter 1
Variational Problems: A Brief Retrospective

Variational and topological methods have proved to be powerful tools in the resolution of concrete nonlinear problems appearing in many disciplines where classical methods may fail. This is the case in particular for *Critical Point Theory*, which has become very successful in recent decades. Here, the aim of critical point theory is to look for critical points of a functional on a manifold, not necessarily in the finite dimensional setting. In general cases, one is interested not only on minima or maxima (that can be found by more direct arguments), but also on saddle points whose existence is usually proved by min-max procedures.

Let M be a Riemannian manifold modeled on a (real) Hilbert space $(E, \langle\cdot,\cdot\rangle)$, and let $\Phi \in C^1(M, \mathbb{R})$. A critical point of Φ on M is a point in M such that the constrained gradient of Φ, denoted by $\nabla^M \Phi$ (whose image lies in the dual space $(TM)^*$), vanishes. More precisely, if $M = \{u \in E : \Psi(u) = 0\}$, with $\Psi \in C^1(E, \mathbb{R})$ such that $\nabla\Psi(u) \neq 0$ for all $u \in M$, then a critical point of Φ on M is a $z \in M$ such that $\nabla\Phi(z) = \lambda\nabla\Psi(z)$, for some Lagrange multiplier $\lambda \in \mathbb{R}$. And if $M = E$, one finds $\nabla\Phi(z) = 0$, namely a stationary point of Φ.

1.1 Some Fundamental Results in the Finite Dimensional Case

Consider the case of a functional Φ on a finite-dimensional closed (compact, without boundary) Riemannian manifold M, an abstract theory relating the number of critical points of Φ to the topological properties of M, was first developed by Lusternik and Schnirelman, and, independently by Morse. In order to find critical points of Φ on M, one of the main ingredient is the deformation of level sets $\Phi^a := \{\xi \in M : \Phi(\xi) \leq a\}$, for $a \in \mathbb{R}$. Here, a deformation (on M) is a continuous map $\eta : [0, 1] \times M \to M$ so that $\eta(0, \cdot) = \mathrm{Id}$ and, for each $t \in [0, 1]$, $\eta(t, \cdot)$ is a

© The Author(s), under exclusive license to Springer Nature Singapore Pte Ltd. 2025

Y. Ding, T. Xu, *Localization Approaches in Strongly Indefinite Problems*, Series in Contemporary Mathematics 6, https://doi.org/10.1007/978-981-97-9539-0_1

homeomorphism. The following statements describe the variational principle under its simplest form.

Lemma 1.1 (Deformation Lemma) *Suppose that the strip $\{\xi \in M : a \leq \Phi(\xi) \leq b\}$ does not contain any critical point of Φ on M. Then Φ^a is a retract by deformation of Φ^b.*

Corollary 1.2 *Under the hypotheses of Lemma 1.1, the homotopy and homology of Φ^a and Φ^b are the same.*

From these statements, we can immediately conclude an existence criterion. Roughly speaking, if Φ^b is topologically different from Φ^a, then Φ possesses at least one critical point in the strip $\{\xi \in M : a \leq \Phi(\xi) \leq b\}$. In practice, one may consider a family of subsets of M, denoted by $\mathscr{F}$, which is invariant under deformations (that is, $\eta(t, U) \in \mathscr{F}$ for all $U \in \mathscr{F}$ and $t \in [0, 1]$ whenever η is a deformation on M) and set

$$c = \inf_{U \in \mathscr{F}} \sup_{\xi \in U} \Phi(\xi).$$

Then, via the Deformation lemma, it follows that there exists a critical point $z \in M$ of Φ such that $\Phi(z) = c$.

The preceding procedure gives rise to the existence of several different critical points, if we can define different families of subsets stable under deformations. This can be illustrated by the following elementary example.

Example 1.3 Let us consider a smooth function $\Phi : S^1 \to \mathbb{R}$, for which we may simply embed its domain S^1 into the xOy-plane as the unit circle centered at the origin. By employing the stereographic projection

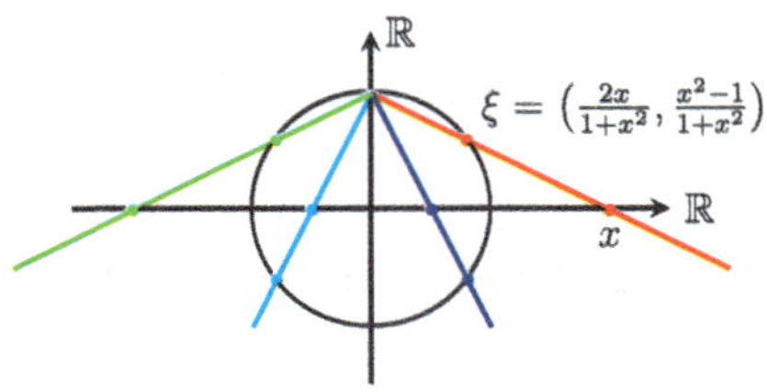

we can use the composed function

$$f(x) = \Phi\Big(\frac{2x}{1 + x^2}, \frac{x^2 - 1}{1 + x^2}\Big)$$

defined on the real line to visualize the shape of Φ. And we only need to keep in mind that the North Pole is treated as the infinity, denoted by $\{\infty\}$. Without loss of generality, we may simply assume that the North Pole is not a critical point of Φ, and suppose that f is characterized by the following graph

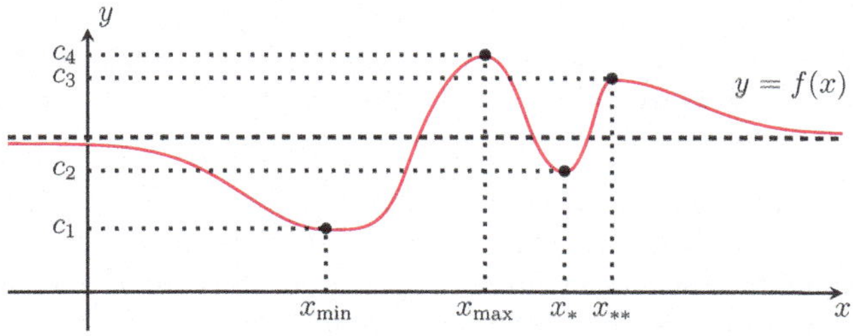

then for some suitable family $\mathscr{F}$ containing subsets of S^1, let us consider the min-max value

$$c = \inf_{U \in \mathscr{F}} \sup_{\xi \in U} \Phi(\xi). \tag{1.1}$$

And we may immediately have the following characterizations

(1) By choosing $\mathscr{F}$ as the collection of all single-point sets, that is $\mathscr{F} = \{\{\xi\} : \xi \in S^1\}$, we find that $\mathscr{F}$ keeps invariant under deformations on S^1. In this case, the min-max value in (1.1) gives us the value c_1, corresponding to the global minimum of Φ, which is obviously inherited form the global minimum of f.

(2) Let us take

$$\mathscr{F} = \left\{ \eta([0,1], \Phi^a) \;\middle|\; \begin{array}{l} a \in \mathbb{R} \text{ so that the level set } f^a \text{ of the} \\ \text{composed function } f \text{is not connected,} \\ \text{and } \eta \text{ is a deformation on } S^1 \end{array} \right\}.$$

Of course the descriptions on the right-hand side is chosen so as to remain invariant under the deformations on S^1. Notice that, for $a \in [c_2, c_4)$, a level set Φ^a itself is included in $\mathscr{F}$ since the identity map $\mathrm{Id} : S^1 \to S^1$ is also a deformation. Hence, it follows that the output of the min-max value in (1.1) is c_2, a local minimum (clearly not the global minimum) of Φ.

(3) The characterization of c_3 is similar as above, and one can use the topological properties of level sets of f to define

$$\mathscr{F} = \left\{ \eta([0,1], \Phi^a) \;\middle|\; \begin{array}{l} a \in \mathbb{R} \text{ so that } f^a \text{ has nontrivial} \\ \text{connected complement, i.e., } \mathbb{R} \setminus f^a \neq \emptyset \\ \text{and is connected in } \mathbb{R}, \text{ and } \eta \\ \text{is a deformation on } S^1 \end{array} \right\}.$$

Then we find $\Phi^a \in \mathscr{F}$ for $a \in [c_3, c_4)$ and $\Phi^a \notin \mathscr{F}$ for $a < c_3$ or $a \geq c_4$. In this way, the min-max value in (1.1) characterizes the value c_3.

(4) By choosing $\mathscr{F} = \{S^1\}$, it follows directly that $\mathscr{F}$ is preserved under deformations since it is exactly the total space S^1 itself. And then, the min-max value in (1.1) gives us nothing but the value c_4, that is the global maximum of Φ.

In the Lusternik-Schnirelman theory, an existence criterion for critical points of Φ in M is achieved by classifying the closed sets that are contractible within M. This classification helps in understanding the topology of the manifold M and provides insights into the critical points of Φ. More precisely, one has:

Theorem 1.4 *The number critical points of* $\Phi \in C^2(M, \mathbb{R})$ *is at least* $\mathrm{Cat}(M)$, *where* $\mathrm{Cat}(M)$ *is the Lusternik-Schnirelman category, namely the minimal integer* k *such that* $M = A_1 \cup A_2 \cup \cdots \cup A_k$ *with* A_i *closed and contractible in* M.

And an important follow-up conclusion is

Theorem 1.5 *Any even* C^2 *functional* Φ *has at least* n *pairs of critical points on the sphere* $\{\xi \in \mathbb{R}^n : |\xi| = R\}$.

A different direction was taken by Morse, who used homology groups to count the critical points of Φ on M. The Morse theory is sharper than the Lusternik-Schnirelman one, but requires the assumption that any possible critical point is non-degenerate. For example, if M is a 2-dimensional torus in $\mathbb{R}^3$, the Lusternik-Schnirelman theory ensures that any functional Φ has on M at least 3 critical points (the minimum, the maximum and one saddle point), while, if all the critical points are non-degenerate, the Morse theory yields the existence of 4 critical points (the minimum, the maximum and two saddle points). We emphasize here that it is possible to find a functional which has exactly 3 critical points on the torus. A nice explicit construction of such a functional is the following: consider the parameterize of the torus as $\mathbb{R}^2/\pi\mathbb{Z}^2$, then take $\Phi(x, y) = \sin(x)\sin(y)\sin(x + y)$. The three critical points are the origin, $(\frac{\pi}{3}, \frac{\pi}{3})$ and $(\frac{2\pi}{3}, \frac{2\pi}{3})$, among which the origin is a degenerate saddle point. Therefore the theorem of Lusternik and Schnirelman is the best possible in general situations.

1.2 In Infinite Dimensional Spaces—Compactness Enters

The extension of critical point theory to an infinite dimensional setting requires a main new ingredient: a compactness condition, which is a fundamental tool to carry out the deformations. The most frequently used formulation of the compactness is due to Palais and Smale and requires that

any sequence $\{u_k\} \subset M$ *such that* $\Phi(u_k) \to c$ *and* $\nabla^M \Phi(u_k) \to 0$
has a converging subsequence.

And we will refer to this as the $(PS)_c$-condition for a smooth functional Φ on a Riemannian Hilbert manifold M. Clearly, this condition is true for the usual situation of a finite dimensional compact Riemannian manifold. Let us mention here that the $(PS)_c$-condition allows for certain standard finite dimensional results to be extended to the infinite dimensional case. For instance, the $(PS)_c$-condition implies that, if Φ is bounded from below, then it either achieves its greatest lower bound or M contains an infinite sequence of points $\{u_k\}$ such that $\Phi(u_k)$ converge to the greatest lower bound while the norm of $\nabla^M\Phi(u_k)$ is bounded away from zero.

To present the deformation result, let us denote $K_c = \{z \in M : \Phi(z) = c, \nabla^M\Phi(z) = 0\}$ the set of critical points of Φ at level c. Then a remarkable consequence of the $(PS)_c$-condition is that:

Lemma 1.6 *If Φ satisfies the $(PS)_c$-condition and $K_c = \emptyset$. Then there exists $\varepsilon > 0$ such that the level set $\Phi^{c-\varepsilon}$ is a retract by deformation of $\Phi^{c+\varepsilon}$.*

Remark 1.7 In fact, in the infinite dimensional setting, there are technical hypotheses for the deformation lemma to be true. For instance, it must be the case that, for the value c with $K_c = \emptyset$, c is not a limit of the set $\{a \in \mathbb{R} : \Phi \text{ has constant value } a \text{ on some component of } M\}$. Here, in Lemma 1.6, the $(PS)_c$-condition is used to rule out such a possibility. One may also use the following hypothesis instead of the $(PS)_c$-condition:

there exists $\varepsilon > 0$ small such that the norm of $\nabla^M\Phi$ is bounded away from zero in the set $\{u \in M : c - \varepsilon < \Phi(u) \le c + \varepsilon\}$.

And then the conclusion of Lemma 1.6 holds true.

From Lemma 1.6, one deduces a result analogous to Theorem 1.4:

Theorem 1.8 *Let $\Phi \in C^2(M, \mathbb{R})$, with M being a connected Riemannian Hilbert manifold without boundary, and suppose that*

(1) $\inf_{u\in M} \Phi(u) > -\infty$;
(2) *$(PS)_c$-condition holds for all $c \in \mathbb{R}$.*

Then Φ has at least $\mathrm{Cat}(M)$ *critical points on M.*

The basic idea in the above theorem is that critical points are obstructions to collapsing a manifold down to a point via the flow associated to the negative of the gradient. Thus critical points are the places in the manifold where topological complexity arises and this particular complexity is well-measured by Lusternik-Schnirelman category. In fact, to prove Theorem 1.8, let us mention the following observation: Define

$$c_j = \inf\{a \in \mathbb{R} : \mathrm{Cat}_M(\Phi^a) \ge j\}$$

where, for a closed subset $B \subset M$, $\mathrm{Cat}_M(B)$ stands for the minimal integer l such that $B \subset A_1 \cup A_2 \cup \cdots \cup A_l$ with A_i closed and contractible in M, we can easily find that

(1) $c_1 = \inf\{\Phi(u) : u \in M\}$;
(2) $c_j \le c_{j+1}$;
(3) $c_j \le \sup\{\Phi(u) : u \in M\}$;
(4) c_j is only defined for $j \le \mathrm{Cat}(M)$.

And, by Lemma 1.6, we have c_j is a critical value of Φ (actually, this is to say that $K_{c_j} \neq \emptyset$). If $c_i \neq c_j$ when $i \neq j$, then the result follows as we obtain distinguished critical values of Φ. If we have $1 \leq i < j \leq \mathrm{Cat}(M)$ so that $-\infty < c = c_i = \cdots = c_j < \infty$, then an interesting conclusion is that $\mathrm{Cat}_M(K_c) \geq j - i + 1$, which means that the number of points in K_c is at least $j - i + 1$. Therefore, in any case, the values $c_1 \leq c_2 \leq \cdots \leq c_{\mathrm{Cat}(M)}$ are critical values of Φ and there are at least $\mathrm{Cat}(M)$ critical points. We will not expound further on the detailed proof of Theorem 1.8, but rather refer the reader to [8, Chapter II Section 5].

If M is the unit sphere in a Hilbert space and Φ is even and satisfies (1)–(2) in Theorem 1.8, then an application of this result leads to an extension of Theorem 1.5, yielding:

Theorem 1.9 *Let $\Phi \in C^2(M, \mathbb{R})$ satisfy* (1)-(2) *of Theorem 1.8, with $M = \{u \in E : \|u\| = 1\}$. Then Φ has infinitely many pairs of critical points on M, each giving rise to solutions of the generalized eigenvalue problem $\nabla\Phi(u) = \lambda u$.*

For an exposition of these topics, see for instance [3, 7, 8].

1.3 Indefinite Functionals—The Way of Unboundedness

The development of critical point theorems is intricately linked to the exploration of differential equations, where critical point theory serves as a fundamental tool for proving the existence of solutions within this mathematical framework.

Example 1.10 Let us consider the linear elliptic boundary value problem

$$\begin{cases} -\Delta u(x) = f(x), & x \in \Omega \\ u(x) = 0, & x \in \partial\Omega \end{cases} \tag{1.2}$$

where Ω denotes a bounded domain in $\mathbb{R}^n$ whose boundary $\partial\Omega$ is smooth. Suppose $f \in C(\overline{\Omega})$, then a function u is a classical solution of (1.2) means that $u \in C^2(\Omega) \cap C(\overline{\Omega})$ and satisfies the equation together with the boundary condition. And, for such a solution, multiplying (1.2) by $\varphi \in C_c^\infty(\Omega, \mathbb{R})$ yields

$$\int_\Omega (\nabla u \cdot \nabla\varphi - f\varphi)dx = 0 \tag{1.3}$$

after an integration by parts. Let $H_0^1(\Omega, \mathbb{R})$ denote the subspace of $H^1(\Omega, \mathbb{R})$, which is the completion of $C_c^\infty(\Omega, \mathbb{R})$ with respect to the norm

$$\|u\|_{H_0^1} = \left(\int_\Omega |\nabla u|^2 dx\right)^{1/2}.$$

If $u \in H_0^1(\Omega, \mathbb{R})$ satisfies (1.3) for all $\varphi \in C_c^\infty(\Omega, \mathbb{R})$, then u is said to be a weak solution of (1.2). Clearly, any classical solution of (1.2) is a weak solution, while under slightly stronger hypotheses on f (say for instance, f is Hölder continuous) the converse is also true. Now, one may consider the functional

$$\Phi(u) = \frac{1}{2}\int_\Omega |\nabla u|^2 dx - \int_\Omega f u \, dx$$

defined on $H_0^1(\Omega, \mathbb{R})$. It is easy to verify that Φ is smooth and its first variation is

$$\nabla\Phi(u)[v] = \int_\Omega \big(\nabla u \cdot \nabla v - f v\big) dx$$

for $v \in H_0^1(\Omega, \mathbb{R})$. Thus u is a critical point of Φ if and only if u is a weak solution of (1.2). Moreover, we have

(∗) By Hölder inequality and the Sobolev embedding $H_0^1(\Omega, \mathbb{R}) \hookrightarrow L^2(\Omega, \mathbb{R})$, we find that

$$\left|\int_\Omega f u \, dx\right| \le C_{\Omega, f} \|u\|_{H_0^1},$$

where $C_{\Omega, f} > 0$ is a constant depending only on the domain Ω and the function f. This suggests that

$$\Phi(u) \ge \frac{1}{2}\|u\|_{H_0^1}^2 - C_{\Omega, f}\|u\|_{H_0^1}, \quad \forall u \in H_0^1(\Omega, \mathbb{R}).$$

This statement actually implies the lower boundedness and the validity of $(PS)_c$-condition for the functional Φ, with $c \in \mathbb{R}$. Specifically, given a sequence $\{u_k\} \subset H_0^1(\Omega, \mathbb{R})$ such that $\Phi(u_k) \to c$ and $\nabla\Phi(u_k) \to 0$, it follows that $\{u_k\}$ is bounded. Consequently, u_k weakly converges to some u_0 (up to a subsequence if necessary). This leads to the relation

$$\|u_k\|_{H_0^1}^2 = \int_\Omega f u_k \, dx + o_k(1) \to \int_\Omega f u_0 \, dx = \|u_0\|_{H_0^1}^2, \quad \text{as } k \to \infty$$

ensuring the convergence of u_k to u_0 in the Hilbert space $H_0^1(\Omega, \mathbb{R})$ under the norm topology. Therefore, by applying Theorem 1.8, a solution of (1.2) can be obtained, while the uniqueness can be established directly through the linearity of the problem.

In more general settings, particularly when addressing nonlinear differential equations, the functionals under consideration may not exhibit boundedness from above or below even modulo finite dimensional subspaces or submanifolds. Such functionals are termed as "*indefinite*" due to their lack of clear bounded behavior, and may not possess any local maxima or minima other than trivial ones. To illustrate this, let's examine the following nonlinear problem as a starting point:

Example 1.11 Consider a second order differential equation for a scalar function u in the form

$$\begin{cases} -u''(x) = u(x)^3, & x \in (0, \pi) \\ u(0) = u(\pi) = 0 \end{cases} \tag{1.4}$$

and the associated Euler-Lagrange functional is given by

$$\Phi(u) = \frac{1}{2}\int_0^{\pi} |u'|^2 dx - \frac{1}{4}\int_0^{\pi} u^4 dx \tag{1.5}$$

for $u \in H_0^1((0, \pi), \mathbb{R})$, where $u' \equiv du/dx$. It is not difficult to show that Φ is differentiable on $H_0^1((0, \pi), \mathbb{R})$ and has $u \equiv 0$ as a local minimum. But for any other nontrivial function $u \in H_0^1((0, \pi), \mathbb{R})$ and $t \in \mathbb{R}$, we can deduce that

$$\Phi(tu) = \frac{t^2}{2}\int_0^{\pi} |u'|^2 dx - \frac{t^4}{4}\int_0^{\pi} u^4 dx \to -\infty$$

as $|t| \to +\infty$. This implies Φ is not bounded from below. Furthermore, for each $k \in \mathbb{N}$, we find the function $u_k(x) := \sin(kx)$ belongs to $H_0^1((0, \pi), \mathbb{R})$ and

$$\Phi(u_k) = \frac{\pi}{4}k^2 - \frac{3\pi}{32} \to +\infty$$

as $k \to +\infty$. So Φ is not bounded from above. Thus it is not obvious that Φ possesses any critical point other than the trivial one $u \equiv 0$.

As a second example of an indefinite functional, emphasizing a stronger sense of indefiniteness, let us progress from the problem (1.4) to a subsequent situation.

Example 1.12 Consider a system of two coupled scalar equations

$$\begin{cases} -u''(x) = v(x)^3, & x \in (0, \pi) \\ -v''(x) = u(x)^3, & x \in (0, \pi) \\ u(0) = u(\pi) = v(0) = v(\pi) = 0 \end{cases}$$

and its corresponding functional

$$\Phi(u, v) = \int_0^{\pi} u'v' \, dx - \frac{1}{4}\int_0^{\pi} (u^4 + v^4) dx \tag{1.6}$$

defined for $(u, v) \in H_0^1((0, \pi), \mathbb{R}) \times H_0^1((0, \pi), \mathbb{R})$. To see its indefinite nature, we shall evaluate the functional in (1.6) by using the pairs (u_k, u_k) and $(u_k, -u_k)$, where $u_k(x) = \sin(kx)$ and $k \in \mathbb{N}$. And we have

$$\Phi(u_k, u_k) = \frac{\pi}{2}k^2 - \frac{3\pi}{16} \to +\infty$$

and

$$\Phi(u_k, -u_k) = -\frac{\pi}{2}k^2 - \frac{3\pi}{16} \to -\infty$$

as $k \to +\infty$. This shows that Φ is not bounded from above or below. And, more specifically, if we set

$$E^+ = \text{span}\{(u_k, u_k) : k \in \mathbb{N}\} \quad \text{and} \quad E^- = \text{span}\{(u_k, -u_k) : k \in N\},$$

we observe that E^+ and E^- are infinite-dimensional subspaces of $H_0^1((0, \pi), \mathbb{R}) \times H_0^1((0, \pi), \mathbb{R})$ on which Φ is not bounded from above and not bounded from below, respectively.

In the previous example, we refer to the functional in (1.6) as "*strongly indefinite*". And the emphasis here is that the quadratic part is positive definite on E^+ and negative definite on E^-, with both subspaces being infinite-dimensional. This characteristic is absent in (1.5). Our upcoming chapter will further present more strongly indefinite variational problems, serving as the primary research focus of this monograph.

1.4 Lack of Compactness

In addition to the indefinite nature of variational problems, there are cases where the Palais-Smale condition may not hold. This situation is relevant for many problems in Mathematical Physics that yield differential equations over all of $\mathbb{R}^n$, and for problems in Differential Geometry that give rise to differential equations with conformal invariance. In the sequel, we collect two important examples characterizing the lack of compactness.

Example 1.13 Let $\Phi : E \to \mathbb{R}$ be a smooth functional defined on a Hilbert space $(E, \langle\cdot, \cdot\rangle)$. It is worth noting that there could exist a sequence $\{u_k\} \subset E$ such that $\Phi(u_k) \to c$ and $\nabla\Phi(u_k) \to 0$ while $\|u_k\| \to +\infty$, as $k \to +\infty$. This phenomenon can occur even in the finite-dimensional Euclidean spaces, and the following explicit construction is due to Brezis and Nirenberg. For $E = \mathbb{R}^2$, the functional $\Phi(x, y) = x^2 + (1 - x)^3 y^2$ has a unique critical, which is the origin. And, since $\Phi(x, y) = x^2 + y^2 + o(x^2 + y^2)$ as $(x, y) \to (0, 0)$, it follows that $(0, 0)$ is a strict local

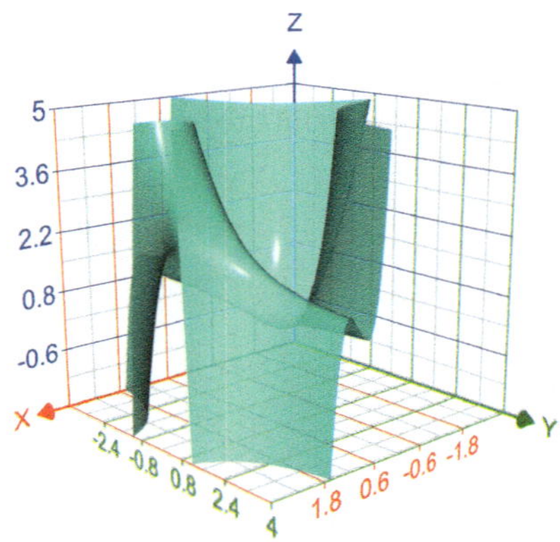

Fig. 1.1 The functional $\Phi(x, y) = x^2 + (1 - x)^3 y^2$

minimum of Φ. Furthermore, this Φ is indefinite since $\Phi(0, t) = t^2 \to +\infty$ and $\Phi(t, t) = t^2 + (1 - t)^3 t^2 \to -\infty$ as $t \to +\infty$. All these can be visualized by Fig. 1.1 below. The particularity of this functional lies in the fact that

$$\Phi\Big(1 + \frac{1}{k}, \sqrt{\frac{2}{3}}k\Big) = 1 + o_k(1) \quad \text{and} \quad \nabla\Phi\Big(1 + \frac{1}{k}, \sqrt{\frac{2}{3}}k\Big) = (0, 0) + o_k(1)$$

as $k \to +\infty$. This suggests that the $(PS)_c$-condition fails at the level $c = 1$.

Example 1.14 Another example of breaking the compactness comes from a famous problem in Differential Geometry, known as the Yamabe problem. Let (M, g) be a closed Riemannian manifold of dimension $m \geq 3$. The Yamabe problem is concerned with finding metrics of constant scalar curvature in the conformal class of g. This problem leads to a semi-linear elliptic PDE for the conformal factor. Specifically, a conformal metric of the form $u^{\frac{4}{m-2}} g$ has constant scalar curvature $\lambda \in \mathbb{R}$ if and only if

$$-\frac{4(m-1)}{m-2}\Delta_g u + \mathrm{Scal}_g u = \lambda u^{\frac{m+2}{m-2}}, \tag{1.7}$$

where $\Delta_g = \mathrm{div}_g \nabla_g$ is the Laplace-Beltrami operator with respect to the metric g and Scal_g denotes the scalar curvature of g. The constant λ can be restricted to the values -1, 1 and 0 depending on whether the *Yamabe invariant*

$$Y(M, [g]) := \inf_{\tilde{g} \in [g]} \Big(\mathrm{Vol}(M, \tilde{g})^{\frac{2-m}{m}} \int_M \mathrm{Scal}_{\tilde{g}} d\,\mathrm{vol}_{\tilde{g}} \Big)$$

of (M, g) has negative sign, positive sign or vanishes, respectively. By considering $\lambda = 1$, we find every solution of (1.7) is a critical point of the functional

$$\Phi(u) = \frac{1}{2}\int_M \left(\frac{4(m-1)}{m-2}|\nabla_g u|^2 + \mathrm{Scal}_g u^2\right) d\,\mathrm{vol}_g - \frac{m-2}{2m}\int_M u^{\frac{2m}{m-2}} d\,\mathrm{vol}_g$$

defined on the Sobolev space $H^1(M, \mathbb{R})$. For a given $\varepsilon > 0$, we define a function $u_\varepsilon : \mathbb{R} \to \mathbb{R}$ by

$$u_\varepsilon(x) = \left(\frac{\sqrt{4m(m-1)}\varepsilon}{\varepsilon^2 + |x|^2}\right)^{\frac{m-2}{2}}$$

Then the function u_ε satisfies the elliptic PDE

$$-\frac{4(m-1)}{m-2}\Delta u_\varepsilon = u_\varepsilon^{\frac{m+2}{m-2}} \quad \text{in } \mathbb{R}^m.$$

And it is well-known that

$$\frac{4(m-1)}{m-2}\int_{\mathbb{R}^m} |\nabla u_\varepsilon|^2 dx = \int_{\mathbb{R}^m} u_\varepsilon^{\frac{2m}{m-2}} dx \equiv [m(m-1)]^{\frac{m}{2}} \omega_m \tag{1.8}$$

and that

$$\int_{\mathbb{R}^m} u_\varepsilon^2 dx = O(\varepsilon^2) \tag{1.9}$$

as $\varepsilon \to 0$, where ω_m stands for the volume of the unit sphere (S^m, g_{S^m}). Let $r_0 > 0$ be smaller than the injectivity radius of (M, g), and η be a smooth cut-off function such that $0 \le \eta(x) \le 1$, $\eta(x) = 1$ for $|x| \le r_0/2$ and $\eta(x) = 0$ for $|x| \ge r_0$. For a point $p \in M$, we define the function $\tilde{u}_\varepsilon$ on M by $\tilde{u}_\varepsilon(\xi) = \eta(\exp_p^{-1}(\xi))u_\varepsilon(\exp_p^{-1}(\xi))$ where $\exp_p : \mathbb{R}^m \cong T_pM \to M$, $x \mapsto \exp_p(x)$ is the exponential map. Then, through a direct computation, we obtain

$$\Phi(\tilde{u}_\varepsilon) = \frac{[m(m-1)]^{\frac{m}{2}}\omega_m}{m} + o_\varepsilon(1)$$

and

$$\nabla\Phi(\tilde{u}_\varepsilon)[v] = o_\varepsilon(1)\|v\|_{H^1}$$

for all $v \in H^1(M, \mathbb{R})$. However, the sequence $\{\tilde{u}_\varepsilon\}$ does not exhibit any compactness in $H^1(M, \mathbb{R})$ since it weakly converges to 0, as indicated by the estimates in (1.8) and (1.9). The resolution of the Yamabe problem stands as a milestone in the realms of variational methods, geometric analysis, and partial differential equation theory. Here, we will refrain from delving too deeply into it. For an extensive discussion and resolution of the Yamabe problem, we direct readers to the following references: [1, 4, 6].

A significant insight in modern variational calculus is the understanding that the breakdown of the Palais-Smale condition does not always signify a dead end. Instead, a more refined analysis of the functional's characteristics may necessitate the development of new variational methods to address and overcome such challenges. The pioneering work in this direction was undertaken by T. Aubin [1] who observed that in the context of the Yamabe problem, the Palais-Smale condition can hold up to specific energy levels related to the optimal constant in the Sobolev inequality. Another important technique for addressing non-compactness was introduced by Sacks and Uhlenbeck [5] in their investigation of harmonic maps. They introduced the method of blowing up the singularities near points of concentration, a strategy that has since been observed in problems related to prescribed curvature problems of Rellich type [2], Yang-Mills equations [10], closed geodesics problems [9] and many other situations. Here, the objective of this monograph is to present a new variational framework for functionals with strong indefiniteness that involve issues of non-compactness. We provide examples demonstrating the applicability of our abstract results to several important problems in Physics, Differential Geometry, and related fields.

References

1. Aubin, T.: Équations différentielles non linéaires et problème de Yamabe concernant la courbure scalaire. J. Math. Pures Appl. **55**, 269–296 (1976)
2. Brezis, H., Coron, J.M.: Multiple solutions of H-systems and Rellich's conjecture. Commun. Pure Appl. Math. **37**, 149–187 (1984)
3. Krasnoselski, M.A.: Topological Methods in the Theory of Nonlinear Integral Equations. Pergamon Press, Oxford (1964)
4. Lee, J.M., Parker, T.: The Yamabe problem. Bull. Am. Math. Soc. **17**, 37–91 (1987)
5. Sacks, J., Uhlenbeck, K.: On the existence of minimal immersions of 2-spheres. Ann. Math. **113**, 1–24 (1981)
6. Schoen, R.: Conformal deformation of a Riemannian metric to constant scalar curvature. J. Differ. Geom. **20**, 479–495 (1984)
7. Schwartz, J.T.: Nonlinear Functional Analysis. Gordon & Breach, New York (1969)
8. Struwe, M.: Variational Methods and Applications to Nonlinear Partial Differential Equations and Hamiltonian Systems. Springer, Berlin (1990)
9. Tanaka, K.: Periodic solutions for singular Hamiltonian systems and closed geodesics on non-compact Riemannian manifolds. AIHP Anal. non-linéaire **17**(1), 1–33 (2000)
10. Taubes, C.H.: Path-connected Yang-Mills moduli spaces. J. Differ. Geom. **19**, 337–392 (1984)

Chapter 2
Strongly Indefinite Problems: Examples and Motivations

In this initial chapter, we will provide an overview of the primary nonlinear variational problems that will be comprehensively explored in the subsequent chapters. Additionally, a concise outline of the abstract setting will be presented to set the stage for the detailed analysis and discussion to follow.

2.1 Semiclassical Standing Waves of Nonlinear Dirac Equations on $\mathbb{R}^3$

Let P be a free classical particle in $\mathbb{R}^3$ with spin $\frac{1}{2}$ whose motion is to be studied in special relativity. Denoted by m its mass, by T its energy and by $p = \frac{mv}{\sqrt{1-v^2/c^2}}$ its momentum, we have

$$T = \sqrt{c^2p^2 + m^2c^4}.$$

where c stands for the speed of light. In quantum mechanics, P is described by a state function $\psi(t, x)$ defined on $\mathbb{R} \times \mathbb{R}^3$, and energy as well as momentum are replaced by the differential operators

$$T \mapsto i\hbar\frac{\partial}{\partial t} \quad \text{and} \quad p \mapsto -i\hbar\nabla,$$

respectively. Then, the state function ψ becomes a solution of the following free equation

$$i\hbar\frac{\partial\psi}{\partial t} = \sqrt{-c^2\hbar^2\Delta + m^2c^4}\,\psi$$

© The Author(s), under exclusive license to Springer Nature Singapore Pte Ltd. 2025
Y. Ding, T. Xu, *Localization Approaches in Strongly Indefinite Problems*, Series in Contemporary Mathematics 6, https://doi.org/10.1007/978-981-97-9539-0_2

involving the 3-dimensional Laplacian $\Delta = \frac{\partial^2}{\partial x_1^2} + \frac{\partial^2}{\partial x_2^2} + \frac{\partial^2}{\partial x_3^2}$. Mathematically, we let

$$L = -ic\hbar \sum_{k=1}^{3} \alpha_k \frac{\partial}{\partial x_k} + mc^2\beta$$

be a possible square root of $-c^2\hbar^2\Delta + m^2c^4$. It is obvious that the coefficients α_k and β of L satisfy the conditions

$$\alpha_k\alpha_l + \alpha_l\alpha_k = 2\delta_{kl} \quad \text{and} \quad \alpha_k\beta + \beta\alpha_k = 0.$$

Writing these as complex 4×4-matrices, we obtain

$$\beta = \begin{pmatrix} I & 0 \\ 0 & -I \end{pmatrix}, \quad \alpha_k = \begin{pmatrix} 0 & \sigma_k \\ \sigma_k & 0 \end{pmatrix}, \quad k = 1, 2, 3,$$

with

$$\sigma_1 = \begin{pmatrix} 0 & 1 \\ 1 & 0 \end{pmatrix}, \quad \sigma_2 = \begin{pmatrix} 0 & -i \\ i & 0 \end{pmatrix}, \quad \sigma_3 = \begin{pmatrix} 1 & 0 \\ 0 & -1 \end{pmatrix}.$$

Since L is a formal differential operator, we can construct from L various operators in the basic Hilbert space $L^2(\mathbb{R}^3, \mathbb{C}^4)$ consisting of all $\mathbb{C}^4$-valued functions $u = (u_1, u_2, u_3, u_4)$ such that

$$|u|_2 = \left(\int_{\mathbb{R}^3} |u|^2 dx \right)^{1/2} < \infty, \quad |u|^2 = \sum_{k=1}^{4} |u_k|^2.$$

The associated inner product is

$$(u, v)_2 = \int_{\mathbb{R}^3} u \cdot \bar{v} dx, \quad u \cdot \bar{v} = \sum_{k=1}^{4} u_k \bar{v}_k.$$

In particular, we have L with domain $C_c^\infty(\mathbb{R}^3, \mathbb{C}^4)$ is densely defined and, since one verifies that $\beta = \beta^*$, $\alpha_k = \alpha_k^*$, we see the linear operator L is self-adjoint.

For a particle in a static field with potential $M(x)$, the Dirac operator is given by

$$L = -ic\hbar \sum_{k=1}^{3} \alpha_k \frac{\partial}{\partial x_k} + mc^2\beta + M(x) \tag{2.1}$$

where $M(x)$ should be understood as the multiplication operator by a 4×4 symmetric matrix function. Without aiming at generality, for a real function $M \in L^\infty(\mathbb{R}^3)$, the operator L in (2.1) is still a densely defined self-adjoint operator in $L^2(\mathbb{R}^3, \mathbb{C}^4)$.

In quantum mechanics, nonlinear couplings appear in the self-interacting scalar theories; see [15, 16, 24] and more recently [7, 13, 14, 18, 19, 33, 43]. This leads us to a nonlinear Dirac equation of the form

$$i\hbar\partial_t\psi = -ic\hbar\sum_{k=1}^{3}\alpha_k\partial_k\psi + mc^2\beta\psi + M(x)\psi - G_\psi(x,\psi), \quad t \in \mathbb{R},\ x \in \mathbb{R}^3, \tag{2.2}$$

where the nonlinear coupling $G(x, \psi)$ describes a self-interaction. Usually, in Quantum electrodynamics nonlinear Dirac equations have to satisfy symmetry constraints, in particular the Poincaré covariance. Nonlinear Dirac equations modeling Bose-Einstein condensates break this symmetry, and often the nonlinearity is a *power-type* function that depends only on the local condensate density (see [21–23] for more background from physics).

By assuming that G satisfies $G(x, e^{i\theta}\psi) = G(x, \psi)$, for all $\theta \in [0, 2\pi]$, we are led to study stationary solutions of (2.2) which may be regarded as "particle-like solution" (see [35]): they propagate without changing their shape and thus have a soliton-like behavior. The stationary solutions of equation (2.2) are found by the ansatz

$$\psi(t,x) = e^{\frac{i\theta t}{\hbar}}u(x);$$

then $u : \mathbb{R}^3 \to \mathbb{C}^4$ satisfies the equation

$$-ic\hbar\sum_{k=1}^{3}\alpha_k\partial_k u + mc^2\beta\, u + \big(M(x)+\theta\big)u = G_u(x,u).$$

Such a u will be called a standing wave. A particular interest is given to the so-called semiclassical states that are standing waves existing for $\hbar \to 0$. Normalizing $c = 1$ and setting $\varepsilon = \hbar$, $a = m$ and $V(x) = M(x) + \theta$, we are finally led to

$$-i\varepsilon\sum_{k=1}^{3}\alpha_k\partial_k u + a\beta u + V(x)u = G_u(x,u) \tag{2.3}$$

It is convenient to make the change of variables $x \mapsto \varepsilon x + x_0$, where $x_0 \in \mathbb{R}^3$ will be chosen in an appropriate way, that leads to

$$-i\alpha\cdot\nabla u + a\beta u + V(\varepsilon x + x_0)u = G_u(\varepsilon x + x_0, u) \tag{2.4}$$

where we denote $\alpha = (\alpha_1, \alpha_2, \alpha_3)$ and $\alpha \cdot \nabla = \sum_{k=1}^{3} \alpha_k \partial_k$ for simplicity. Solutions of (2.4) can be obtained as critical points of the corresponding Euler functional

$$\Phi_\varepsilon(u) = \frac{1}{2}\int_{\mathbb{R}^3} \left[-i\alpha\cdot\nabla u\cdot\bar{u} + a\beta u\cdot\bar{u} + V(\varepsilon x + x_0)u\cdot\bar{u}\right]dx - \int_{\mathbb{R}^3} G(\varepsilon x + x_0, u)dx$$

defined on $H^{1/2}(\mathbb{R}^3, \mathbb{C}^4)$ (for the definition of the Sobolev space $H^{1/2}(\mathbb{R}^3, \mathbb{C}^4)$, see Chap. 4).

To prove existence and localization of solutions of (2.4), one of the main difficulties is the lack of compactness. This is the same for the elliptic equations on $\mathbb{R}^n$. Since the Sobolev embedding $H^{1/2}(\mathbb{R}^3, \mathbb{C}^4) \hookrightarrow L^p(\mathbb{R}^3, \mathbb{C}^4)$ is not compact for all $p \in [2, 3]$, Φ_ε does not satisfy, in general, the Palais-Smale (PS, for short) compactness condition. This is the case when V is constant and G is independent of the spatial variables. To overcome this difficulty a usual strategy is to apply the P.L. Lions Concentration-Compactness principle. Roughly, suppose that G is independent of the spatial variables and that $\lim_{|x|\to\infty} V(x) = V_\infty$. Let us consider the *limit functional*

$$\Phi_\infty(u) = \frac{1}{2}\int_{\mathbb{R}^3} \left[- i\alpha \cdot \nabla u \cdot \bar{u} + a\beta u \cdot \bar{u} + V_\infty u \cdot \bar{u}\right]dx - \int_{\mathbb{R}^3} G(u)dx$$

which has a ground state critical level γ_∞, the lowest nontrivial critical level of Φ_∞. Generally, using the Concentration-Compactness principle, existence results are found by imposing conditions that permit to compare the critical levels of Φ_ε with γ_∞. For instance, if $V(x)$ is a real-valued function such that $V(x) < V_\infty$ for all $x \in \mathbb{R}^3$, then it readily follows that the ground state critical level of Φ_ε is lower than the corresponding level γ_∞ of Φ_∞. Since it is possible to prove that the PS condition holds at levels lower than γ_∞, this yields the existence of a solution to (2.4). See for instance [10, 12]. There is also an additional difficulty for the nonlinear Dirac equations which does not present in its elliptic counterpart: The functional Φ_ε is strongly indefinite, i.e., its Morse index and co-index at any critical point are infinite. Thus the standard variational methods such as Morse theory and its variants are not directly applicable. In general, the problem of finding critical points for such functionals seems widely open and is one of challenging problems in calculus of variations.

It is very natural to ask the question whether there are approaches that give rise to existence results for nonlinear Dirac equations, which do not require the preceding comparison procedure and the global assumption on V.

Motivated by this question, we will deal in Chap. 4 with problems on $\mathbb{R}^3$ whose model is (2.4). Our approach will prove that if x_0 is a strict local minimum, local maximum or a isolated saddle point of the potential function V, then (2.4) has for ε small a solution concentrates around x_0. This kind of solutions are called *spike layers* or simply *spikes*. Form the physical point of view, spikes are important since they show that nonlinear Dirac equations of the type (2.4) are not dispersive and the energy is localized in packets.

2.2 A Coupled System of Multi-Component Incongruent Diffusion

A system of Reaction-Diffusion (RD) equations comprises reaction terms and diffusion terms, i.e. the typical form is as follows:

$$\partial_t z = \text{div}_x\big(\mathbf{D}\nabla_x z\big) + M(x)z + f(x, z) \tag{2.5}$$

where $z(t, x) \in \mathbb{R}^m$, $m > 1$, is a state variable and describes density or concentration of multi-component substances, populations at position $x \in \mathbb{R}^N$ at time t. The first term on the right-hand side describes the “diffusion”, including $\mathbf{D}$ as a matrix of diffusion coefficients D_{ij} (the diagonal elements of $\mathbf{D}$ describe the main-term diffusion rate and the off-diagonal elements express the cross-terms diffusion which was suggested firstly in 1932 see [34]). Creation and killing in the reaction process (birth, death, etc) are described by the scalar field $M(x)$ which could be a matrix-valued function, and the nonlinear part, $f(x, z)$ is a smooth function $f : \mathbb{R}^N \times \mathbb{R}^m \to \mathbb{R}^m$ called reaction kinetics and describes processes really “change” the present z, i.e. something happens to it (coupling actions, chemical reactions), not just diffuse in the space.

Multi-component diffusion is more complicated than is often realized. In general, systems of RD equations allow for much more complex behavior than a scalar RD equation does. For example, a ternary system (two solutes in a solvent) may has four coefficients in the diffusion matrix, not just two. And the diffusion coefficients can be large or small and be positive or negative, thus having a substantial effect on flows of matter. Meanwhile, the interacting reaction terms are of interest and lead to interesting behavior. For instance, oscillating phenomena can evolve—as these oscillations can spread in space via diffusion and instabilities may develop spatial phenomena like pattern formation can be observed. Here for a detailed survey, we would refer the readers to [31].

On the macroscopic level, fluxes of chemical components (species) are due to convection and molecular fluxes, where the latter essentially refers to diffusive transport. The almost exclusively employed constitutive “law” to model diffusive fluxes within continuum mechanical models is Fick’s law (which requires positive diffusion coefficients), stating that the flux of a chemical component is proportional to the gradient of the concentration of this species, directed against the gradient. This leads to positive coefficients D_{ii} in $\mathbf{D}$, however, thermodynamic conditions do not require the diagonal of $\mathbf{D}$ to be positive. Contrary to popular belief, there are systems with negative D_{ii} for a particular choice of the solvent component. Typical example is acetic acid + chloroform with water being chosen as solvent, a totally unexpected result is the negative chloroform main term diffusion coefficient, that is a negative D_{22} value (see [30, 44] for the Chemistry details). This behavior has been interpreted as a thermodynamic effect caused by the diffusion of “salted-out” chloroform down the water concentration gradient produced by the chloroform gradient under conditions of constant acetic acid concentration.

Motivated exactly by the unusual phenomenon of negative main term diffusion, let us consider the multi-component nonlinear RD system (2.5) in the situation $z = (u, v) \in \mathbb{R}^{2K}$, some $K \geq 1$, and

$$\mathbf{D} = \begin{pmatrix} D_u & 0 \\ 0 & D_v \end{pmatrix} = \varepsilon^2 \mathcal{J}, \quad \text{where} \quad \mathcal{J} = \begin{pmatrix} \mathrm{Id} & 0 \\ 0 & -\mathrm{Id} \end{pmatrix} \tag{2.6}$$

and $\varepsilon > 0$ being a parameter characterizing the pace of the diffusion process (for simplicity we have ignored the cross-term diffusions). To the best of our knowledge negative main term diffusion, the phenomenon in which one species to be driven from lower to higher concentrations, has generally been neglected in the study of RD systems. The existence problem will be discussed in Chap. 5, in particular, we will show the effect of external potentials in the existence, localization and multiplicity of solutions to (2.5).

We suppose the RD system (2.5)–(2.6) is in variational form, that is, it is the Euler-Lagrange equation of some energy functional. This happens when f is the derivatives of some function $F(x, z)$. In this case, let us also remark that the RD system (2.5)–(2.6) can be viewed as an infinite-dimensional Hamiltonian system:

$$\beta \frac{dz}{dt} = \nabla \mathcal{H}(z), \quad \beta = \begin{pmatrix} 0 & \mathrm{Id} \\ -\mathrm{Id} & 0 \end{pmatrix}, \tag{2.7}$$

where $\mathcal{H}$ is some energy functional on a real Hilbert space $\mathscr{H} \subset L^2(\mathbb{R}^N, \mathbb{R}^{2K})$, and the matrix β can be regarded as a skew-symmetric operator. Such an interesting feature makes variational method applicable, especially, the techniques developed for variational problems with strongly indefinite structure can be employed.

2.3 Bifurcation from the Spectrum: The Spinorial Brezis-Nirenberg Problem

Let E be a Hilbert space, let $F : \mathbb{R} \times E \to E$ be a smooth mapping and suppose that $F(\lambda, 0) = 0$ for all $\lambda \in \mathbb{R}$. If there exists λ_0 with the property that the equation $F(\lambda, u) = 0$ has a sequence of solutions (λ_n, u_n), with $u_n \neq 0$ and such that $(\lambda_n, u_0) \to (\lambda_0, 0)$ as $n \to \infty$, we say that λ_0 is a bifurcation point (for the equation $F = 0$). If $F(\lambda, u) = \lambda u - \nabla K(u)$ and $\nabla K : E \to E$ is a compact operator, a theorem by Krasnoselski [25] ensures that every eigenvalue of $\nabla^2 K(0)$ is a bifurcation point.

When dealing with non-compact problems, the preceding approach cannot be carried out anymore and one needs to use a functional approach, in particular, an additional localization procedure arises so that the compactness can be recovered. A typical example is given by the spinorial Brezis-Nirenberg problem: Let (M, g, σ) be an m-dimensional closed spin manifold, $m \geq 2$, where g is a Riemannian metric and $\sigma : P_{Spin}(M) \to P_{SO}(M)$ is a fixed spin structure on M. We denote $\mathbb{S}(M) = P_{Spin}(M) \times_\rho \mathbb{S}_m \to M$ the spinor bundle. It is a complex vector bundle associated to $P_{Spin}(M) \to M$ via the spin representation $\rho : Spin(m) \to \mathrm{Aut}(\mathbb{S}_m)$. The (Atiyah-Singer) Dirac operator $D_g : C^\infty(M, \mathbb{S}(M)) \to C^\infty(M, \mathbb{S}(M))$ is defined by

$$D_g := \mathfrak{M} \circ \nabla : C^\infty(M, \mathbb{S}(M)) \xrightarrow{\nabla} C^\infty(M, T^*M \otimes \mathbb{S}(M))$$

$$\cong C^\infty(M, TM \otimes \mathbb{S}(M)) \xrightarrow{\mathfrak{M}} C^\infty(M, \mathbb{S}(M))$$

where ∇ is the canonical lift of the Levi-Civita connection on $P_{SO}(M)$ via the double covering $P_{Spin}(M) \to P_{SO}(M)$ and $\mathfrak{M}$ is the Clifford multiplication. The spinorial Brezis-Nirenberg problem is read as

$$D_g \psi = \lambda \psi + |\psi|_g^{2^*-2} \psi \quad \text{on } M, \tag{2.8}$$

where $2^* = \frac{2m}{m-1}$, $\lambda \in \mathbb{R}$ and $\psi \in C^1(M, \mathbb{S}(M))$ is a spinor. The main characteristic of the problem is that the exponent 2^* is critical in the sense of the Sobolev embedding $H^{\frac{1}{2}}(M, \mathbb{S}(M)) \hookrightarrow L^q(M, \mathbb{S}(M))$ for $1 \le q \le 2^*$, which is compact only if $q < 2^*$.

The problem above arises in differential geometry: For any metric $\tilde{g} = e^{2u} g \in [g]$, let us denote the smallest (i.e. first) positive eigenvalue of the corresponding Dirac operator $D_{\tilde{g}}$ by $\lambda_1^+(\tilde{g})$. Then the *Bär-Hijazi-Lott invariant* of (M, g, σ) is a positive real number $\lambda_{min}^+(M, g, \sigma)$ defined by (see for instance [2])

$$\lambda_{min}^+(M, g, \sigma) := \inf_{\tilde{g} \in [g]} \lambda_1^+(\tilde{g}) \operatorname{Vol}(M, \tilde{g})^{\frac{1}{m}}.$$

It is clear that the expression on the right-hand-side is chosen so as to remain conformally invariant, in a similar way as the Yamabe invariant. In a similar way as for the Yamabe problem, the Bär-Hijazi-Lott invariant cannot be greater than that of the sphere: for $m \ge 2$,

$$\lambda_{min}^+(M, g, \sigma) \le \lambda_{min}^+(S^m, g_{S^m}, \sigma_{S^m}) = \frac{m}{2} \omega_m^{\frac{1}{m}} \tag{2.9}$$

where σ_{S^m} stands for the unique spin structure on S^m. The proof relies on a suitable cut-off argument performed on Killing spinors on the gluing of a sphere with large radius to the manifold, see [3, 4] for details. Then we are led to the following open problem (see [20]):

Open Problem *Whether or not* (2.9) *is a strict inequality when* (M, g, σ) *is not conformally equivalent to the standard sphere* $(S^m, g_{S^m}, \sigma_{S^m})$ *of the same dimension?*

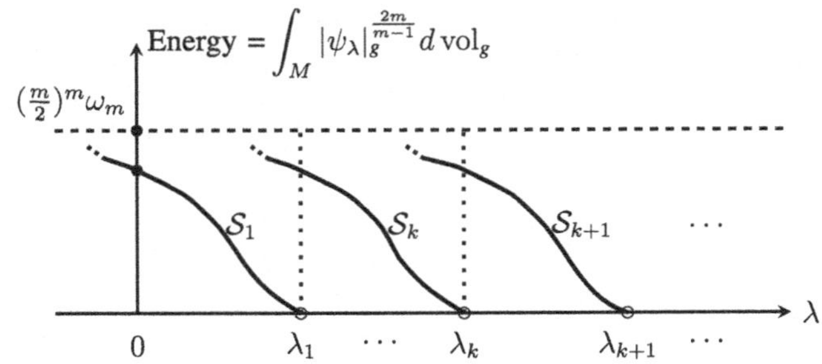

Fig. 2.1 Ground-state energy branches bifurcating from eigenvalues

And it is sufficient to find non-trivial solutions of

$$D_g\psi = |\psi|_g^{\frac{2}{m-1}}\psi \quad \text{with} \int_M |\psi|_g^{\frac{2m}{m-1}} d\,\mathrm{vol}_g < \Big(\frac{m}{2}\Big)^m \omega_m. \tag{2.10}$$

Evidently, the spinorial Brezis-Nirenberg problem (2.8) is a linear perturbation of (2.10). And the existence and bifurcation results of (2.8) for $\lambda \in \mathbb{R}$ are of particular geometric interests since the limiting behavior of a solution (λ, ψ_λ) as $\lambda \to 0$ may shed some light on the above open problem.

The energy functional associated to (2.8) is given by

$$\mathcal{L}_\lambda(\psi) = \frac{1}{2}\int_M (D_g\psi, \psi)_g d\,\mathrm{vol}_g - \frac{\lambda}{2}\int_M |\psi|_g^2 d\,\mathrm{vol}_g - \frac{1}{2^*}\int_M |\psi|_g^{2^*} d\,\mathrm{vol}_g,$$

which is strongly indefinite because the spectrum $\mathrm{spec}(D_g)$ consists of an infinite sequence of eigenvalues $\cdots < \lambda_{-1} < \lambda_0 \le 0 < \lambda_1 < \lambda_2 < \cdots$ with $|\lambda_k| \to \infty$ as $|k| \to \infty$. In Chap. 6, we will present a global variational approach that works in all dimensions $m \ge 2$ and also $\lambda \in \mathrm{spec}(D_g)$. By setting $\mathcal{S}_k$, $k = 1, 2, \ldots,$ the energy branches bifurcating from the k-th positive eigenvalue of D_g, we can visualize in Fig. 2.1 below the existence and bifurcation diagram in case (2.9) holds strictly, otherwise the branch $\mathcal{S}_1$ may be ended at the point $\big(0, (\frac{m}{2})^m\omega_m\big)$.

2.4 Conformal Embeddings of $S^2 \to \mathbb{R}^3$ with Prescribed Mean Curvature

One of the fundamental topics in classical differential geometry concerns the question of whether a smooth Riemannian manifold can be isometrically immersed/embedded into Euclidean spaces. It has been conjectured by Schläfli in 1873 that every m-dimensional smooth Riemannian manifold (M^m, g) admits a local isometric embedding into $\mathbb{R}^{N_m}$, with $N_m = \frac{m(m+1)}{2}$. Several important achievements have been made towards this problem in the last century. In one of his outstanding papers, J. Nash [32] showed the existence of a global isometric embedding of any m-dimensional smooth Riemannian manifold in $\mathbb{R}^N$ with $N = 3N_m + 4m$ in the case of compact M and $N = (m+1)(3N_m + 4m)$ in the case of non-compact M. Following Nash, we mention the book "Partial Differential Relations" published in 1986 by Gromov, which contains various problems related to the isometric embedding of Riemannian manifolds. Furthermore, in this book, Gromov showed that $N = N_m + 2m + 3$ is enough for the compact case.

In low dimensions, if one asks the more specific question of whether a given 2-dimensional Riemann surface can be isometrically immersed/embedded into Euclidean 3-space, not too much is known. Besides some general local existence results (see for example [28, 37] for real analytic metrics and for smooth metrics under suitable curvature assumptions), non-existence results seem to be easier to come by: for instance, Hilbert's classical result that the hyperbolic plane does not admit an isometric immersion into $\mathbb{R}^3$, and the fact that a compact non-positively curved Riemann surface cannot be isometrically immersed into $\mathbb{R}^3$.

In the classical differential geometry of surfaces, the Gauß-Codazzi-Mainardi equations consist of a pair of related equations, and it incorporates the extrinsic curvature (or mean curvature) of the surface. The equations are precisely the integrability conditions to obtain an immersed surface in $\mathbb{R}^3$ (locally or on a simply connected domain). With the development of Spin Geometry and the theory of Dirac operators, an equivalent albeit simpler expression becomes well-known, namely the existence of a special spinor field called *generalized Killing spinor field* (see for instance [27, 36, 40–42] and references therein). This expression is the so-called *spinorial Weierstrass representation*, and can be understood as a geometrically invariant way for the representation of surfaces in Euclidean 3-space.

In late 1990s, T. Friedrich has written a very nice article in which the spinorial Weierstrass representation is beautifully explained, see [17, Theorem 13]. Eventually, it turns out that there exists a one-to-one correspondence between the existence of an isometric immersion $(\widetilde{M}^2, g) \hookrightarrow (\mathbb{R}^3, g_{\mathbb{R}^3})$ with mean curvature H and the existence of a spinor field φ with constant length satisfying the following Dirac equation

$$D_g\varphi = H\varphi, \quad |\varphi|_g \equiv 1 \tag{2.11}$$

where $\widetilde{M}^2$ is the universal covering of M^2, D_g and $|\cdot|_g$ are respectively the Dirac operator and the hermitian metric on the spinor bundle $\mathbb{S}(M^2)$ over (M^2, g).

Following this idea, the situation is much clearer for simply connected surfaces. And one may naturally ask the following question:

Question 1 *Given an arbitrary smooth function $H : S^2 \to \mathbb{R}$, what is the condition on H so that there is an isometric immersion $S^2 \hookrightarrow \mathbb{R}^3$ realizing H as its mean curvature?*

Or it is equivalent to consider the following question via the spinorial Weierstrass representation:

Question 2 *What are necessary and sufficient conditions on H so that there is a solution to Eq.* (2.11)*?*

It seems interesting to see that Question 2 can be investigated via a conformal change of (2.11). That is, we may begin with the standard Riemann sphere (S^2, g_{S^2}) and consider the following nonlinear partial differential equation involving the Dirac operator $D_{g_{S^2}}$:

$$D_{g_{S^2}}\psi = H|\psi|^2_{g_{S^2}}\psi \quad \text{on } (S^2, g_{S^2}). \tag{2.12}$$

Suppose we have a spinor field ψ that satisfies (2.12) and that never vanishes. Then one may introduce a conformal metric $g = |\psi|^4_{g_{S^2}} g_{S^2}$ on S^2. Through the well-known formula for the Dirac operator under a conformal change of the metrics (see for instance [20, Proposition 1.3.10]), we obtain a new spinor field $\varphi := \iota(|\psi|^{-1}_{g_{S^2}}\psi)$ which satisfies

$$D_g\varphi = H\varphi, \qquad |\varphi|_g \equiv 1,$$

where $\iota : (\mathbb{S}(S^2), g_{S^2}) \to (\mathbb{S}(S^2), g)$ is a fiberwise isometry. And thus we find an isometric immersion $(S^2, g) \hookrightarrow (\mathbb{R}^3, g_{\mathbb{R}^3})$ that realizes H as prescribed mean curvature.

We will make use of the above spinorial tool in the study of Question 2. From this point of view, it seems natural to consider a two-step procedure:

- find a non-trivial spinor field ψ satisfying Eq. (2.12),
- show that the zero set under $|\psi|_{g_{S^2}} : S^2 \to [0, \infty)$ is empty, i.e. $\psi^{-1}(0) = \emptyset$.

However this is not an easy task. In general, even if one already knows there exists a non-trivial solution ψ to (2.12), it is still difficult to know whether ψ possesses zeros. This is due to the lack of an adequate replacement for the maximum principle developed for second order elliptic partial differential operators. Let us mention a further question besides Question 1 and 2:

Question 3 *Assume that a function $H : S^2 \to \mathbb{R}$ is given such that it is the mean curvature of an immersion. Under which condition is this immersion an embedding? And when does it have self-intersections?*

These questions will be discussed in Chap. 7 together with more general results dealing with a so-called localized concentration-compactness alternative property of the equation (2.12).

2.5 Other Problems with Localization

There are several further strongly indefinite problems whose main feature is that they possess solutions localizing around points or at manifolds.

2.5.1 Singularly Perturbed Elliptic Problems

An important example is given by Hamiltonian elliptic systems on Riemannian manifolds: Let (M, g) be a smooth closed Riemannian m-manifold, $m \geq 3$. We consider the singularly perturbed nonlinear elliptic problem

$$\begin{cases} -\varepsilon^2 \Delta_g v + v = f_1(u), & \\ & \text{on } M \\ -\varepsilon^2 \Delta_g u + u = f_2(v), & \\ u, v > 0, & \end{cases} \tag{2.13}$$

where $\Delta_g = \mathrm{div}_g \nabla_g$ is the Laplace-Beltrami operator on (M, g) and f_1, f_2 fall into a typical class of superlinear functions. Beside the existence of a solution for a fixed value of ε, we are interested in the blow-up profile (the localized behavior) of these solutions as ε goes to zero. Particularly, we are looking for a local geometric quantity of (M, g) which affects the location of a global maximum point of a solution of (2.13) as $\varepsilon \searrow 0$.

Problems like (2.13) arise in the study of some Reaction-Diffusion systems with chemical of biological motivation. The simplest example of a Hamiltonian elliptic system is the widely-studied Lane-Emden system, i.e., when $f_1 = v^p$, $f_2 = u^q$ in (2.13). Even for this system, important open problems subsist (see for instance [8]). Hamiltonian elliptic systems are very usual in biology, more specifically in models from population dynamics (see [31]) whose stationary states satisfy systems of type (2.13). An important difficulty in the study of Hamiltonian elliptic systems is the fact that the Euler functional is strongly indefinite. Our goal is to get a general existence result for small ε and to show a localized concentration phenomenon in (2.13) in terms of the geometric background of (M, g). These topics will be discussed in Chap. 8 Sect. 8.1.

2.5.2 Coupled Nonlinear Dirac Systems

Recently, see [5, 11], it has been proved that there exist solutions of the uncoupled nonlinear Dirac equation (2.3) concentrating around non-degenerate critical points of the external potential function V. It is also natural to consider more general relativistic descriptions of spin $1/2$ particles. For instance, one can consider a stationary nonlinear Maxwell-Dirac system like

$$\begin{cases} \alpha \cdot (i\varepsilon\nabla + Q(x)\mathbf{A})u - a\beta u - \theta u - Q(x)A_0 u = G_u(x, u), \\ -\Delta A_k = 4\pi Q(x)(\alpha_k u) \cdot \bar{u} \quad k = 0, 1, 2, 3, \end{cases} \tag{2.14}$$

where A_0 and $\mathbf{A} = (A_1, A_2, A_3)$ models the electromagnetic field and Q stands for the pointwise charge distribution.

Similar to (2.3), one can show that solutions of (2.14) possess a localized concentration behavior. A new feature of this case is that the linear external potentials A_0, $\mathbf{A}$ and Q do not effect the concentration. Roughly, one proves that the spatial distribution of the nonlinear self-interaction G will play the role of precipitating factor, to govern the location of the concentration phenomenon.

The results dealing with localized concentration behavior for stationary nonlinear Maxwell-Dirac system and related problems will be discussed in Chap. 8 Sect. 8.2.

2.6 The Abstract Setting

The problems discussed above can be studied by means of a common abstract setting. Let E be a real Hilbert space with inner product $\langle\cdot,\cdot\rangle$ and norm $\|\cdot\|$. We consider in E the nonlinear equation of the form

$$Lz = N(z), \quad z \neq 0, \tag{2.15}$$

where $L : (E, \langle\cdot,\cdot\rangle) \to (E, \langle\cdot,\cdot\rangle)$ is an invertible continuous self-adjoint linear operator and N is a superlinear potential operator with potential Ψ, that is, $N = \nabla\Psi$. Let $\sigma(L)$ denote the spectrum of L. Motivated by the preceding discussions, we set E^+ and E^- the invariant subspaces of L corresponding to $\sigma(L) \cap (0, +\infty)$ and $\sigma(L) \cap (-\infty, 0)$, respectively, and assume that there exists $\delta > 0$ such that

$$\langle Lz, z\rangle \geq \delta\|z\|^2, \quad \forall z \in E^+ \tag{2.16}$$

and

$$\langle Lz, z\rangle \leq -\delta\|z\|^2, \quad \forall z \in E^-. \tag{2.17}$$

Here, we call L *strongly indefinite* if both the dimensions of $E^\pm$ are infinite. Formally, (2.15) is the Euler-Lagrange equation of the functional

$$\Phi(z) = \frac{1}{2}\langle Lz, z\rangle - \Psi(z), \tag{2.18}$$

where z varies in E. Then Φ is neither bounded from above nor below, even modulo subspaces of finite dimension or codimension.

The basic common structure of strongly indefinite problems has been previously emphasized, among others, by Amann [1] and Benci-Rabinowitz [6]. Since these problems have been the target of investigation for decades and it is not our intention to give a complete bibliography, we shall refer the interested reader to the basic references [9, 26, 29, 38].

From a variational point of view, the aforementioned problems are typically solved by minimizing an associated constrained problem. This formulation was developed by Szulkin and Weth [39]. The advantage of this kind of approach is to give the most simplified characterization of the critical values for a indefinite functional. As the corresponding functionals of the problems discussed above are of strongly indefinite type, the direct method of the calculus of variation fails, but the reduction techniques and Nehari manifold techniques typically work. Hence these problems share many similarities.

However, only a few statements can be proven simultaneously for all of these problems. First of all, let us mention for instance that both problems (2.3) and (2.8) concern with a family of differential equations (parameterized by $\varepsilon > 0$ in (2.3) and by $\lambda \in \mathbb{R}$ in (2.8)). In our applications in Chap. 4, by some truncation arguments, the functional associated to (2.3) will satisfy the Palais-Smale compactness condition. By employing the Nehari manifold technique in [39], one does not need to worry about the existence of a ground state solution. In this regard, the main target is to study the localized concentration behavior of the solutions and the corresponding energies as $\varepsilon \to 0$. Since it is hopeless to formulate the ground state solutions of (2.3) explicitly, one possible way to characterize the qualitative properties of the limiting behavior is to compute approximations of those ground state energies by using some suitable test functions. These test functions are varying synchronously with ε. Being differently, the analysis in the spinorial Brezis-Nirenberg problem is more involved due to the appearance of the conformally invariant nonlinearity in (2.8) and the number one issue is to show the existence result. Standard methods developed in [9, 26, 29, 38, 39] etc. do not apply here due to the lack of compactness. To attack this problem, it is sufficient to deal with only one functional (i.e. λ is fixed) and the main target here is to establish a priori energy estimate for the associated functional and to show the lowest minimax level stays below some threshold for non-compactness. Similar to the case (2.3), the energy estimate is based on a careful choice of some test functions. Since λ is fixed, the test functions should be constructed independently. Because of these differences, most analytical properties have to be proven for each problem separately. We consider it hence as remarkable that our abstract approaches on the energy estimate can be extended to all such problems, and the list could continue.

Remarks on the Exposition

In order to limit the monograph to a reasonable length, we will only give the outline of the proofs which are based on arguments already employed. This will be mainly the case in the last chapters.

References

1. Amann, H.: Saddle points and multiple solutions of differential equations. Math. Z. **169**(2), 127–166 (1979)
2. Ammann, B.: A variational problem in conformal spin geometry. Habilitationsschrift, Universität Hamburg (2003)

3. Ammann, B.: A spin-conformal lower bound of the first positive Dirac eigenvalue. Differ. Geom. Appl. **18**(1), 21–32 (2003)
4. Ammann, B., Grossjean, J.-F., Humbert, E., Morel, B.: A spinorial analogue of Aubin's inequality. Math. Z. **260**, 127–151 (2008)
5. Bartsch, T., Xu, T.: Strongly localized semiclassical states for nonlinear Dirac equations. Discrete Contin. Dyn. Syst. **41**(1), 29–60 (2021). Special issue celebrating 25 years of excellence
6. Benci, V., Rabinowitz, P.: Critical point theorems for indefinite functionals. Invent. Math. **52**, 241–273 (1979)
7. Cazenave, T., Vázquez, L.: Existence of localized solutions for a classical nonlinear Dirac field. Commun. Math. Phys. **105**(1), 35–47 (1986)
8. de Figueiredo, D.G.: Nonlinear elliptic systems. Anais Acad. Brasil. Cienc. **72**(4), 453–469 (2000)
9. Ding, Y.H.: Variational Methods for Strongly Indefinite Problems. Interdiscip. Math. Sci., vol. 7. World Scientific Publ., Singapore (2007)
10. Ding, Y.H.: Semi-classical ground states concentrating on the nonlinear potential for a Dirac equation. J. Differ. Equ. **249**, 1015–1034 (2010)
11. Ding, Y.H., Xu, T.: Localized concentration of semiclassical states for nonlinear Dirac equations. Arch. Ration. Mech. Anal. **216**(2), 415–447 (2015)
12. Ding, Y.H., Lee, C., Ruf, B.: On semiclassical states of a nonlinear Dirac equation. In: Proceedings of the Royal Society of Edinburgh: Section A Mathematics 143.04 (2013), pp. 765–790
13. Esteban, M.J., Séré, E.: Stationary states of the nonlinear Dirac equation: a variational approach. Commun. Math. Phys. **171**(2), 323–350 (1995)
14. Esteban, M.J., Séré, E.: An overview on linear and nonlinear Dirac equations. Current developments in partial differential equations (Temuco, 1999). Discrete Contin. Dyn. Syst. **8**(2), 381–397 (2002)
15. Finkelstein, R., LeLevier, R., Ruderman, M.: Nonlinear spinor fields. Phys. Rev. **83**(2), 326–332 (1951)
16. Finkelstein, R., Fronsdal, C., Kaus, P.: Nonlinear spinor field. Phys. Rev. **103**(5), 1571–1579 (1956)
17. Friedrich, T.: On the spinor representation of surfaces in Euclidean 3-*space*. J. Geom. Phys. **28**(1–2), 143–157 (1998)
18. Fushchich, W., Zhdanov, R.: Symmetry and exact solutions of nonlinear spinor equations. Phys. Rep. **172**, 123–174 (1989)
19. Fushchich, W., Zhdanov, R.: Symmetries and Exact Solutions of Nonlinear Dirac Equations, 384 pp. Mathematical Ukraina Publisher, Kyiv (1997)
20. Ginoux, N.: The Dirac Spectrum. Lecture Notes in Math., vol. 1976. Springer, Berlin (2009)
21. Haddad, L.H., Carr, L.D.: The nonlinear Dirac equation in Bose-Einstein condensates: foundation and symmetries. Physica D **238**(15), 1413–1421 (2009)
22. Haddad, L.H., Carr, L.D.: The nonlinear Dirac equation in Bose-Einstein condensates: II. Relativistic soliton stability analysis. New J. Phys. **17**, 063034, 22 pp. (2015)
23. Haddad, L.H., Weaver, C.M., Carr, L.D.: The nonlinear Dirac equation in Bose-Einstein condensates: I. Relativistic solitons in armchair nanoribbon optical lattices. New J. Phys. **17**, 063033, 23 pp. (2015)
24. Ivanenko, D.D.: Notes to the theory of interaction via particles. Zh.Éksp. Teor. Fiz. **8**, 260–266 (1938)
25. Krasnoselski, M.A.: Topological Methods in the Theory of Nonlinear Integral Equations. Pergamon Press, Oxford (1964)
26. Kryszewski, W., Szulkin, A.: Generalized linking theorem with an application to semilinear Schrödinger equation. Adv. Differ. Equ. **3**, 441–472 (1998)
27. Kusner, R., Schmitt, N.: Representation of surfaces in space. arXiv:dg-ga/9610005 (1996)

28. Lin, C.S.: The local isometric embedding in $\mathbb{R}^3$ of two-dimensional Riemannian manifolds with gaussian curvature changing sign cleanly. Commun. Pure Appl. Math. **39**(6), 867–887 (1986)
29. Mawhin, J., Willem, M.: Critical Point Theory and Hamiltonian Systems. Applied Mathematical Sciences, vol. 74. Springer, New York (1989)
30. Miller, D.G., Vitagliano, V., Sartorio, R.: Some comments on multicomponent diffsuion: negative main term diffusion coefficients, Second law constraints, Solvent choices, and Reference frame transformations. J. Phys. Chem. **90**, 1509–1519 (1986)
31. Murray, J.: Mathematical Biology, II: Spatial Models and Biomedical Applications. Springer, Berlin (2003)
32. Nash, J.: The imbedding problem for Riemannian manifolds. Ann. Math. **63**(1), 20–63 (1956)
33. Ng, W.K., Parwani, R.R.: Nonlinear dirac equations. Symm. Integr. Geom. Method. Appl. **5**, 023 (2009)
34. Onsager, L., Fuoss, R.M.: Irreversible processes in electrolytes. Diffusion, conductance and viscous flow in arbitrary mixtures of strong electrolytes. J. Phys. Chem. **36**, 2689–2778 (1932)
35. Ranada, A.F.: Classical nonlinear Dirac field models of extended particles. In: Barut, A.O. (ed.) Quantum Theory, Groups, Fields and Particles. Reidel, Amsterdam (1983)
36. Roth, J.: Spinorial characterizations of surfaces into 3-homogeneous manifolds. J. Geom. Phys. **60**, 1045–1061 (2010)
37. Spivak, M.D.: A Comprehensive Introduction to Differential Geometry. Publish or Perish, Houston (1970)
38. Struwe, M.: Variational Methods and Applications to Nonlinear Partial Differential Equations and Hamiltonian Systems. Springer, Berlin (1990)
39. Szulkin, A., Weth, T.: The method of Nehari manifold. In: Handbook of Nonconvex Analysis and Applications, pp. 597–632. Int. Press, Somerville, (2010)
40. Taimanov, I.A.: Surfaces of revolution in terms of solitons. Ann. Global Anal. Geom. **15**(5), 419–435 (1997)
41. Taimanov, I.A.: Modified Novikov-Veselov equation and differential geometry of surfaces. Am. Math. Soc. Transl. Ser. **179**(2), 133–151 (1997)
42. Taimanov, I.A.: The Weierstrass representation of closed surfaces in $\mathbb{R}^3$. Funct. Anal. Appl. **32**(4), 258–267 (1998)
43. Toyama, F.M., Hosono, Y., Ilyas, B., Nogami, Y.: Reduction of the nonlinear Dirac equation to a nonlinear Schrödinger equation with a correction term. J. Phys. A **27**(9), 3139–3148 (1994)
44. Vitagliano, V., Sartorio, R., Scala, S., Spaduzzi, D.: Diffusion in a ternary system and the critical mixing point. J. Solution Chem. 7(8), 605–622 (1978)

Chapter 3
Localized Energy Estimates for Strongly Indefinite Functionals

In this chapter we will prove some abstract results on the energy inequalities of strongly indefinite functionals on a Hilbert space E, whose norm and scalar product will be denoted, respectively, by $\|\cdot\|$ and $\langle\cdot,\cdot\rangle$. We anticipate that most of the results we will discuss could be carried out in a Banach spaces.

3.1 A Quick Review on Critical Point Theory

The classical Calculus of Variations deals with finding minima of functionals $\Phi : X \to \mathbb{R}$ that are bounded below. The basic idea of the direct method is to consider a minimizing sequence $\Phi(u_n) \to \inf \Phi$, to find a convergent subsequence $u_{n_k} \to u$, and to show that $\Phi(u) = \inf \Phi$. In order to make this work the space X should have a topology which is rather weak for the existence of a convergent subsequence, and rather strong so that Φ is lower semicontinuous. In many applications the functional is not bounded below and instead of a minimizer one is interested in saddle points. This is the concern of the Calculus of Variations in the Large or Critical Point Theory, which has undergone an enormous development in the last century due to the work of mathematicians like Morse, Lusternik, Schnirelman, Palais, Smale, Rabinowitz, Ambrosetti, Lions, Struwe, Witten, Floer and many others, with applications to problems from analysis, geometry and mathematical physics.

Motivated by several applications, for instance to finite- and infinite-dimensional Hamiltonian systems, nonlinear Schrödinger equations and nonlinear Dirac equations, we were led to consider C^1-functionals $\Phi : E \to \mathbb{R}$ defined on the product $E = X \oplus Y$ of Banach spaces X and Y with $\dim X = \infty$ and $\dim Y = \infty$, and

© The Author(s), under exclusive license to Springer Nature Singapore Pte Ltd. 2025
Y. Ding, T. Xu, *Localization Approaches in Strongly Indefinite Problems*, Series in Contemporary Mathematics 6, https://doi.org/10.1007/978-981-97-9539-0_3

we need to work with the weak topology on Y in order to gain compactness. The functionals typically have the form

$$\Phi(u) = \frac{1}{2}\left(\|x\|^2 - \|y\|^2\right) - \Psi(u) \quad \text{for } u = x + y \in X \oplus Y. \tag{3.1}$$

Since $\nabla\Psi : E \to E^*$ is not completely continuous and the Palais-Smale condition does not hold in general. It makes applications of Leray-Schauder degree type arguments rather subtle. On the other hand the functional $\Psi : E \to \mathbb{R}$ is weakly sequentially lower semicontinuous and $\nabla\Psi : E \to E^*$ is weakly sequentially continuous. It turns out that the product topology

$$\mathcal{T} = (\text{weak topology on } Y) \times (\text{norm topology on } X)$$

is well suited for certain arguments because $\Phi : (E, \mathcal{T}) \to \mathbb{R}$ is sequentially upper semicontinuous, and $\nabla\Phi : (E, \mathcal{T}) \to (E^*, \text{weak* topology})$ is continuous. Given a finite-dimensional subspace $F \subset X$ the unit ball of $F \oplus Y$ is $\mathcal{T}$-compact, and given a bounded sequence $(u_n = x_n + y_n)_n$ the Y-part $(y_n)_n$ $\mathcal{T}$-converges (up to a subsequence). When one wants to develop critical point theory with this topology on E one needs to construct deformations on E which are $\mathcal{T}$-continuous. Deformations are usually obtained by integrating vector fields which in turn are constructed with the help of partitions of unity. So one needs to construct these in a $\mathcal{T}$-Lipschitz continuous way. A more difficult situation occurs when one is interested in "normalized solutions", that is critical points of Φ constrained to the unit sphere $SE = \{u \in E : \|u\| = 1\}$ or to other finite-codimensional submanifolds X of E.

In this section we will outline some typical results in critical point theory for strongly indefinite problems, referring to [3, 8] for detailed proofs and more complete results. Let X, Y be Banach spaces with Y being separable and reflexive, and set $E = X \oplus Y$. By abuse of notation, we consistently write $\|\cdot\|$ for the norms on X, Y, E and their dual spaces. Let $\mathcal{S} \subset Y^*$ be a dense subset, $\langle\cdot,\cdot\rangle : Y \times Y^* \to \mathbb{R}$ be the duality pairing and $\mathcal{D} = \{d_s : s \in \mathcal{S}\}$ be the associated family of semi-metrics on $Y \cong Y^{**}$ defined by

$$d_s(y_1, y_2) = |\langle y_1 - y_2, s\rangle| \quad \text{for } y_1, y_2 \in Y.$$

We consider $\mathcal{P}$ be the family of semi-norms on E consisting of all semi-norms

$$p_s : E = X \oplus Y \to \mathbb{R}, \quad p_s(x + y) = \|x\| + |s(y)|, \qquad s \in \mathcal{S}.$$

Thus $\mathcal{P}$ induces the product topology on E given by the $\mathcal{D}$-topology on Y and the norm topology on X. We simply denote this topology as $\mathcal{T}_{\mathcal{P}}$. And evidently, $\mathcal{T}_{\mathcal{P}}$ is contained in the product of norm topology in X and the weak topology in Y. In particular, $u_n = x_n + y_n \xrightarrow{\mathcal{T}_{\mathcal{P}}} u = x + y$ provided that $x_n \to x$ and $y_n \rightharpoonup y$.

For $a, b \in \mathbb{R}$ we write $\Phi^b := \{u \in E : \Phi(u) \le b\}$, $\Phi_a := \{u \in E : \Phi(u) \ge a\}$, and $\Phi_a^b := \Phi_a \cap \Phi^b$. Denoted by $\mathcal{T}_w$ the weak-topology of E and $\mathcal{T}_{w^*}$ the weak*-topology of E^*, for a functional $\Phi \in C^1(E, \mathbb{R})$, our basic hypothesis is:

(Φ_0) $\Phi : (E, \mathcal{T}_\mathcal{P}) \to \mathbb{R}$ is upper semi-continuous, that is, Φ_a is $\mathcal{P}$-closed for every $a \in \mathbb{R}$; and $\nabla\Phi : (\Phi_a, \mathcal{T}_\mathcal{P}) \to (E^*, \mathcal{T}_{w^*})$ is continuous for every $a \in \mathbb{R}$.

In our applications (Φ_0) holds because the following result applies.

Proposition 3.1 (Bartsch and Ding [3]) *Consider a functional* $\Phi \in C^1(E, \mathbb{R})$ *of the form* (3.1) *such that*

(i) $\Psi \in C^1(E, \mathbb{R})$ *is bounded from below;*
(ii) $\Psi : (E, \mathcal{T}_w) \to \mathbb{R}$ *is sequentially lower semicontinuous, that is,* $u_n \rightharpoonup u$ *in* E *implies* $\Psi(u) \le \liminf \Psi(u_n)$*;*
(iii) $\nabla\Psi : (E, \mathcal{T}_w) \to (E^*, \mathcal{T}_{w^*})$ *is sequentially continuous.*
(iv) $\nu : E \to \mathbb{R}$, $\nu(u) = \|u\|^2$, *is* C^1 *and* $\nabla\nu : (E, \mathcal{T}_w) \to (E^*, \mathcal{T}_{w^*})$ *is sequentially continuous.*

Then Φ *satisfies* (Φ_0). *Moreover, for any countable dense subset* $\mathcal{S}_0 \subset \mathcal{S}$, Φ *satisfies* (Φ_0) *with* $\mathcal{P}$ *replaced by* $\mathcal{P}_0 := \{p_s \in \mathcal{P} : s \in \mathcal{S}_0\}$ *and* $\mathcal{T}_\mathcal{P}$ *by* $\mathcal{T}_{\mathcal{P}_0}$ *associated to* $\mathcal{P}_0$.

The following theorem is a $\mathcal{P}$-continuous version of the non-critical interval theorem in critical point theory.

Theorem 3.2 (Bartsch and Ding [3]) *Consider* $a, b \in \mathbb{R}$ *with* $a < b$ *so that* Φ_a *is* $\mathcal{P}$*-closed and* $\nabla\Phi : (\Phi_a^b, \mathcal{T}_\mathcal{P}) \to (E^*, \mathcal{T}_{w^*})$ *is continuous. Suppose moreover that*

$$\alpha := \inf\{\|\nabla\Phi(u)\| : u \in \Phi_a^b\} > 0 \tag{3.2}$$

Then there exists a deformation $\eta : [0, 1] \times \Phi^b \to \Phi^b$ *with the properties:*

(i) η *is continuous with either the* $\mathcal{P}$*-topology or the norm topology on* Φ^b*;*
(ii) for each t *the map* $u \mapsto \eta(t, u)$ *is a homeomorphism of* Φ^b *onto* $\eta(t, \Phi^b)$ *with the* $\mathcal{P}$*-topology or with the norm topology;*
(iii) $\eta(0, u) = u$ *for all* $u \in \Phi^b$*;*
(iv) $\eta(t, \Phi^c) \subset \Phi^c$ *for all* $c \in [a, b]$ *and all* $t \in [0, 1]$*;*
(v) $\eta(1, \Phi^b) \subset \Phi^a$*;*
(vi) each point $u \in \Phi^b$ *has a* $\mathcal{P}$*-neighborhood* U *in* Φ^b *so that the set* $\{v - \eta(t, v) : v \in U, 0 \le t \le 1\}$ *is contained in a finite-dimensional subspace of* E.
(vii) if a finite group G *acts isometrically on* E *and if* Φ *is* G*-invariant, then* η *is equivariant in* u.

By $P^X : E = X \oplus Y \to X$ we denote the continuous projection onto X along Y, and by $P^Y := I - P^X : E \to Y$ the complementary projection.

Theorem 3.3 (Bartsch and Ding [3]) *Consider $a, b \in \mathbb{R}$ with $a < b$ so that Φ_a is $\mathcal{P}$-closed and $\nabla\Phi : (\Phi_a^b, \mathcal{T}_\mathcal{P}) \to (E^*, \mathcal{T}_{w^*})$ is continuous. Suppose moreover that*

$$\alpha := \inf\{(1 + \|u\|)\|\nabla\Phi(u)\| : u \in \Phi_a^b\} > 0 \tag{3.3}$$

and

$$\text{there exists } \gamma > 0 \text{ with } \|u\| < \gamma\|P^X u\| \text{ for all } u \in \Phi_a^b. \tag{3.4}$$

Then there exists a deformation $\eta : [0, 1] \times \Phi^b \to \Phi^b$ with the properties (i)–(vii) from Theorem 3.2.

Next we introduce a infinite-dimensional version of linking. It is well-known that linking is essentially a finite-dimensional concept depending on degree theory or methods from algebraic topology. Here we extend it in a rather general and simple way. We need some notations. Given a subset $A \subset Z$ of a locally convex topological vector space we write $L(A) := \overline{\text{span}(A)}$ for the smallest closed linear subspace containing A, and we write ∂A for the boundary of A in $L(A)$. For a linear subspace $F \subset Z$ we set $A_F := A \cap F$. Finally let $I = [0, 1]$.

Definition 3.4 (Bartsch and Ding [3]) Given two subsets $Q, S \subset Z$ with $S \cap \partial Q = \emptyset$, we say that Q finitely links with S if for any finite-dimensional linear subspace $F \subset Z$ with $F \cap S \neq \emptyset$, and any continuous deformation $h : I \times Q_F \to F + L(S)$ with $h(0, u) = u$ for all u, and $h(I \times \partial Q_F) \cap S = \emptyset$ there holds $h(t, Q_F) \cap S \neq \emptyset$ for all $t \in I$.

Example 3.5 We present three examples of finite linking. In all cases the proof of the finite linking property is not difficult and based on a Brouwer degree argument.

(a) Given an open subset $\mathcal{O} \subset Z$, $u_0 \in \mathcal{O}$, and $u_1 \in Z \setminus \overline{\mathcal{O}}$, then $Q = \{tu_1 + (1 - t)u_0 : t \in I\}$ finitely links with $S := \partial\mathcal{O}$.
(b) Suppose Z is the topological sum $Z = Z_1 \oplus Z_2$ of two linear subspaces, $\mathcal{O} \subset Z_1$ is open and $u_0 \in \mathcal{O}$. Then $Q = \overline{\mathcal{O}}$ finitely links with $S = \{u_0\} \times Z_2$.
(c) Given $Z = Z_1 \oplus Z_2$ as in b), two open subsets $\mathcal{O}_1 \subset Z_1$, $\mathcal{O}_2 \subset Z_2$, and $u_1 \in \mathcal{O}_1$, $u_2 \in Z_2 \setminus \overline{\mathcal{O}_2}$. Then $Q = \overline{\mathcal{O}_1} \times \{tu_2 : t \in I\}$ finitely links with $S = \{u_1\} \times \partial\mathcal{O}_2$.

Now we come back to our functional $\Phi : E \to \mathbb{R}$. If $Q \subset E$ finitely links with $S \subset E$ we set

$$\Gamma_{Q,S} := \{h \in C(I \times Q, E) : h \text{ satisfies } (h_1) - (h_5)\}$$

where

(h_1) $h : I \times (Q, \mathcal{T}_\mathcal{P}) \to (E, \mathcal{T}_\mathcal{P})$ is continuous;
(h_2) $h(0, u) = u$ for all $u \in Q$;
(h_3) $\Phi(h(t, u)) \leq \Phi(u)$ for all $t \in I$, $u \in Q$;
(h_4) $h(I \times \partial Q) \cap S = \emptyset$

(h_5) each $(t, u) \in I \times Q$ has a $\mathcal{P}$-open neighborhood W such that the set $\{v - h(s, v) : (s, v) \in W \cap (I \times Q)\}$ is contained in a finite-dimensional subspace of E.

Recall that $(u_n)_n$ is a $(PS)_c$-sequence if $\Phi(u_n) \to c$ and $\|\nabla\Phi(u_n)\| \to 0$ as $n \to \infty$. We say that $(u_n)_n$ is a $(C)_c$-sequence if $\Phi(u_n) \to c$ and $(1+\|u_n\|)\nabla\Phi(u_n) \to 0$ as $n \to \infty$. Then our critical point theorem reads as

Theorem 3.6 (Bartsch and Ding [3]) *Suppose Φ satisfies (Φ_0) with $\mathcal{P}$ countable, and let $Q, S \subset E$ be such that Q is $\mathcal{P}$-compact and Q finitely links with S. If $\sup\Phi(\partial Q) \le \inf\Phi(S)$ then there exists a $(PS)_c$-sequence for*

$$c := \inf_{h\in\Gamma_{Q,S}} \sup_{u\in Q} \Phi(h(1, u)) \in [\inf\Phi(S), \sup\Phi(Q)].$$

If $c = \inf\Phi(S)$ and if for all $\delta > 0$ the set $S^\delta := \{u \in E : dist_{\|\cdot\|}(u, S) \le \delta\}$ is $\mathcal{P}$-closed then there exists a $(PS)_c$-sequence $(u_n)_n$ with $u_n \to S$ in norm.

Proof The inequality $c \le \sup\Phi(Q)$ is obvious. In order to see $c \ge \inf\Phi(S)$ we first observe that $h(I \times \partial Q) \cap S = \emptyset$ for every $h \in \Gamma_{Q,S}$ by (h_3). Since Q is $\mathcal{P}$-compact there exists a finite-dimensional subspace F containing $\{u-h(t, u) : (t, u) \in I\times Q\}$. Consequently $h(I \times Q_F) \subset F$. Since Q finitely links with S there exists $u \in Q$ with $h(1, u) \in S$ which implies $\sup_{u\in Q}\Phi(h(1, u)) \ge \inf\Phi(S)$ as claimed.

Assume that $\|\nabla\Phi(u)\| \ge \alpha$ for all $u \in \Phi_{c-\varepsilon}^{c+\varepsilon}$, some $\alpha, \varepsilon > 0$. We can take η to be the deformation from Theorem 3.2 for $a := c - \varepsilon$, $b := c + \varepsilon$. Now we choose $h \in \Gamma_{Q,S}$ with $\sup\Phi(h(1, Q)) < c + \varepsilon$ and define $g : I \times Q \to E$ by $g(t, u) := \eta(t, h(t, u))$. Then $g(0, u) = u$ for all u and g satisfies $(h_1) - (h_4)$. Moreover, (h_5) follows from the equality $u - g(t, u) = (u - h(t, u)) + (h(t, u) - \eta(t, h(t, u)))$. Thus $g \in \Gamma_{Q,S}$ which leads to the contradiction $c \le \sup_{u\in Q}\Phi(g(1, u)) \le c - \varepsilon$.

We have seen that there exists a $(PS)_c$-sequence. Now suppose $c = \inf\Phi(S)$. If there does not exist a $(PS)_c$-sequence converging to S in norm then there exist $\varepsilon > 0$, $\delta > 0$, and $\alpha > 0$ so that $\|\nabla\Phi(u)\| \ge \alpha$ for all $u \in S^\delta \cap \Phi_{c-\varepsilon}^{c+\varepsilon}$. For such u we choose $w(u) \in E$ with $\|w(u)\| \le 2$ and $\nabla\Phi(u)[w(u)] > \|\nabla\Phi(u)\|$. We then choose a $\mathcal{P}$-open neighborhood $N(u)$ of u so that $\nabla\Phi(v)[w(u)] > \|\nabla\Phi(u)\| \ge \alpha$ for all $v \in N(u) \cap \Phi_{c-\varepsilon}$. For $u \in \Phi_{c-\varepsilon}^{c+\varepsilon} \setminus S^\delta$ we put $N(u) := E \setminus S^\delta$ and $w(u) := 0$. Lastly, for $u \in E \setminus \Phi_{c-\varepsilon}$ we set $N(u) := E \setminus \Phi_{c-\varepsilon}$ and $w(u) := 0$. Then $W := \bigcup_{u\in\Phi^{c+\varepsilon}} N(u)$ is $\mathcal{P}$-open. Let $\{U_j : j \in J\}$ be a $\mathcal{P}$-locally finite $\mathcal{P}$-open refinement of $\{N(u) : u \in \Phi^{c+\varepsilon}\}$ and $\{\pi_j : j \in J\}$ a $\mathcal{P}$-locally $\mathcal{P}$-Lipschitz partition of unity subordinated to the covering $\{U_j : j \in J\}$. For $j \in J$ with $U_j \cap S^\delta \ne \emptyset$ we choose u_j with $U_j \subset N(u_j)$ and define $w_j := w(u_j)$. For $j \in J$ with $U_j \cap S^\delta = \emptyset$ we set $w_j := 0$. Then the vector field $f(u) := -\sum_{j\in J}\pi_j(u)w_j$ satisfies $\|f(u)\| \le 2$ for all $u \in W$, and it satisfies $\nabla\Phi(u)[f(u)] \ge \alpha$ for all $u \in S^\delta$. Let us denote $\varphi^t(u)$ the associated flow as in [8, Chapter 3].

By construction we have that, if $u \in \Phi^{c+\alpha\delta/4} \setminus S^{\delta/2}$ then $\varphi^t(u) \notin S$ for all $t \ge 0$. Moreover, if $u \in S^{\delta/2}$ we have $\varphi^t(u) \in S^\delta$ for $0 \le t \le \delta/4$, hence $\Phi(\varphi^{\delta/2}(u)) \le \Phi(u) - \alpha\delta/4 < c$. Now take any $h \in \Gamma_{Q,S}$ with $\sup\Phi(h(1, u)) \le c + \alpha\delta/4$. Our considerations yield $\varphi^{\delta/2}(h(1,u)) \notin S$ for all $u \in Q$. This contradicts the linking condition on Q and S because $\varphi^{\delta/2} \circ h(t,\,.) \in \Gamma_{Q,S}$. □

Similarly, the finitely linking yields also a $(C)_c$-sequence. We need the additional assumption:

(Φ_+) there exists $\zeta > 0$ such that $\|u\| < \zeta\|P^X u\|$ for all $u \in \Phi_0$.

Theorem 3.7 *Suppose Φ satisfies (Φ_0) and (Φ_+). Let $Q, S \subset E$ be such that Q is $\mathcal{P}$-compact and Q finitely links with S. If $\kappa := \inf \Phi(S) > 0$ and $\sup \Phi(\partial Q) \le \kappa$ then Φ has a $(C)_c$-sequence with $\kappa \le c \le \sup \Phi(Q)$.*

Proof Repeating the arguments of the first two paragraphs of the proof of Theorem 3.6 with the application of Theorem 3.2 replaced by Theorem 3.3 yields the desired conclusion. □

As a corollary of Theorems 3.6 we obtain an improvement of a very useful critical point theorem of Kryszewski and Szulkin [11].

Theorem 3.8 (Bartsch and Ding [3]) *Consider a functional $\Phi : E \to \mathbb{R}$ satisfying (Φ_0) with $\mathcal{P}$ countable, e.g. Φ is as in Proposition 3.1. Suppose there exist $R > r > 0$ and $e \in X$, $\|e\| = 1$ such that we have for $S := \{u \in X : \|u\| = r\}$, $Q = \{v+te \in E : v \in Y,\ \|v\| < R,\ 0 < t < R\}$: $\inf \Phi(S) \ge \Phi(0) \ge \sup \Phi(\partial Q)$. Then there exists a $(PS)_c$-sequence for*

$$c := \inf_{h\in\Gamma_{Q,S}} \sup_{u\in Q} \Phi(h(1,u)) \in [\inf \Phi(S), \sup \Phi(Q)].$$

If $c = \inf \Phi(S)$ then there exists a $(PS)_c$-sequence $(u_n)_n$ with $u_n \to S$ in norm.

The original theorem in [11] deals with the case where E is a Hilbert space, Φ is as in Proposition 3.1, and $\inf \Phi(S) > \Phi(0) \ge \sup \Phi(\partial Q)$. The additional information on the Palais-Smale sequence in the case $c = \inf \Phi(S)$ has not been obtained in [11]. It is however important in applications when $c = \Phi(0)$ in order to construct a nontrivial critical point. If the stronger hypothesis $\inf \Phi(S) > \Phi(0)$ holds then $c > \Phi(0)$. This is sufficient to deduce the existence of a nontrivial critical point.

We have also the following consequence of Theorem 3.7.

Theorem 3.9 *Let Φ satisfy (Φ_0) and (Φ_+), and suppose there exist $R > r > 0$ and $e \in X$, $\|e\| = 1$ such that for $S := \{u \in X : \|u\| = r\}$, $Q = \{v + te \in E : v \in Y,\ \|v\| < R, 0 < t < R\}$ we have $\kappa := \inf \Phi(S) > 0$ and $\sup \Phi(\partial Q) \le \kappa$ then Φ has a $(C)_c$-sequence with $\kappa \le c \le \sup \Phi(Q)$.*

3.2 A First Glimpse on the Localized Energy Inequality

In this and in the subsequent section we will discuss some localized energy inequalities at ground state critical levels of a class of strongly indefinite functionals. The specific feature of these functionals is that they are perturbative in nature.

In what follows, on a Hilbert space E, we shall denote by $\mathscr{L}(E) = \mathscr{L}(E, E)$ the space of bounded linear maps from E to E, endowed with the uniform operator norm, and by $\mathscr{L}_s(E)$ the same space endowed with the strong operator topology. The adjoint of $A \in \mathscr{L}(E)$ is denoted by A^*, and by self-adjointness we mean $A = A^*$. Let G be a Lie group and let $\mathscr{T} : G \to U(E)$ be a representation of G on the group of the unitary linear transformations on E. We set $\mathscr{G} = \mathscr{T}(G)$ and sometimes, when no ambiguity is possible, we will use $\mathscr{G}$ instead of G to stand for the Lie group.

Definition 3.10 A subset $M \subset E$ is called $\mathscr{G}$-invariant if $g(M) = M$ for every $g \in \mathscr{G}$. A functional Φ on E is called $\mathscr{G}$-invariant if $\Phi \circ g = \Phi$ for every $g \in \mathscr{G}$. A map h from E to E is called $\mathscr{G}$-equivariant if $h \circ g = g \circ h$ for every $g \in \mathscr{G}$.

Let $\{A_\varepsilon\}_{\varepsilon>0} \subset \mathscr{L}(E)$ be a family of $\mathscr{G}$-equivariant self-adjoint operators. Let $\{\Psi_\varepsilon\}_{\varepsilon>0} \subset C^2(E, \mathbb{R})$ be a $\mathscr{G}$-invariant family, and set $N_\varepsilon := \nabla\Psi_\varepsilon : E \to E$. Considering a given splitting $E = X \oplus Y$ of E into $\mathscr{G}$-invariant orthogonal subspaces X and Y with associated bounded projections P^X and P^Y, we write $z^X := P^X z$ and $z^Y := P^Y z$ for $z \in E$. Then we are interested in localizing critical points of the functionals

$$\Phi_\varepsilon : E \to \mathbb{R}, \quad \Phi_\varepsilon(z) := \frac{1}{2}\left(\|z^X\|^2 - \|z^Y\|^2\right) + \frac{1}{2}\langle A_\varepsilon z, z\rangle - \Psi_\varepsilon(z)$$

for ε small. Let $A_0 \in \mathscr{L}(E)$ be a $\mathscr{G}$-equivariant self-adjoint operator and Ψ_0 be a $\mathscr{G}$-invariant C^2 functional, and set $N_0 := \nabla\Psi_0 : E \to E$. Consider

$$\Phi_0 : E \to \mathbb{R}, \quad \Phi_0(z) := \frac{1}{2}\left(\|z^X\|^2 - \|z^Y\|^2\right) + \frac{1}{2}\langle A_0 z, z\rangle - \Psi_0(z),$$

as singular limit functional.

Since we are interested in the situation when ε is small, by setting $\mathcal{E} = [0, 1]$, we will be concerned with $\{\Phi_\varepsilon\}_{\varepsilon\in\mathcal{E}} := \{\Phi_0\} \cup \{\Phi_\varepsilon\}_{\varepsilon\in(0,1]}$. Now we collect some hypotheses on $\{\Phi_\varepsilon\}_{\varepsilon\in\mathcal{E}}$ which we will impose in the various results:

- $(A1)$ There exists $\theta \in (0, 1)$ such that $\sup_{\varepsilon\in(0,1]} \|A_\varepsilon\| \le \theta$.
- $(A2)$ $A_\varepsilon \to A_0$ in $\mathscr{L}_s(E)$ as $\varepsilon \to 0$.
- $(N1)$ For each $\varepsilon \in \mathcal{E}$, Ψ_ε is non-negative and convex, and $N_\varepsilon : (E, \mathcal{T}_w) \to (E, \mathcal{T}_w)$ is sequentially continuous.
- $(N2)$ For each $z \in E$, $N_\varepsilon(z) \to N_0(z)$ in E as $\varepsilon \to 0$.
- $(N3)$ There exists $\kappa \in C(\mathbb{R}^+, \mathbb{R}^+)$ ($\mathbb{R}^+ = [0, \infty)$), independent of ε, such that

$$\left|\nabla^2\Psi_\varepsilon(z)[v, w]\right| \le \kappa(\|z\|) \cdot \|v\| \cdot \|w\|$$

for $z, v, w \in E$ and $\varepsilon \in \mathcal{E}$.

$(N4)$ For all $\varepsilon \in \mathcal{E}$ and $z \in E \setminus \{0\}$, $\widehat{\Psi}_\varepsilon(z) := \frac{1}{2}\nabla\Psi_\varepsilon(z)[z] - \Psi_\varepsilon(z) > 0$, and $\widehat{\Psi}_\varepsilon : (E, \mathcal{T}_w) \to \mathbb{R}$ is sequentially lower semi-continuous.

$(N5)$ Given arbitrarily $\varepsilon \in \mathcal{E}$, for every $z \in E \setminus \{0\}$ and $w \in E$, it holds that

$$\nabla^2\Psi_\varepsilon(z)[z + w, z + w] - 2\nabla\Psi_\varepsilon(z)[w] > \nabla\Psi_\varepsilon(z)[z].$$

$(I1)$ there exist $\rho, \tau > 0$, both independent of $\varepsilon \in \mathcal{E}$, such that $\Phi_\varepsilon\big|_{B_\rho^X} \geq 0$ and $\Phi_\varepsilon\big|_{S_\rho^X} \geq \tau$ where $B_\rho^X := B_\rho \cap X = \{z \in X : \|z\| \leq \rho\}$ and $S_\rho^X := \partial B_\rho^X = \{z \in X : \|z\| = \rho\}$.

$(I2)$ for any $e \in X \setminus \{0\}$ set $E_e = \mathbb{R}^+ e \oplus Y$, either $\sup_{z \in E_e} \Phi_0(z) = +\infty$ or $\Phi_0(z) \to -\infty$ as $z \in E_e$ and $\|z\| \to \infty$ (here $\mathbb{R}^+ = [0, \infty)$).

Remark 3.11 Condition $(A2)$ and $(N2)$ are quite natural in singular perturbation problems. Both are satisfied in the scaled equations when the parameter varies. Condition $(N3)$ and $(N4)$ are rather harmless, also $(N5)$ holds under rather general assumptions on the nonlinearity. They are technical assumptions which are deeply related to the proof of our results, and they will be easily checked if growth conditions on the nonlinearity are given.

Definition 3.12 A $\mathscr{G}$-invariant functional $\Phi \in C^1(E, \mathbb{R})$ is said to satisfy the $\mathscr{G}$-weak $(C)_c$-condition (or $(PS)_c$-condition) if for each $(C)_c$-sequence (or $(PS)_c$-sequence) $\{z_n\}$ there exists correspondingly $\{g_n\} \subset \mathscr{G}$ such that $\{g_n z_n\}$ possesses a subsequence converge weakly to a point in $E \setminus \{0\}$.

3.2.1 An Infinite-Dimensional Reduction

Observe that $\Psi_\varepsilon(z) \geq 0$ for all $z \in E$ and that $(N1)$ yields

$$\nabla^2\Psi_\varepsilon(z)[w, w] \geq 0 \quad \text{for any } w \in E.$$

This can be seen by the fact $\Psi_\varepsilon \in C^2(E, \mathbb{R})$ and Ψ_ε is convex for each $\varepsilon \in \mathcal{E}$. Note that by $(I1)$ we have $\Phi_\varepsilon(0) \geq 0$, which implies $\Psi_\varepsilon(0) = 0$. And condition $(N3)$ yields

$$\Psi_\varepsilon(z) = \int_0^1 \int_0^t \nabla^2\Psi_\varepsilon(sz)[z, z]\, ds dt \leq C(\kappa, \|z\|)\|z\|^2, \quad \forall z \in E \tag{3.5}$$

where $C(\kappa, \|z\|) > 0$ is a constant depending only on the function κ and $\|z\|$.

Consider $\varepsilon \in \mathcal{E}$ being fixed, we define the nonlinear functional $\phi_v : Y \to \mathbb{R}$ by

$$\phi_v(w) = \Phi_\varepsilon(v + w) \quad \text{for } v \in X.$$

Notice that $(A1)$ and $(A2)$ implies $\sup_{\varepsilon\in\mathcal{E}} \|A_\varepsilon\| \le \theta < 1$, we infer

$$\phi_v(w) \le \frac{1+\theta}{2}\|v\|^2 - \frac{1-\theta}{2}\|w\|^2. \tag{3.6}$$

Moreover, by taking derivatives, we have

$$\begin{aligned}\nabla^2\phi_v(w)[z,z] &= -\|z\|^2 + \langle A_\varepsilon z, z\rangle - \nabla^2\Psi_\varepsilon(v+w)[z,z] \\ &\le -(1-\theta)\|z\|^2\end{aligned} \tag{3.7}$$

for any $w, z \in Y$.

As a consequence of (3.6) and (3.7), ϕ_v is strictly concave and $\phi_v(w) \to -\infty$ as $\|w\| \to \infty$. It follows from the weak upper semi-continuity of ϕ_v that there is unique strict maximum point $h_\varepsilon(v)$ for ϕ_v, which we can easily confirm to be the only critical point of ϕ_v on Y. The uniquely defined map $h_\varepsilon : X \to Y$ can be seen as a reduction of Φ_ε on X satisfying

$$\Phi_\varepsilon\big(v+h_\varepsilon(v)\big) = \phi_v\big(h_\varepsilon(v)\big) = \max_{w\in Y}\phi_v(w) = \max_{w\in Y}\Phi_\varepsilon(v+w). \tag{3.8}$$

It follows from (3.8)

$$\begin{aligned}0 &\le \Phi_\varepsilon\big(v+h_\varepsilon(v)\big) - \Phi_\varepsilon(v) \\ &= -\frac{1}{2}\|h_\varepsilon(v)\|^2 + \frac{1}{2}\langle A_\varepsilon(v+h_\varepsilon(v)), v+h_\varepsilon(v)\rangle - \Psi_\varepsilon\big(v+h_\varepsilon(v)\big) \\ &\quad - \frac{1}{2}\langle A_\varepsilon v, v\rangle + \Psi_\varepsilon(v) \\ &\le -\frac{1}{2}\|h_\varepsilon(v)\|^2 + \frac{\theta}{2}\|h_\varepsilon(v)\|^2 + \frac{\theta}{2}\|v\|^2 + \frac{\theta}{2}\|v\|^2 + \Psi_\varepsilon(v)\end{aligned}$$

for $v \in X$. Hence

$$\|h_\varepsilon(v)\|^2 \le \frac{2\theta}{1-\theta}\|v\|^2 + \frac{2}{1-\theta}\Psi_\varepsilon(v)$$

and the boundedness of Ψ_ε (see (3.5)) implies that of h_ε. If $v \in X$ and $g \in \mathcal{G}$, we have by invariance of Φ_ε and by (3.8)

$$\begin{aligned}\Phi_\varepsilon\big(gv+h_\varepsilon(gv)\big) &= \Phi_\varepsilon\big(v+g^{-1}h_\varepsilon(gv)\big) \le \Phi_\varepsilon\big(v+h_\varepsilon(v)\big) \\ &= \Phi_\varepsilon\big(gv+gh_\varepsilon(v)\big) \le \Phi_\varepsilon\big(gv+h_\varepsilon(gv)\big).\end{aligned}$$

Therefore, we conclude

$$\Phi_\varepsilon\big(gv+gh_\varepsilon(v)\big) = \Phi_\varepsilon\big(gv+h_\varepsilon(gv)\big)$$

which together with (3.8) implies that $g \circ h_\varepsilon = h_\varepsilon \circ g$, i.e. h_ε is $\mathscr{G}$-equivariant.

Next we define $\pi : X \times Y \to Y$ by

$$\pi(v, w) = P^Y \circ \nabla \Phi_\varepsilon(v + w)$$

where we have identified E^* with E via the isomorphism induced from the Riesz representation theorem. Observe that, for every $v \in X$, we derive from the definition of h_ε:

$$0 = \nabla \phi_v\big(h_\varepsilon(v)\big)[w] = \nabla \Phi_\varepsilon\big(v + h_\varepsilon(v)\big)[w] \quad \text{for any } w \in Y.$$

This implies

$$\pi\big(v, h_\varepsilon(v)\big) = 0 \quad \text{for } v \in X. \tag{3.9}$$

Notice that $\partial_w \pi(v, w) = P^Y \circ \nabla^2 \Phi_\varepsilon(v + w)\big|_Y$ is a bounded linear operator on Y. And from (3.7), we infer $\partial_w \pi(v, w)$ is an isomorphism with

$$\big\| \partial_w \pi(v, w)^{-1} \big\| \le \frac{1}{1 - \theta} \quad \forall v \in X. \tag{3.10}$$

Therefore (3.9) and (3.10) together with the implicit function theorem yield the uniquely defined map $h_\varepsilon : X \to Y$ is C^1 smooth with

$$\nabla h_\varepsilon(v) = -\partial_w \pi\big(v, h_\varepsilon(v)\big)^{-1} \circ \partial_v \pi\big(v, h_\varepsilon(v)\big), \quad \forall v \in X$$

where $\partial_v \pi(v, w) = P^Y \circ \nabla^2 \Phi_\varepsilon(v + w)\big|_X$.

Now set

$$I_\varepsilon = \Phi_\varepsilon^{red} : X \to \mathbb{R}, \quad I_\varepsilon(v) = \Phi_\varepsilon\big(v + h_\varepsilon(v)\big),$$

where justifies the reduction process. We have $I_\varepsilon \in C^1(X, \mathbb{R})$ is $\mathscr{G}$-invariant. And we can conclude from the above arguments that:

Proposition 3.13 *Suppose* $(A1)$–$(A2)$, $(N1)$ *and* $(N3)$ *are satisfied. Then* $I_\varepsilon \in C^1(X, \mathbb{R})$ *for each* $\varepsilon \in \mathcal{E}$ *and critical points of* I_ε *and* Φ_ε *are in one-to-one correspondence via the injective map* $v \mapsto v + h_\varepsilon(v)$ *from* X *to* E*. Moreover, if* $\{v_n\} \subset X$ *is a* $(C)_c$*-sequence (or* $(PS)_c$*-sequence) of* I_ε*, then* $\{v_n + h_\varepsilon(v_n)\}$ *is a* $(C)_c$*-sequence (or* $(PS)_c$*-sequence) of* Φ_ε.

Remark 3.14 The second part of the above proposition may seen to be not so obvious, however, by taking the derivative of I_ε we have

$$\begin{aligned} \nabla I_\varepsilon(v)[w] &= \nabla \Phi_\varepsilon\big(v + h_\varepsilon(v)\big)\big[w + \nabla h_\varepsilon(v)[w]\big] \\ &= \nabla \Phi_\varepsilon\big(v + h_\varepsilon(v)\big)[w + y] \end{aligned}$$

for all $v, w \in X$ and $y \in Y$. And hence $\|\nabla I_\varepsilon(v)\|_{X^*} = \big\|\nabla\Phi_\varepsilon\big(v + h_\varepsilon(v)\big)\big\|_{E^*}$ which implies nothing but the second conclusion in Proposition 3.13. Let us mention here that a reduction of a strongly indefinite functional to a functional on X is well known under strong differentiability condition, see for example [12, 13]. In [12, 13] a reduction in two steps has been performed: first to X and then to a Nehari manifold on X. However, since we are interested in the geometric situation $(I2)$, the so-called Nehari manifold is not necessarily defined for all directions in X, this context requires somewhat different arguments.

3.2.2 Asymptotic Approximation at the Ground State Level

In order to state our next results, we shall give another observation. Due to the fact $h_\varepsilon(v)$ is the unique critical point of ϕ_v on Y that, by setting $z = w - h_\varepsilon(v)$ for $w \in Y$ and $l(t) := \Phi_\varepsilon\big(v + h_\varepsilon(v) + tz\big)$, we have $l(1) = \Phi_\varepsilon(v + w)$, $l(0) = \Phi_\varepsilon\big(v + h_\varepsilon(v)\big)$ and $l'(0) = 0$. So, by $l(1) - l(0) = \int_0^1 (1 - s)l''(s)\,ds$, we deduce

$$
\begin{aligned}
&\Phi_\varepsilon(v + w) - \Phi_\varepsilon\big(v + h_\varepsilon(v)\big)\\
&= \int_0^1 (1 - s)\nabla^2\Phi_\varepsilon\big(v + h_\varepsilon(v) + sz\big)[z, z]\,ds\\
&= -\int_0^1 (1 - s)\big(\|z\|^2 - \langle A_\varepsilon z, z\rangle\big)\,ds - \int_0^1 (1 - s)\nabla^2\Psi_\varepsilon\big(v + h_\varepsilon(v) + sz\big)[z, z]\,ds.
\end{aligned}
$$

Consequently, we have

$$
\begin{aligned}
&\Phi_\varepsilon\big(v + h_\varepsilon(v)\big) - \Phi_\varepsilon(v + w)\\
&= \frac{1}{2}\|z\|^2 - \frac{1}{2}\langle A_\varepsilon z, z\rangle + \int_0^1 (1 - s)\nabla^2\Psi_\varepsilon\big(v + h_\varepsilon(v) + sz\big)[z, z]\,ds
\end{aligned}
\tag{3.11}
$$

for all $v \in X$ and $w \in Y$.

Lemma 3.15 *Suppose* (A1)–(A2) *and* (N1)–(N3). *Then* $h_\varepsilon(v) \to h_0(v)$ *in* Y *as* $\varepsilon \to 0$ *for* $v \in X$.

Proof For ease of notations, set $z_\varepsilon = v + h_\varepsilon(v)$, $w = v + h_0(v)$ and $v_\varepsilon = z_\varepsilon - w$. It sufficient to show $\|v_\varepsilon\| \to 0$ as $\varepsilon \to 0$.

Taking into account that

$$
\Phi_\varepsilon(z) = \Phi_0(z) + \frac{1}{2}\langle (A_\varepsilon - A_0)z, z\rangle - \big(\Psi_\varepsilon(z) - \Psi_0(z)\big), \quad \forall z \in E
$$

we infer

$$\begin{aligned}&\big(\Phi_\varepsilon(z_\varepsilon)-\Phi_\varepsilon(w)\big)+\big(\Phi_0(w)-\Phi_0(z_\varepsilon)\big)\\&=\frac{1}{2}\left\langle (A_\varepsilon-A_0)z_\varepsilon, z_\varepsilon\right\rangle-\frac{1}{2}\left\langle (A_\varepsilon-A_0)w, w\right\rangle+\big(\Psi_0(z_\varepsilon)-\Psi_0(w)\big)\\&-\big(\Psi_\varepsilon(z_\varepsilon)-\Psi_\varepsilon(w)\big).\end{aligned} \tag{3.12}$$

Remark that

$$\Psi_0(z_\varepsilon)-\Psi_0(w)=\nabla\Psi_0(w)[v_\varepsilon]+\int_0^1(1-s)\nabla^2\Psi_0(w+sv_\varepsilon)[v_\varepsilon, v_\varepsilon]\,ds, \tag{3.13}$$

$$\Psi_\varepsilon(z_\varepsilon)-\Psi_\varepsilon(w)=\nabla\Psi_\varepsilon(w)[v_\varepsilon]+\int_0^1(1-s)\nabla^2\Psi_\varepsilon(w+sv_\varepsilon)[v_\varepsilon, v_\varepsilon]\,ds, \tag{3.14}$$

and by (3.11)

$$\begin{aligned}&\Phi_\varepsilon(z_\varepsilon)-\Phi_\varepsilon(w)\\&=\frac{1}{2}\|v_\varepsilon\|^2-\frac{1}{2}\left\langle A_\varepsilon v_\varepsilon, v_\varepsilon\right\rangle+\int_0^1(1-s)\nabla^2\Psi_\varepsilon\big(z_\varepsilon-sv_\varepsilon\big)[v_\varepsilon, v_\varepsilon]\,ds,\end{aligned} \tag{3.15}$$

$$\begin{aligned}&\Phi_0(w)-\Phi_0(z_\varepsilon)\\&=\frac{1}{2}\|v_\varepsilon\|^2-\frac{1}{2}\left\langle A_0 v_\varepsilon, v_\varepsilon\right\rangle+\int_0^1(1-s)\nabla^2\Psi_0\big(w+sv_\varepsilon\big)[v_\varepsilon, v_\varepsilon]\,ds.\end{aligned} \tag{3.16}$$

We derived from (3.12)–(3.16) and the fact Ψ_ε is convex for all $\varepsilon\in\mathcal{E}$ that

$$\begin{aligned}&\|v_\varepsilon\|^2-\frac{1}{2}\left\langle (A_\varepsilon+A_0)v_\varepsilon, v_\varepsilon\right\rangle\\&\le\frac{1}{2}\left\langle (A_\varepsilon-A_0)z_\varepsilon, z_\varepsilon\right\rangle-\frac{1}{2}\left\langle (A_\varepsilon-A_0)w, w\right\rangle+\nabla\Psi_0(w)[v_\varepsilon]-\nabla\Psi_\varepsilon(w)[v_\varepsilon]\\&=\frac{1}{2}\left\langle (A_\varepsilon-A_0)v_\varepsilon, v_\varepsilon\right\rangle+\left\langle (A_\varepsilon-A_0)w, v_\varepsilon\right\rangle+\nabla\Psi_0(w)[v_\varepsilon]-\nabla\Psi_\varepsilon(w)[v_\varepsilon].\end{aligned}$$

This implies

$$\|v_\varepsilon\|^2-\left\langle A_\varepsilon v_\varepsilon, v_\varepsilon\right\rangle\le\left\langle (A_\varepsilon-A_0)w, v_\varepsilon\right\rangle+\left\langle N_0(w)-N_\varepsilon(w), v_\varepsilon\right\rangle,$$

and, from $(A2)$ and $(N2)$, we have

$$(1-\theta)\|v_\varepsilon\|^2\le o(1)\|v_\varepsilon\|,$$

and thereby the proof is completed. □

As a corollary of Lemma 3.15, we shall give a first relationship between I_ε for $\varepsilon > 0$ and I_0 that is:

Corollary 3.16 *Suppose* $(A1)$–$(A2)$ *and* $(N1)$–$(N3)$ *are satisfied.* $I_\varepsilon(v) \to I_0(v)$ *as* $\varepsilon \to 0$ *for* $v \in X$.

Proof As in the proof of Lemma 3.15, we set $z_\varepsilon = v + h_\varepsilon(v)$, $w = v + h_0(v)$ and $v_\varepsilon = z_\varepsilon - w$ for $v \in X$.

Recall the definition of I_ε for $\varepsilon \in \mathcal{E}$, by virtue of Lemma 3.15, we only need to show that $\Psi_\varepsilon(z_\varepsilon) \to \Psi_0(w)$ as $\varepsilon \to 0$. This can be seen by the fact

$$\langle A_\varepsilon z_\varepsilon, z_\varepsilon\rangle = \langle A_0 w, w\rangle + \langle (A_\varepsilon - A_0)w, w\rangle + O(\|v_\varepsilon\|)$$

and $\|v_\varepsilon\| = o(1)$ as $\varepsilon \to 0$.

From (3.5), we have

$$\Psi_\varepsilon(z_\varepsilon) = \int_0^1 \nabla\Psi_\varepsilon(tz_\varepsilon)[z_\varepsilon]\,dt = \int_0^1 \int_0^t \nabla^2\Psi_\varepsilon(sz_\varepsilon)[z_\varepsilon, z_\varepsilon]\,ds dt, \tag{3.17}$$

$$\Psi_0(w) = \int_0^1 \nabla\Psi_0(tw)[w]\,dt = \int_0^1 \int_0^t \nabla^2\Psi_0(sw)[w, w]\,ds dt. \tag{3.18}$$

By virtue of $(N3)$, we easily conclude the family $\{f_\varepsilon\}$, where

$$f_\varepsilon : [0, 1] \to \mathbb{R}, \quad f_\varepsilon(t) = \nabla\Psi_\varepsilon(tz_\varepsilon)[z_\varepsilon],$$

is uniformly bounded and equicontinuous. Then, by Arzelà-Ascoli theorem, the family $\{f_\varepsilon\}$ is compact in $C[0, 1]$. Notice that $z_\varepsilon \to w$ in E as $\varepsilon \to 0$, a standard argument shows $f_\varepsilon(t) \to f_0(t)$ pointwise on $[0, 1]$ as $\varepsilon \to 0$. Hence we have f_ε converges to f_0 in the $C[0, 1]$ topology as ε shrinks. This together with (3.17) and (3.18), we see that $\Psi_\varepsilon(z_\varepsilon)$ converges to $\Psi_0(w)$ as $\varepsilon \to 0$. □

Next we shall give geometric structures of I_ε for $\varepsilon \neq 0$. Recall that we assumed

$$c_0 = \inf_{e \in X\setminus\{0\}} \sup_{z \in E_e} \Phi_0(z), \tag{3.19}$$

to be a critical value for Φ_0, then our result will be:

Proposition 3.17 *Under the assumptions* $(A1)$–$(A2)$, $(N1)$–$(N5)$ *and* $(I1)$–$(I2)$, *for* $\varepsilon > 0$ *small enough,* I_ε *possesses the mountain-pass structure:*

(1) $I_\varepsilon(0) = 0$ *and there exist* $r > 0$ *and* $\tau > 0$ *(both independent of* ε*) such that* $I_\varepsilon|_{S_r^X} \geq \tau$.
(2) *there exists* $v_0 \in X$ *(independent of* ε*) such that* $\|v_0\| > r$ *and* $I_\varepsilon(v_0) < 0$.

Moreover, if Φ_ε satisfy the $\mathscr{G}$-weak $(C)_c$-condition (or $(PS)_c$-condition) for all $c >$ 0, then

$$c'_\varepsilon = \inf_{\nu \in \Gamma_\varepsilon} \max_{t \in [0,1]} I_\varepsilon\big(\nu(t)\big) \tag{3.20}$$

is a critical value for I_ε, where

$$\Gamma_\varepsilon = \big\{\nu \in C([0,1], X) : \nu(0) = 0,\ I_\varepsilon\big(\nu(1)\big) < 0\big\}.$$

Before giving the proof of Proposition 3.17, we shall give some equivalent characterizations of the critical value of Φ_0 defined in (3.19) which are essential in our proofs. Set

$$c'_0 = \inf_{\nu \in \Gamma_0} \max_{t \in [0,1]} I_0\big(\nu(t)\big),$$

and

$$c''_0 = \inf_{e \in X \setminus \{0\}} \sup_{t \geq 0} I_0(te),$$

where $\Gamma_0 := \big\{\nu \in C([0,1], X) : \nu(0) = 0,\ I_0(\nu(1)) < 0\big\}$.

Lemma 3.18 *Suppose* $(A1)$–$(A2)$, $(N1)$–$(N5)$ *and* $(I1)$–$(I2)$ *are satisfied. If* $c_0 < \infty$ *is a critical value of* Φ_0*, then* $c_0 = c'_0 = c''_0$.

Proof Obviously, by $(I2)$ and $c_0 < \infty$, it follows from the definition of I_0 that $c'_0 \leq c''_0 \leq c_0$. So, in what follows, we shall prove $c_0 \leq c'_0$.

Claim 3.1 If $v \in X \setminus \{0\}$ satisfies $\nabla I_0(v)[v] = 0$, then $\nabla^2 I_0(v)[v, v] < 0$.

In order to verify Claim 3.1, we fist do some basic calculations. Recall that $h_0(v)$ is the unique critical point of ϕ_v on Y for $v \in X$, we have

$$-\langle h_0(v), y\rangle + \big\langle A_0\big(v + h_0(v)\big), y\big\rangle - \nabla\Psi_0\big(v + h_0(v)\big)[y] = 0 \quad \forall y \in Y. \tag{3.21}$$

Set $z = v + h_0(v)$ and $w = h_0'(v)v - h_0(v)$, then

$$\begin{aligned}
\nabla I_0'(v)[v] &= \|v\|^2 - \langle h_0(v), \nabla h_0(v)[v]\rangle + \big\langle A_0\big(v + h_0(v)\big), v + \nabla h_0(v)[v]\big\rangle \\
&\quad - \nabla\Psi_0\big(v + h_0(v)\big)\big[v + \nabla h_0(v)[v]\big] \\
&= \|v\|^2 + \big\langle A_0\big(v + h_0(v)\big), v\big\rangle - \nabla\Psi_0\big(v + h_0(v)\big)[v] \\
&= \|v\|^2 - \Big\langle h_0(v), z^Y + y\Big\rangle + \big\langle A_0\big(v + h_0(v)\big), z + y\big\rangle \\
&\quad - \nabla\Psi_0\big(v + h_0(v)\big)[z + y] \\
&= \nabla\Phi_0(z)[z + y]
\end{aligned} \tag{3.22}$$

for all $y \in Y$. Since (3.21) is valid for all $v \in X$, by taking derivative with respect to v, we deduce

$$\begin{aligned}
0 \equiv &- \langle -\nabla h_0(v)[v], y\rangle + \big\langle A_0\big(v + \nabla h_0(v)[v]\big), y\big\rangle \\
&- \nabla^2\Psi_0\big(v + h_0(v)\big)\big[v + \nabla h_0(v)[v], y\big]
\end{aligned} \tag{3.23}$$

for all $y \in Y$. So, choose $y = z^Y + w = \nabla h_0(v)[v]$ in (3.23), we infer

$$\begin{aligned}
\nabla I_0(v)[v, v] &= \|v\|^2 + \langle A_0(z + w), v\rangle - \nabla^2\Psi_0(z)[z + w, v] \\
&= \|v\|^2 - \|z^Y + w\|^2 - \langle A_0(z + w), z + w\rangle \\
&\quad - \nabla^2\Psi_0(z)[z + w, z + w] \\
&= \nabla^2\Phi_0(z)[z + w, z + w].
\end{aligned}$$

Taking into account that $\nabla\Phi_0(z)[z] = \nabla I_0(v)[v] = 0$ (which follows from (3.22)), we can conclude

$$\begin{aligned}
\nabla^2 I_0(v)[v, v] &= \nabla^2\Phi_0(z)[z + w, z + w] \\
&= \nabla^2\Phi_0(z)[z, z] + 2\nabla^2\Phi_0(z)[z, w] + \nabla^2\Phi_0(z)[w, w] \\
&= \|z^X\|^2 - \|z^Y\|^2 + \langle A_0 z, z\rangle - \nabla^2\Psi_0(z)[z, z] \\
&\quad + 2\Big(-\Big\langle z^Y, w\Big\rangle + \langle A_0 z, w\rangle - \nabla^2\Psi_0(z)[z, w]\Big) \\
&\quad + \Big(-\|w\|^2 + \langle A_0 w, w\rangle - \nabla^2\Psi_0(z)[w, w]\Big) \\
&= \big(\nabla\Psi_0(z)[z] - \nabla^2\Psi_0(z)[z, z]\big) + 2\big(\nabla\Psi_0(z)[w] - \nabla^2\Psi_0(z)[z, w]\big) \\
&\quad - \nabla^2\Psi_0(z)[w, w] - \|w\|^2 + \langle A_0 w, w\rangle \\
&< 0
\end{aligned}$$

due to $(N5)$ and $z \neq 0$.

Let $v \in X \setminus \{0\}$, we find the function $t \mapsto I_0(tv)$ has at most one nontrivial critical point $t = t(v) > 0$ which (if exists) will be the maxima point. Denoted by

$$\mathscr{M} = \{t(v)v : v \in X \setminus \{0\},\ t(v) < +\infty\},$$

we have $\mathscr{M} \neq \emptyset$ since c_0 is a critical value of Φ_0. We also observe that

$$c_0'' = \inf_{z \in \mathscr{M}} I_0(z),$$

moreover, by (I_2) and $\mathscr{M} \neq \emptyset$, we infer $\Gamma_0 \neq \emptyset$.

Claim 3.2 $c_0'' = c_0$.

Let $e \in \mathscr{M}$, then $\Phi_0'(e + h_0(e))\big|_{E_e} = 0$ (recall $E_e = \mathbb{R}^+ e \oplus Y$). Hence $c_0 \leq \max_{z \in E_e} \Phi_0(z) = I_0(e)$, which implies $c_0 \leq c_0''$.

Claim 3.3 $c_0'' \leq c_0'$.

We only need to show that given $\nu \in \Gamma_0$ there exists $\bar{t} \in [0, 1]$ such that $\nu(\bar{t}) \in \mathscr{M}$. Assuming contrarily we have $\nu([0, 1]) \cap \mathscr{M} = \emptyset$. As a result of $(I1)$,

$$\nabla I_0(\nu(t))[\nu(t)] > 0 \quad \text{for } t > 0 \text{ small.}$$

Since the function $t \mapsto \nabla I_0(\nu(t))[\nu(t)]$ is continuous and $\nabla I_0(\nu(t))[\nu(t)] \neq 0$ for all $t \in (0, 1]$, we have

$$\nabla I_0(\nu(t))[\nu(t)] > 0 \quad \text{for all } t \in [0, 1].$$

Then we find by $(N4)$

$$\begin{aligned} I_0(\nu(t)) &= \frac{1}{2}\nabla I_0(\nu(t))[\nu(t)] + \widehat{\Psi}_0\big(\nu(t) + h_0(\nu(t))\big) \\ &\geq \frac{1}{2}\nabla I_0(\nu(t))[\nu(t)] > 0 \end{aligned}$$

for all $t \in (0, 1]$ which is absurd.

Combing Claims 3.1–3.3, we have the assertion proved. □

Proof of Proposition 3.17 Since we have $I_\varepsilon(v) \geq \Phi_\varepsilon(v)$ for all $v \in X$, (1) follows easily from (I_1).

To check (2), let $w = w^X + w^Y \in E = X \oplus Y$ be a critical point of Φ_0 with $\Phi_0(w) = c_0$. Then, by virtue of Proposition 3.13, we have $w^Y = h_0(w^X)$. A direct consequence of Lemma 3.18 is

$$c_0 = I_0(w^X) = \max_{t \geq 0} I_0(tw^X),$$

and, by $(I2)$, we can conclude there exists $t_0 > 0$ (large enough) such that

$$I_0\big(t_0 w^X\big) < -1.$$

As a result of Corollary 3.16, we have

$$\begin{aligned} I_\varepsilon\big(t_0 w^X\big) &= I_0\big(t_0 w^X\big) + o(1) \\ &\leq -\frac{1}{2} + o(1) \end{aligned}$$

as $\varepsilon \to 0$. Therefore, there is $\varepsilon_0 > 0$ such that $I_\varepsilon\big(t_0 w^X\big) < 0$ for all $\varepsilon \in (0, \varepsilon_0]$.

As a consequence of the mountain-pass structure, we obtain a $(C)_{c'_\varepsilon}$-sequence for I_ε denoted by $\{v_\varepsilon^n\}_{n=1}^\infty$. By Proposition 3.13 and $\mathscr{G}$-weak $(C)_{c'_\varepsilon}$-condition for Φ_ε, we conclude that there exists $v_\varepsilon \neq 0$ such that $\nabla I_\varepsilon(v_\varepsilon) = 0$. Moreover, from $(N4)$, we have

$$\begin{aligned} c'_\varepsilon &= \lim_{n\to\infty} \Big(I_\varepsilon(v_\varepsilon^n) - \frac{1}{2}\nabla I_\varepsilon(v_\varepsilon^n)[v_\varepsilon^n]\Big) \\ &= \lim_{n\to\infty} \Big(\Phi_\varepsilon\big(v_\varepsilon^n + h_\varepsilon(v_\varepsilon^n)\big) - \frac{1}{2}\nabla\Phi_\varepsilon\big(v_\varepsilon^n + h_\varepsilon(v_\varepsilon^n)\big)\big[v_\varepsilon^n + h_\varepsilon(v_\varepsilon^n)\big]\Big) \\ &= \lim_{n\to\infty} \Big(\frac{1}{2}\nabla\Psi_\varepsilon\big(v_\varepsilon^n + h_\varepsilon(v_\varepsilon^n)\big)\big[v_\varepsilon^n + h_\varepsilon(v_\varepsilon^n)\big] - \Psi_\varepsilon\big(v_\varepsilon^n + h_\varepsilon(v_\varepsilon^n)\big)\Big) \qquad (3.24) \\ &= \lim_{n\to\infty} \widehat{\Psi}_\varepsilon\big(v_\varepsilon^n + h_\varepsilon(v_\varepsilon^n)\big) \geq \widehat{\Psi}_\varepsilon\big(v_\varepsilon + h_\varepsilon(v_\varepsilon)\big) \\ &= I_\varepsilon(v_\varepsilon) - \frac{1}{2}\nabla I_\varepsilon(v_\varepsilon)[v_\varepsilon] = I_\varepsilon(v_\varepsilon) > 0. \end{aligned}$$

Now set

$$c''_\varepsilon := \inf_{e\in X\setminus\{0\}} \sup_{t\geq 0} I_\varepsilon(te),$$

and

$$c_\varepsilon := \inf_{e\in X\setminus\{0\}} \sup_{z\in E_e} \Phi_\varepsilon(z). \qquad (3.25)$$

Let us repeat the proof of Claim 3.1 in Lemma 3.18, from which we can conclude: let $e \in X \setminus \{0\}$, the function $t \mapsto I_\varepsilon(te)$ has at most one nontrivial critical point $t = t(e) > 0$ which (if exists) will be the maximum point.

Noting that (3.24) and $v_\varepsilon \in X \setminus \{0\}$ is a critical point of I_ε implies

$$c''_\varepsilon \leq \sup_{t\geq 0} I_\varepsilon(tv_\varepsilon) = I_\varepsilon(v_\varepsilon) \leq c'_\varepsilon < \infty.$$

And on the other hand, it is not difficult to check $c'_\varepsilon \le c''_\varepsilon$. Hence we have $c'_\varepsilon = c''_\varepsilon$. Meanwhile, $c_\varepsilon = c''_\varepsilon$ is much obvious since the definition of h_ε implies

$$I_\varepsilon(te) = \Phi_\varepsilon\big(te + h_\varepsilon(te)\big) = \max_{w\in Y} \Phi_\varepsilon(te+w)$$

and therefore

$$\sup_{t\ge 0} I_\varepsilon(te) = \sup_{t\ge 0}\max_{w\in Y} \Phi_\varepsilon(te+w) = \sup_{z\in Ee} \Phi_\varepsilon(z).$$

By taking infimum with respect to $e \in X \setminus \{0\}$, we have $c_\varepsilon = c'_\varepsilon = c''_\varepsilon$.

Due to the above observation, if we have proved

$$I_\varepsilon(v_\varepsilon) \ge c''_\varepsilon, \tag{3.26}$$

then we can conclude $I_\varepsilon(v_\varepsilon) = c'_\varepsilon$ immediately from (3.24).

In fact, let us set

$$\mathscr{M}_\varepsilon := \big\{t(v)v :\ v \in X\setminus\{0\},\ 0 < t(v) < \infty \text{ such that } \nabla I_\varepsilon(t(v)v)[v] = 0\big\},$$

we infer that

$$c''_\varepsilon = \inf_{z\in\mathscr{M}_\varepsilon} I_\varepsilon(z).$$

Since $v_\varepsilon \in \mathscr{M}_\varepsilon$, (3.26) is obviously valid. And the proof is thereby completed. □

As a by-product of the proof of Proposition 3.17 we have

Lemma 3.19 *Let $\varepsilon \in (0, \varepsilon_0]$ be such that Proposition 3.17 is valid. Then $c_\varepsilon = c'_\varepsilon = c''_\varepsilon$ characterize the ground state energy of Φ_ε.*

In what follows, we shall show the asymptotic behaviour of the critical values found in Proposition 3.17.

Lemma 3.20 *Let $\varepsilon \in (0, \varepsilon_0]$ be such that Proposition 3.17 is valid. Then $c_\varepsilon \le c_0 + o(1)$ as $\varepsilon \to 0$.*

Proof Again, Let $w = w^X + w^Y \in E = X \oplus Y$ be the critical point of Φ_0 such that $\Phi_0(w) = I_0(w^X) = c_0$. Set $t_0 > 0$ such that $I_0(t_0 w^X) \le -1$. By virtue of Lemmas 3.18 and 3.19, it sufficient to prove

$$I_\varepsilon(tw^X) = I_0(tw^X) + o(1) \quad \text{uniformly in } t \in [0, t_0] \tag{3.27}$$

as $\varepsilon \to 0$.

To this end, we only need to show the family $\{f_\varepsilon\} \subset C[0, t_0]$

$$f_\varepsilon(t) := I_\varepsilon(tw^X)$$

is uniformly bounded and equicontinuous. This can be seen from Corollary 3.16 that if $\{f_\varepsilon\}$ is compact in the $C[0, t_0]$ topology then (3.27) is valid.

Clearly, $f_\varepsilon \in C^1$ and the uniformly boundedness of $\{f_\varepsilon\}$ and $\{f'_\varepsilon\}$ on $[0, t_0]$ comes easily from $(A1)$–$(A2)$ and $(N3)$. So, by Arzelà-Ascoli theorem, we have $\{f_\varepsilon\}$ is compact in the $C[0, t_0]$. And therefore, by (3.27), we conclude

$$\begin{aligned} c_\varepsilon &\le \sup_{t\ge 0} I_\varepsilon(tw^X) = \sup_{t\in[0,t_0]} I_\varepsilon(tw^X) = \sup_{t\in[0,t_0]} I_0(tw^X) + o(1) \\ &= \sup_{t\ge 0} I_0(tw^X) + o(1) = I_0(w^X) + o(1) \\ &= c_0 + o(1) \end{aligned}$$

as $\varepsilon \to 0$. □

Now, combining Proposition 3.17, Lemmas 3.19 and 3.20 we summarize the following result which together with Proposition 3.13 gives a localized energy inequality.

Theorem 3.21 (Ding and Xu [9]) *Under the assumptions of Proposition 3.17, for $\varepsilon > 0$ small, I_ε possesses a nontrivial critical value which can be characterized by*

$$c_\varepsilon = \inf_{e\in X\setminus\{0\}} \sup_{t\ge 0} I_\varepsilon(te).$$

Moreover, $c_\varepsilon \le c_0 + o(1)$ as $\varepsilon \to 0$.

3.2.3 A Further Multiple Existence Result

The critical value in Theorem 3.21 or its equivalent expression (3.25) can be interpreted as a two-step procedure: first to maximize the functional Φ_ε on $E_e = \mathbb{R}^+ e \oplus Y$ and then to minimize along very direction e in $X\setminus\{0\}$. It is natural to ask the question whether there are critical points of Φ_ε that corresponding to certain higher critical levels, which are not explicitly characterized by minimax schemes. To give a trial resolution on this problem, compared with the $\mathscr{G}$-weak compact condition in Definition 3.12, we need first of all a stronger compactness:

Definition 3.22 We say that a $\mathscr{G}$-invariant functional $\Phi \in C^1(E, \mathbb{R})$ satisfies the $\mathscr{G}$-$(PS)_c$-condition if every $(PS)_c$-sequence has a subsequence which converges after an accordingly $\mathscr{G}$-action:

$$\left.\begin{aligned} \Phi(z_n) &\to c \\ \nabla\Phi(z_n) &\to 0 \end{aligned}\right\} \Rightarrow g_n z_n \to z \in E \text{ along a subsequence for some } \{g_n\} \subset \mathscr{G}.$$

Then, in addition to the condition (Φ_0) mentioned in Proposition 3.1, we will assume that

(Φ1) *There exists* $r > 0$ *with* $\rho := \inf \Phi(S_r^X) > \Phi(0) = 0$ *where* $S_r^X := \{z \in X : \|z\| = r\}$;
(Φ2) *There exists a finite dimensional subspace* $X_0 \subset X$ *and* $R > r$ *such that for* $E_0 := X_0 \oplus Y$ *and* $B_0 := \{z \in E_0 : \|z\| \leq R\}$ *there holds* $d := \sup \Phi(E_0) < \infty$ *and* $\sup \Phi(E_0 \setminus B_0) < \inf \Phi(B_r^X)$, *where* $B_r^X := \{z \in X : \|z\| \leq r\}$.

In the case in which Φ is even, i.e. Φ admits a $\mathbb{Z}_2$-action, let us introduce the notations as was introduced in [3, 8]: for $c \in \mathbb{R}$, denoted by $\mathcal{M}(\Phi^c)$ the collection of maps $h : \Phi^c \to E$ satisfying

(*i*) *h is* $\mathcal{T}_\mathcal{P}$*-continuous and odd;*
(*ii*) $h(\Phi^a) \subset \Phi^a$ *for all* $a \in [\rho, d]$;
(*iii*) *each* $z \in \Phi^c$ *has a* $\mathcal{T}_\mathcal{P}$*-open neighborhood* $O \subset E$ *such that* $(\mathrm{id} - h)(O \cap \Phi^c)$ *is contained in a finite dimensional linear subspace.*

Then a pseudo-index of Φ^c can be defined by

$$\psi(c) := \min \left\{ \mathrm{gen}(h(\Phi^c) \cap S_r^X) : h \in \mathcal{M}(\Phi^c) \right\} \in \mathbb{N} \cup \{\infty\} \tag{3.28}$$

where $\mathrm{gen}(\cdot)$ denotes the classical $\mathbb{Z}_2$-genus.

Let us emphasize here the situation that Φ may admits a group action of $\mathcal{G}$ in addition to the $\mathbb{Z}_2$-action is a considerable issue in our abstract settings. If $\mathcal{G}$ is a compact group, then a generalized index theory can be employed as was introduced by Rabinowitz [14] (see also [2] for related material). However, in general, the group $\mathcal{G}$ may be non-compact and quotient space $E/\mathcal{G}$ will no longer be a linear space or a manifold. Therefore classical results can not be applied. One way to treat such $\mathcal{G} \times \mathbb{Z}_2$-invariant problem is to separate these two group actions. For such, in the sequel, let us introduce the following concept

Definition 3.23 Let E be a Hilbert space (or Banach space) associated with a group action of $\mathcal{G} \times \mathbb{Z}_2$. We say that $\mathcal{G}$ is separated with respect to $\mathbb{Z}_2$ if and only if, for all closed $\mathbb{Z}_2$-invariant subset $A \subset E$, the $\mathcal{G}$-orbit $\mathcal{G}(A) := \{gz : g \in \mathcal{G}, z \in A\}$ is closed and $\mathrm{gen}(A) = \mathrm{gen}(\mathcal{G}(A))$.

Let us remark that, for a separated $\mathcal{G}$-action, there holds $\mathcal{G}(z_1) \cap \mathcal{G}(z_2) = \emptyset$ for any $\mathcal{G}$-distinct elements z_1 and z_2. Hence, for a closed $\mathbb{Z}_2$-invariant subset A, $\mathcal{G}(A)$ is the disjoint union of all the $\mathcal{G}$-orbits of its elements.

Then our general existence result comes as follows

Theorem 3.24 (Ding and Xu [10]) *Let* Φ *be a even functional satisfying* (Φ0)-(Φ2). *If* Φ *also admits another separated group action of* $\mathcal{G}$ *(not necessarily compact) and satisfies the* $\mathcal{G}$-$(PS)_c$*-condition for every* $c \in [\rho, d]$. *Then* Φ *has* *at least* $n := \dim X_0$ *pairs of* $\mathscr{G}$*-distinct critical points. In particular, if* Φ *has only finitely many* $\mathscr{G}$*-distinct critical points in* Φ_ρ^d*, then*

$$c_i = \inf\big\{c : \psi(c) \geq i\big\} \in [\rho, d\,], \quad i = 1, 2, \dots, n \tag{3.29}$$

are critical values satisfying $\rho \leq c_1 < c_2 < \cdots < c_n \leq d$.

Remark 3.25 Theorem 3.24 generalizes Theorem 4.6 of [3] (see also [8, p. 31]) in the sense that Φ is allowed to carry a $\mathscr{G}$-invariant property for some non-compact group action. This generalization permits us to treat such as RD systems of the form (2.5)–(2.6).

Remark 3.26 Theorem 3.24 holds true for more general classes of symmetries, typical example is Φ admits a group action of $\mathscr{G} \times G$ where G is some compact group and $\mathscr{G}$ is separated. Typical examples of G include the case of a finite group such as $\mathbb{Z}_p$ for a prime number p and a compact Lie group such as S^1, $SU(2)$, et al. In all these cases, there exists an index $i : \{A \subset E : A \text{ is } G\text{-invariant}\} \to \mathbb{N} \cup \{\infty\}$ for G-action satisfying the monotonicity, sub-additivity, super-variance as well as a dimension property: $i(F \setminus \{0\}) = c \cdot \dim F$ for any finite-dimensional G-invariant linear subspace $F \subset E$. We refer the readers to [2, 4, 6] for a discussion of group actions, index theories, examples and applications.

Proof of Theorem 3.24 For a subset $\mathcal{S} \subset E$ and $\sigma > 0$, let us denote the σ-neighborhood of $\mathcal{S}$ as

$$U_\sigma(\mathcal{S}) := \big\{z \in E : \inf_{w \in \mathcal{S}} \|z - w\| < \sigma\big\},$$

we remind the reader the following definition of (PS)-attractor (see [8, Chapter 3]):

Definition 3.27 A subset $\mathscr{A} \subset E$ is said to be a $(PS)_c$-attractor for Φ if any $(PS)_c$-sequence approaches the σ-neighborhood of $\mathscr{A} \cap \Phi_{c-\delta}^{c+\delta}$ for any $\sigma, \delta > 0$:

$$\left.\begin{aligned} \Phi(z_n) &\to c \\ \nabla\Phi(z_n) &\to 0 \end{aligned}\right\} \Rightarrow z_n \in U_\sigma\big(\mathscr{A} \cap \Phi_{c-\delta}^{c+\delta}\big) \text{ for all } n \text{ suitably large.}$$

And given $\mathcal{I} \subset \mathbb{R}$, $\mathscr{A} \subset E$ is said to be a $(PS)_{\mathcal{I}}$-attractor if $\mathscr{A}$ is a $(PS)_c$-attractor for all $c \in \mathcal{I}$.

For later arguments we introduce a comparison function $\psi_l : [0, l] \to \mathbb{N} \cup \{\infty\}$: let $l > 0$ and

$$\mathcal{M}_0(\Phi^l) := \big\{h \in \mathcal{M}(\Phi^l) : h \text{ is a homeomorphism from } \Phi^l \text{ to } h(\Phi^l)\big\},$$

for $c \in [0, l]$, define

$$\psi_l(c) := \min\big\{\operatorname{gen}(h(\Phi^c) \cap S_r^X) : h \in \mathcal{M}_0(\Phi^l)\big\}.$$

Note that due to $\mathcal{M}_0(\Phi^l) \subset \mathcal{M}(\Phi^l) \hookrightarrow \mathcal{M}(\Phi^c)$ via the restriction $h \mapsto h|_{\Phi^c}$ we have $\psi(c) \le \psi_l(c)$ for all $c \in [0, l]$.

Now, we can sketch the proof as follows: If $c_i \in [\rho, d]$ as was defined in (3.29) is not a critical value, then for any sufficiently small $\bar{\delta} > 0$ we have $\inf\{\|\Phi'(z)\| : z \in \Phi_{c_i-\bar{\delta}}^{c_i+\bar{\delta}}\} > 0$. By virtue of the Deformation Theorem (a version developed for strongly indefinite functionals can be found in [8, Theorem 3.2]), we can infer the existence of $\eta \in C([0,1] \times \Phi^{c_i+\delta}, \Phi^{c_i+\delta})$ for some $\delta \in (0, \bar{\delta})$ such that $g := \eta(1, \cdot) \in \mathcal{M}(\Phi^{c_i+\delta})$ and $g(\Phi^{c_i+\delta}) \subset \Phi^{c_i-\delta}$. This is impossible since we can deduce that

$$\begin{aligned}\psi(c_i - \delta) &= \min\{\mathrm{gen}(h(\Phi^{c_i-\delta}) \cap S_r^X) : h \in \mathcal{M}(\Phi^{c_i-\delta})\}\\ &\ge \min\{\mathrm{gen}(h \circ g(\Phi^{c_i+\delta}) \cap S_r^X) : h \in \mathcal{M}(\Phi^{c_i-\delta})\}\\ &\ge \min\{\mathrm{gen}(h(\Phi^{c_i+\delta}) \cap S_r^X) : h \in \mathcal{M}(\Phi^{c_i+\delta})\}\\ &= \psi(c_i + \delta),\end{aligned}$$

and the monotonicity of the $\mathbb{Z}_2$-genus implies $\psi(c_i - \delta) = \psi(c_i + \delta)$.

If Φ has only finitely many $\mathcal{G}$-distinct critical points in Φ_ρ^d, then thanks to the $\mathcal{G}$-(PS)-condition we have

$$\mathcal{A} := \{gz : g \in \mathcal{G} \text{ and } z \in \Phi_\rho^d \text{ such that } \Phi'(z) = 0\}$$

is a $(PS)_{\mathcal{I}}$-attractor with $\mathcal{I} := [\rho, d]$. Moreover, we also have that $\mathcal{A}/\mathcal{G}$ is the critical set of Φ in Φ_ρ^d which is finite. Hence, we deduce that

$$\inf\{\|P^X z - P^X w\| : z, w \in \mathcal{A} \text{ and } P^X z, P^X w \text{ are } \mathcal{G}\text{-distinct}\} > 0.$$

For $\sigma > 0$ small we then have that $U_\sigma(P^X\mathcal{A}) \subset X$ is the union of disjoint σ-$\mathcal{G}$-orbits around the elements of $P^X\mathcal{A}$. This, jointly with the fact the $\mathcal{G}$-action is separated, implies that $\mathrm{gen}(U_\sigma) = \mathrm{gen}(U_\sigma(P^X\mathcal{A})) = \mathrm{gen}(P^X\mathcal{A}) = 1$ where $U_\sigma := U_\sigma(P^X\mathcal{A}) \times Y$. Let $\eta \in C([0,1] \times \Phi^d, \Phi^d)$ be a deformation deduced from Theorem 3.5 a) in [8]. For $\delta > 0$ small enough the map $g := \eta(1, \cdot)$ satisfies

$g(\Phi^{c_i+\delta}) \subset \Phi^{c_i-\delta} \cup U_\sigma$. Let $l = d + 1$ and choose $h_0 \in \mathcal{M}_0(\Phi^l)$ such that $\psi_l(c_i - \delta) = \text{gen}(h_0(\Phi^{c_i-\delta}) \cap S_r^X)$. And consequently,

$$\begin{aligned}
\psi_l(c_i + \delta) &= \min \left\{ \text{gen}(h(\Phi^{c_i+\delta}) \cap S_r^X) : h \in \mathcal{M}_0(\Phi^l) \right\} \\
&\leq \text{gen}(h_0 \circ g(\Phi^{c_i+\delta}) \cap S_r^X) \\
&\leq \text{gen}(h_0(\Phi^{c_i-\delta} \cup U_\sigma) \cap S_r^X) \\
&\leq \text{gen}(h_0(\Phi^{c_i-\delta}) \cap S_r^X) + \text{gen}(h_0(U_\sigma)) \\
&= \text{gen}(h_0(\Phi^{c_i-\delta}) \cap S_r^X) + 1 \\
&= \psi_l(c_i - \delta) + 1.
\end{aligned}$$

It is then immediate that $\psi_l(c_i + \delta) = \psi_l(c_i - \delta) + 1$, otherwise one would get a contradiction to the definition of c_i. And finally, the fact $\psi_l(c_i) \geq i$ implies $\rho \leq c_1 < c_2 < \cdots < c_n \leq d$ which completes the proof. □

3.3 A Refined Energy Estimate

Our next result is to establish an refined approximate expansion for the minimax level c_ε in Theorem 3.21. Particularly, this approximate expansion will be based on a sequence of test elements in Φ_ε. Motivated by the applications mentioned in Chap. 2, we need to cover the cases that the test elements are varying synchronously and non-synchronously with the perturbation parameter in the variational problem.

To this end, let us consider a class of specific situations (some of them might be sub-cases in the framework of Sect. 3.2, but some are not): We take $\mathcal{A} \subset \mathbb{R}$ to be an closed interval and for each $\alpha \in \mathcal{A}$, we equip E accordingly a scalar product $\langle \cdot, \cdot \rangle_\alpha$ and the induce norm $\| \cdot \|_\alpha$. Then we can consider in E the following family of equations

$$L_\alpha z = N_\alpha(z), \quad z \neq 0, \ \alpha \in \mathcal{A}, \tag{3.30}$$

where $L_\alpha : (E, \langle \cdot, \cdot \rangle_\alpha) \to (E, \langle \cdot, \cdot \rangle_\alpha)$ is of strongly indefinite type as was given in (2.16) and (2.17). For the nonlinear potential operator N_α, we will consider in this section the case in which it has the form

(H1) $N_\alpha(z) = \nabla \Psi_\alpha(z)$ and there is a continuous function $\varrho : [0, \infty) \to [0, \infty)$ such that $|||\nabla N_\alpha(z)|||_\alpha \leq \varrho(\|z\|_\alpha)$, where $|||\cdot|||_\alpha$ denotes the operator norm for continuous linear operators on $(E, \langle \cdot, \cdot \rangle_\alpha)$.

(H2) $\lim_{\|z\| \to 0} \Psi_\alpha(z)/\|z\|_\alpha = 0$ and there exists $p > 2$ such that $\langle N_\alpha(z), z \rangle_\alpha \geq p\Psi_\alpha(z)$ for all $z \in E$.

(H3) $\Psi_\alpha \not\equiv 0$ on E and is convex.

(H4) There exist $K > 0$ and $\mu \in (\frac{1}{2}, 1)$ such that $\|N_\alpha(z)\|_\alpha \le K \langle N_\alpha(z), z\rangle_\alpha^\mu + \mu\delta\|z\|_\alpha$ for all $z \in E$, where $\delta > 0$ is the constant in (2.16) and (2.17).

(H5) There exists $\kappa > 1$ such that, for arbitrary $z, w \in E$,

$$\langle \nabla N_\alpha(z)[z+w], z+w\rangle_\alpha - 2\langle N_\alpha(z), w\rangle_\alpha \ge \kappa \langle N_\alpha(z), z\rangle_\alpha .$$

(H6) If there is a $\|\cdot\|_\alpha$-bounded sequence $\{z_j\} \subset E$ such that

$$L_\alpha z_j - N_\alpha(z_j) \to 0 \quad \text{and} \quad \liminf_{j\to\infty} \big\langle N_\alpha(z_j), z_j\big\rangle_\alpha > 0,$$

then there exists $z \ne 0$ with $L_\alpha z = N_\alpha(z)$ in E.

Here the function ϱ and the positive constants δ, p, K, μ and κ are independent of $\alpha \in \mathcal{A}$.

We shall denote $E_\alpha^\pm$ the L_α-invariant subspaces associated to the positive and negative components of $\sigma(L_\alpha)$, which are orthogonal with respect to the scalar product $\langle\cdot,\cdot\rangle_\alpha$. Then we have $E = E_\alpha^+ \oplus E_\alpha^-$, and each element $z \in E$ possesses the representation $z = z_\alpha^+ + z_\alpha^-$ with $z_\alpha^\pm \in E_\alpha^\pm$. The energy functional corresponds to (3.30) is given by

$$\Phi_\alpha(z) = \frac{1}{2}\langle L_\alpha z, z\rangle_\alpha - \Psi_\alpha(z), \quad z \in E,\ \alpha \in \mathcal{A}.$$

It can be seen from (H2) that $\Psi_\alpha(0) = 0$ and the function $\Psi_\alpha(tz)/t^p$, $t > 0$, is nondecreasing in t for all $z \in E$. Combined with the convexity in (H3), it follows that $\Psi_\alpha \ge 0$ (i.e. 0 is the global minimum of Ψ_α) and

$$\Psi_\alpha(z) = 0 \Longrightarrow \Psi_\alpha(tz) = 0 \text{ for all } t \in [0,1].$$

Furthermore, from (H4) and the obvious inequality

$$ab \le \mu a^{\frac{1}{\mu}} + (1-\mu) b^{\frac{1}{1-\mu}}, \quad \text{for } a, b \ge 0$$

we deduce that $\|N_\alpha(z)\|_\alpha \le \epsilon\mu\|N_\alpha(z)\|_\alpha + C_\epsilon(1-\mu)K^{\frac{1}{1-\mu}}\|z\|_\alpha^{\frac{\mu}{1-\mu}} + \mu\delta\|z\|_\alpha$ where $\epsilon > 0$ is arbitrary and $C_\epsilon > 0$ is a constant depending only on ϵ. If ϵ is fixed small, then we have the following boundedness of N_α uniformly in $\alpha \in \mathcal{A}$:

$$\|N_\alpha(z)\|_\alpha \le \frac{C_\epsilon(1-\mu)K^{\frac{1}{1-\mu}}}{1-\epsilon\mu}\|z\|_\alpha^{\frac{\mu}{1-\mu}} + \frac{\mu\delta}{1-\epsilon\mu}\|z\|_\alpha, \quad \text{for all } z \in E. \tag{3.31}$$

Beyond the positiveness and the convexity of Ψ_α, (H5) is stronger than $(N5)$ in Sect. 3.2 but they both yield that $\nabla N_\alpha(z) \ne 0$ when $\Psi_\alpha(z) > 0$. And finally, it is clear that if $N_\alpha : (E, \|\cdot\|_\alpha) \to (E, \|\cdot\|_\alpha)$ is compact then (H6) holds. We emphasize here that the sequence $\{z_j\}$ in (H6) may not be compact in general. Actually such sequence can even converge weakly to 0 in E, and then a concentration-compactness type argument can be employed (see for instance [1, 5, 7, 8, 11] and references therein, where some semilinear problems on $\mathbb{R}^N$ are considered).

In what follows, we are intend to characterize qualitative properties of critical values of the functionals Φ_α. Under the hypotheses imposed above, for each α, the functional Φ_α is a well-defined C^2-functional on E. Recall that 0 lies in a gap of the spectrum of $L_\alpha : (E, \|\cdot\|_\alpha) \to (E, \|\cdot\|_\alpha)$, the Spectral theory asserts that the L_α-invariant subspaces $E_\alpha^\pm$ associated to the positive and negative components of $\sigma(L_\alpha)$ are orthogonal with respect to the scalar product $\langle\cdot,\cdot\rangle_\alpha$. Particularly, $E = E_\alpha^+ \oplus E_\alpha^-$ and each element $z \in E$ possesses the representation $z = z_\alpha^+ + z_\alpha^-$, where $z_\alpha^\pm \in E_\alpha^\pm$. And we easily see that

$$\Phi_\alpha(z) = \frac{1}{2}\left\langle L_\alpha z_\alpha^+, z_\alpha^+\right\rangle_\alpha + \frac{1}{2}\left\langle L_\alpha z_\alpha^-, z_\alpha^-\right\rangle_\alpha - \Psi_\alpha(z),$$

$$\langle \nabla\Phi_\alpha(z), w\rangle_\alpha = \langle L_\alpha z - N_\alpha(z), w\rangle_\alpha ,$$

and

$$\left\langle \nabla^2\Phi_\alpha(z)[\varphi], w\right\rangle_\alpha = \langle L_\alpha\varphi - \nabla N_\alpha(z)[\varphi], w\rangle_\alpha$$

for $z, w, \varphi \in E$.

3.3.1 Generalized Nehari Manifold

We begin with the following result concerning the boundedness of a Palais-Smale sequence of Φ_α.

Lemma 3.28 *For each $\alpha \in \mathcal{A}$, let $\{z_n\} \subset E$ and $R > 0$ such that $\sup_n \Phi_\alpha(z_n) \le R$ and $\nabla\Phi_\alpha(z_n) \to 0$. Then, for large n, $\|z_n\|_\alpha$ is bounded by a constant depending only on R.*

Proof For sufficient large n, we have

$$\begin{aligned}\langle N_\alpha(z_n), z_n\rangle_\alpha &= \langle L_\alpha z_n, z_n\rangle_\alpha - \langle \nabla\Phi_\alpha(z_n), z_n\rangle_\alpha \\ &= 2\Phi_\alpha(z_n) + 2\Psi_\alpha(z_n) + o_n(1)\|z_n\|_\alpha \\ &\le 2R + \frac{2}{p}\langle N_\alpha(z_n), z_n\rangle_\alpha + o_n(1)\|z_n\|_\alpha .\end{aligned}$$

Hence

$$\langle N_\alpha(z_n), z_n\rangle_\alpha \le \frac{p}{p-2}\left(2R + o_n(1)\|z_n\|_\alpha\right)$$

and so, by (H4),

$$\begin{aligned}\delta\|z_n\|_\alpha &\le \|L_\alpha z_n\|_\alpha \le \|\nabla\Phi_\alpha(z_n)\|_\alpha + \|N_\alpha(z_n)\|_\alpha \\ &\le o_n(1) + K\,\langle N_\alpha(z_n), z_n\rangle_\alpha^\mu + \mu\delta\|z_n\|_\alpha \\ &\le o_n(1) + K\frac{p^\mu}{(p-2)^\mu}\big(2R + o_n(1)\|z_n\|_\alpha\big)^\mu + \mu\delta\|z_n\|_\alpha.\end{aligned}$$

Since $\mu < 1$, this implies the boundedness of $\|z_n\|_\alpha$. □

By taking the advantage of the positiveness and convexity of Ψ_α, together with (2.17), we have a very good geometric behavior of the functional Φ_α in the following sense:

(i) Since Ψ_α is non-negative, for a given $\varphi \in E_\alpha^+$, the functional Φ_α^φ defined on E_α^- by

$$\Phi_\alpha^\varphi(w) = \frac{1}{2}\langle L_\alpha\varphi, \varphi\rangle + \frac{1}{2}\langle L_\alpha w, w\rangle - \Psi_\alpha(\varphi + w), \qquad w \in E_\alpha^-$$

is anti-coercive, i.e., $\Phi_\alpha^\varphi(w) \to -\infty$ as $\|w\|_\alpha \to \infty$.

(ii) Since Ψ_α is convex, Φ_α^φ is strictly concave, in other words, the quadratic form

$$\left\langle \nabla^2\Phi_\alpha^\varphi(w)[\,\cdot\,], \cdot\right\rangle_\alpha : E_\alpha^- \times E_\alpha^- \to \mathbb{R}$$

is negative definite.

Combining the above two properties, for each $\alpha \in \mathcal{A}$ and $\varphi \in E_\alpha^+ \setminus \{0\}$, we get a chance to maximize the functional Φ_α on $\mathbb{R}^+\varphi \oplus E_\alpha^-$, where $\mathbb{R}^+ = (0, \infty)$. This procedure follows exactly the same idea in Sect. 3.2.1. Remark that φ is varying in the space $E_\alpha^+ \setminus \{0\}$, we can restrict ourselves to the choice $\varphi \in S_\alpha^+ := \{\varphi \in E_\alpha^+ : \|\varphi\|_\alpha = 1\}$ without changing the maxima of Φ_α on $\mathbb{R}^+\varphi \oplus E_\alpha^-$. By virtue of (3.31) in which we can fix $\epsilon \le 1 - \mu$, we deduce from the second part of (H2) that

$$\Psi_\alpha(z) \le \frac{1}{p}\langle N_\alpha(z), z\rangle_\alpha \le \frac{\delta}{p}\|z\|_\alpha^2 + C\|z\|_\alpha^{\frac{1}{1-\mu}} \tag{3.32}$$

for all $z \in E$, where $C > 0$ is a constant. Indeed, this is a direct consequence of $\frac{\mu}{1-\epsilon\mu} < \frac{\mu}{1-\epsilon} \le 1$. And so, by (2.16) and the fact $\frac{1}{1-\mu} > 2$, we have the existence of $\tau > 0$ such that

$$\max_{z\in\mathbb{R}^+\varphi\oplus E_\alpha^-}\Phi_\alpha(z) \ge \max_{t>0}\Phi_\alpha(t\varphi) \ge \tau \quad \text{for all } \varphi \in S_\alpha^+. \tag{3.33}$$

Therefore, we have found a candidate min-max scheme for Φ_α: first maximize the functional Φ_α on $\mathbb{R}^+\varphi\oplus E_\alpha^-$ and then minimize with respect to $\varphi \in S_\alpha^+$. Specifically, let $P_\alpha : E \to E_\alpha^+$ denote the orthogonal projection onto E_α^+, we have the following variational formulation, which is a variant of Proposition 3.13.

Proposition 3.29 *For each* $\alpha \in \mathcal{A}$,

(1) *there exists* $\chi_\alpha \in C^1(E_\alpha^+, E_\alpha^-)$ *such that for* $\varphi \in E_\alpha^+$

$$w \in E_\alpha^-,\ w \neq \chi_\alpha(\varphi) \Longrightarrow \Phi_\alpha(\varphi + w) < \Phi_\alpha(\varphi + \chi_\alpha(\varphi)),$$

that is $\chi_\alpha(\varphi)$ *is the unique maximizer of* Φ_α^φ *on* E_α^- *and hence*

$$L_\alpha\big(\chi_\alpha(\varphi)\big) = (I - P_\alpha)N_\alpha(\varphi + \chi_\alpha(\varphi)); \tag{3.34}$$

furthermore, for $\varphi \in E_\alpha^+$,

$$\|\chi_\alpha(\varphi)\|_\alpha^2 \le \frac{2}{\delta}\Psi_\alpha(\varphi) \tag{3.35}$$

and

$$|||\nabla \chi_\alpha(\varphi)|||_\alpha \le \frac{1}{\delta}\, |||\nabla N_\alpha(\varphi + \chi_\alpha(\varphi))|||_\alpha. \tag{3.36}$$

(2) *if* $\{\varphi_n\} \subset E_\alpha^+$ *is a Palais-Smale sequence for the reduced functional*

$$J_\alpha = \Phi_\alpha^{red} : E_\alpha^+ \to \mathbb{R}, \quad J_\alpha(\varphi) = \Phi_\alpha(\varphi + \chi_\alpha(\varphi)),$$

then $\{\varphi_n + \chi_\alpha(\varphi_n)\}$ *is a Palais-Smale sequence for* Φ_α, *in particular,* $\|\nabla J_\alpha(\varphi)\|_\alpha = \|\nabla \Phi_\alpha(\varphi + \chi_\alpha(\varphi))\|_\alpha$ *for all* $\varphi \in E_\alpha^+$.

By virtue of Proposition 3.29, we are reduced to consider the critical values of the functional J_α on E_α^+, $\alpha \in \mathcal{A}$. The essence of our analysis is to consider the values

$$\gamma_\alpha = \inf_{\varphi \in E_\alpha^+ \setminus \{0\}} \max_{t>0} J_\alpha(t\varphi), \tag{3.37}$$

which has the same expression as the critical value in Theorem 3.21. Since $J_\alpha(\varphi) \ge \Phi_\alpha(\varphi)$ for all $\varphi \in E_\alpha^+$, together with (3.32), we have the existence of $r > 0$ such that

$$\|\varphi\|_\alpha \le r \Longrightarrow J_\alpha(\varphi) \ge 0 \quad \text{and} \quad \|\varphi\|_\alpha = r \Longrightarrow J_\alpha(\varphi) \ge \tau.$$

It is all clear that if there exists additionally $\phi \in E_\alpha^+$ such that $J_\alpha(\phi) < 0$, then the reduced functional J_α possesses the mountain pass geometry, and hence $\gamma_\alpha \ge \tau > 0$ is the least value of J_α at which there exist non-trivial critical points (as was indicated by Theorem 3.21). This is the case when the nonlinearity $\Psi_\alpha(\varphi)$ vanishes if and only if $\varphi = 0$, see for instance [1, 5]. Unfortunately, since the assumptions (H1)–(H5) allow Ψ_α to vanish on certain part of E, we do not have an apparent linking structure for J_α. Here, our strategy consists of finding critical point of J_α on its Nehari-type manifold $\mathscr{N}_\alpha$ defined as

$$\mathscr{N}_\alpha = \{\varphi \in E_\alpha^+ \setminus \{0\} : \langle \nabla J_\alpha(\varphi), \varphi \rangle_\alpha = 0\}.$$

Different from the usual concept of Nehari manifold, we emphasize here that the set $\mathscr{N}_\alpha$ above is not homeomorphic to the sphere S_α^+ in general. In fact, as we will see later in Chap. 8, we allow that there exists $\varphi \in E_\alpha^+ \setminus \{0\}$ such that $\Psi_\alpha(t\varphi) \equiv 0$ for all $t > 0$. Then, by Proposition 3.29, we have $\chi_\alpha(t\varphi) \equiv 0$, and hence $J_\alpha(t\varphi) = \frac{t^2}{2} \langle L_\alpha \varphi, \varphi \rangle_\alpha$, which suggests that $J_\alpha(t\varphi)$ is a quadratic function and have no critical points for $t \in [0, \infty)$ other than $t = 0$.

Lemma 3.30 *If $\varphi \in \mathscr{N}_\alpha$, then $\Psi_\alpha(\varphi) > 0$ and $\Psi_\alpha(\varphi + \chi_\alpha(\varphi)) > 0$.*

Proof Assume contrarily that $\Psi_\alpha(\varphi) = 0$, then it follows that $N_\alpha(\varphi) = 0$ (since 0 is the global minimum of Ψ_α). And from (3.35), we have $\chi_\alpha(\varphi) = 0$. Hence

$$\langle \nabla J_\alpha(\varphi), \varphi \rangle_\alpha = \langle L_\alpha \varphi - P_\alpha N_\alpha(\varphi), \varphi \rangle_\alpha = \langle L_\alpha \varphi, \varphi \rangle_\alpha > 0,$$

which is a contradiction.

If $\Psi_\alpha(\varphi + \chi_\alpha(\varphi)) = 0$, then it also follows that $N_\alpha(\varphi + \chi_\alpha(\varphi)) = 0$. By (3.34) and the invertibility of L_α on E_α^-, we deduce that $\chi_\alpha(\varphi) = 0$. Thus we get a similar contradiction as above. □

Let us introduce the functional $K_\alpha : E_\alpha^+ \to \mathbb{R}$ by

$$K_\alpha(\varphi) = \langle \nabla J_\alpha(\varphi), \varphi \rangle_\alpha . \tag{3.38}$$

We have $K_\alpha \in C^1(E_\alpha^+, \mathbb{R})$ and its derivative (with respect to the inner product $\langle \cdot, \cdot \rangle_\alpha$ and the induced norm $\| \cdot \|_\alpha$) is given by

$$\langle \nabla K_\alpha(\varphi), w \rangle_\alpha = \langle \nabla J_\alpha(\varphi), w \rangle_\alpha + \left\langle \nabla^2 J_\alpha(\varphi)[w], \varphi \right\rangle_\alpha$$

for all $\varphi, w \in E_\alpha^+$. Also, we have $\mathscr{N}_\alpha = K_\alpha^{-1}(0) \setminus \{0\}$.

Lemma 3.31 *For $\varphi \in E_\alpha^+$,*

$$\langle \nabla K_\alpha(\varphi), \varphi \rangle_\alpha \le 2K_\alpha(\varphi) - (\kappa - 1) \langle N_\alpha(\varphi + \chi_\alpha(\varphi)), \varphi + \chi_\alpha(\varphi) \rangle_\alpha ,$$

where $\kappa > 1$ is the constant in (H5).

Proof Since

$$\begin{aligned}\langle \nabla J_\alpha(\varphi), \phi \rangle_\alpha &= \langle L_\alpha \varphi - P_\alpha N_\alpha(\varphi + \chi_\alpha(\varphi)), \phi \rangle_\alpha \\ &= \langle L_\alpha \varphi, \phi \rangle_\alpha - \langle N_\alpha(\varphi + \chi_\alpha(\varphi)), \phi \rangle_\alpha\end{aligned}$$

for all $\varphi, \phi \in E_\alpha^+$. It follows directly that

$$\left\langle \nabla^2 J_\alpha(\varphi)[\phi], \phi \right\rangle_\alpha = \langle L_\alpha \phi, \phi \rangle_\alpha - \left\langle \nabla N_\alpha(\varphi + \chi_\alpha(\varphi))\left[\phi + \nabla \chi_\alpha(\varphi)[\phi]\right], \phi \right\rangle_\alpha \tag{3.39}$$

By taking derivative with respect to φ in the relation (3.34), we have

$$L_\alpha\left(\nabla \chi_\alpha(\varphi)[\varphi]\right) = (I - P_\alpha)\nabla N_\alpha(\varphi + \chi_\alpha(\varphi))\left[\varphi + \nabla \chi_\alpha(\varphi)[\varphi]\right] \tag{3.40}$$

for $\varphi \in E_\alpha^+$. For ease of notations, we denote $z_\varphi = \varphi + \chi_\alpha(\varphi)$ and $w_\varphi = \nabla \chi_\alpha(\varphi)[\varphi] - \chi_\alpha(\varphi)$. Then we deduce from (3.39) and the relations (3.34) and (3.40) that

$$\begin{aligned}\left\langle \nabla^2 J_\alpha(\varphi)[\varphi], \varphi \right\rangle_\alpha &= \langle L_\alpha \varphi, \varphi \rangle_\alpha - \left\langle \nabla N_\alpha(z_\varphi)[z_\varphi + w_\varphi], \varphi \right\rangle_\alpha \\ &= \langle \nabla J_\alpha(\varphi), \varphi \rangle_\alpha + \left\langle N_\alpha(z_\varphi), \varphi \right\rangle_\alpha - \left\langle \nabla N_\alpha(z_\varphi)[z_\varphi + w_\varphi], \varphi \right\rangle_\alpha \\ &= \langle \nabla J_\alpha(\varphi), \varphi \rangle_\alpha - \left\langle \nabla N_\alpha(z_\varphi)[z_\varphi + w_\varphi], z_\varphi + w_\varphi \right\rangle_\alpha \\ &\quad + 2\left\langle N_\alpha(z_\varphi), w_\varphi \right\rangle_\alpha + \left\langle N_\alpha(z_\varphi), z_\varphi \right\rangle_\alpha + \left\langle L_\alpha w_\varphi, w_\varphi \right\rangle_\alpha \\ &\le \langle \nabla J_\alpha(\varphi), \varphi \rangle_\alpha - (\kappa - 1)\left\langle N_\alpha(z_\varphi), z_\varphi \right\rangle_\alpha - \delta \|w_\varphi\|_\alpha^2\end{aligned}$$

for $\varphi \in E_\alpha^+$, where we used (H5) and (2.17). Hence the assertion follows. □

It follows from the above two lemmas and the second part of (H2) that, for $\varphi \in \mathcal{N}_\alpha$, $\nabla K_\alpha(\varphi) \neq 0$. Hence $\mathcal{N}_\alpha$ is locally a C^1-manifold. Moreover, if $\varphi \in \mathcal{N}_\alpha$ is a constrained critical point of J_α, then φ is also a critical point of J_α on E_α. Indeed, according to the Lagrange multiplier rule, we have the existence of $\lambda \in \mathbb{R}$ such that $\nabla J_\alpha(\varphi) = \lambda \nabla K_\alpha(\varphi)$. Note that $\varphi \in \mathcal{N}_\alpha$, we find $\langle \nabla J_\alpha(\varphi), \varphi \rangle = 0$. Together with Lemma 3.31, this implies $\lambda = 0$ and φ is a critical point of J_α on E_α.

For later use, we also mention that if $\mathcal{N}_\alpha \neq \emptyset$ then $\mathcal{N}_\alpha$ is bounded away from 0, i.e., there exists $r_0 > 0$ such that $\|\varphi\|_\alpha > r_0$ for all $\varphi \in \mathcal{N}_\alpha$. In fact, for a given $\varphi \in S_\alpha^+$, let us consider the function $t \mapsto J_\alpha(t\varphi)$, $t \ge 0$. Then we have

$$\frac{d}{dt} J_\alpha(t\varphi) = \langle \nabla J_\alpha(t\varphi), \varphi \rangle_\alpha = \frac{K_\alpha(t\varphi)}{t}$$

and

$$\frac{d^2}{dt^2}J_\alpha(t\varphi) = \left\langle \nabla^2 J_\alpha(t\varphi)[\varphi], \varphi \right\rangle_\alpha = \frac{1}{t^2}\big(\langle \nabla K_\alpha(t\varphi), t\varphi\rangle_\alpha - K_\alpha(t\varphi)\big).$$

It follows from Lemma 3.31 that if there is $t_\varphi > 0$ such that $K_\alpha(t_\varphi\varphi) = 0$, i.e., $t_\varphi\varphi \in \mathcal{N}_\alpha$, then $\frac{d^2}{dt^2}J_\alpha(t\varphi)\big|_{t=t_\varphi} < 0$. And hence t_φ is a strict local maximum point. We actually have more. In fact, by Lemma 3.30 and 3.31, we can see that $K_\alpha(t\varphi) < 0$ for all $t > t_\varphi$. Therefore, t_φ should be the unique critical point of $J_\alpha(t\varphi)$, $t > 0$, and hence the global maximum point. Now, by (3.33), we find $\max_{t>0} J_\alpha(t\varphi) \geq \tau$, and this implies the existence of $r_0 > 0$ such that $t_\varphi \geq r_0$.

3.3.2 *Refined Upper Bound Expansion Via Test Elements*

We note that the min-max value in (3.37) has the following characterization

$$\gamma_\alpha = \inf_{\varphi\in E_\alpha^+\setminus\{0\}} \max_{t>0} J_\alpha(t\varphi) = \inf_{\varphi\in E_\alpha^+\setminus\{0\}} \max_{z\in\mathbb{R}^+\varphi\oplus E_\alpha^-} \Phi_\alpha(z) = \inf_{\varphi\in\mathcal{N}_\alpha} J_\alpha(\varphi). \tag{3.41}$$

Particularly, $\gamma_\alpha < +\infty$ provided that $\mathcal{N}_\alpha \neq \emptyset$. Our next step is to describe a qualitative property of γ_α for $\alpha \in \mathcal{A}$. Precisely, let us take an arbitrary sequence $\{\alpha_n\} \subset \mathcal{A}$ (it is allowed that $\{\alpha_n\}$ consists of only finite number of distinct values) and consider $\{z_n\} \subset E$ such that

$$c_1 \leq \Phi_n(z_n) \leq c_2 \quad \text{and} \quad \|\nabla\Phi_n(z_n)\|_n \to 0 \quad \text{as } n \to \infty, \tag{3.42}$$

for some fixed constants $c_1, c_2 > 0$, where we write $\Phi_n = \Phi_{\alpha_n}$ and $\|\cdot\|_n = \|\cdot\|_{\alpha_n}$ for simplicity. In this setting, if $\alpha_n \equiv \alpha \in \mathcal{A}$ (that is, α_n equals to a fixed value), then $\{z_n\}$ is nothing but a generic Palais-Smale sequence for Φ_α. When α_n is varying for different n, the situation is more involved since we have a sequence of functionals Φ_n and $\{z_n\}$ is no longer a Palais-Smale sequence. Particularly, the boundedness of z_n needs to be measured by $\|\cdot\|_n$ correspondingly.

Lemma 3.32 *Under* (3.42), *we have:*

(1) $\|z_n\|_n$ *is uniformly bounded in* n;
(2) $\|(I - P_n)z_n - \chi_n(P_n z_n)\|_n = O(\|\nabla\Phi_n(z_n)\|_n)$ *as* $n \to \infty$, *where* $P_n = P_{\alpha_n} : E \to E_{\alpha_n}^+$ *is the orthogonal projection and* $\chi_n = \chi_{\alpha_n} \in C^1(E_{\alpha_n}^+, E_{\alpha_n}^-)$ *is the reduction map;*
(3) $\nabla J_n(P_n z_n) \to 0$ *as* $n \to \infty$, *where* $J_n = J_{\alpha_n}$.

Proof By repeating the arguments in Lemma 3.28, we see the boundedness of $\|z_n\|_n$. To check (2), let us set $\phi_n = P_n z_n + \chi_n(P_n z_n)$ and $w_n = z_n - \phi_n$. Then, we have $w_n = (I - P_n)z_n - \chi_n(P_n z_n) \in E_{\alpha_n}^-$. Thanks to the relation (3.34), we see that

$$0 = \langle \nabla \Phi_n(\phi_n), w_n \rangle_n = \langle L_n \phi_n, w_n \rangle_n - \langle N_n(\phi_n), w_n \rangle_n$$

where, for simplicity, we write $L_n = L_{\alpha_n}$, $N_n = N_{\alpha_n}$ and $\langle \cdot, \cdot \rangle_n = \langle \cdot, \cdot \rangle_{\alpha_n}$. Since

$$\langle \nabla \Phi_n(z_n), w_n \rangle_n = \langle L_n z_n, w_n \rangle_n - \langle N_n(z_n), w_n \rangle_n ,$$

it follows that

$$\langle \nabla \Phi_n(z_n), w_n \rangle_n = \langle L_n w_n, w_n \rangle_n + \langle N_n(\phi_n), w_n \rangle_n - \langle N_n(z_n), w_n \rangle_n . \tag{3.43}$$

By the convexity in (H3), we see that

$$\langle N_n(z_n), w_n \rangle_n - \langle N_n(\phi_n), w_n \rangle_n = \langle \nabla N_n(\phi_n + \theta_n w_n)[w_n], w_n \rangle_n \geq 0$$

for some $\theta_n \in (0, 1)$. Together with (2.17) and (3.43), this implies

$$\|w_n\|_n^2 \leq \frac{1}{\delta} \|\nabla \Phi_n(z_n)\|_n \cdot \|w_n\|_n,$$

and hence $\|w_n\|_n = O(\|\nabla \Phi_n(z_n)\|_n)$ as $n \to \infty$.

To see (3), it follows from (2) and the C^2-smoothness of Φ_n that $\|\nabla \Phi_n(\phi_n)\|_n \to 0$. Finally, since $\|\nabla J_n(P_n z_n)\|_n = \|\nabla \Phi_n(\phi_n)\|_n$, we have $\|\nabla J_n(P_n z_n)\|_n \to 0$ as $n \to \infty$. □

Let $\mathscr{N}_n = \mathscr{N}_{\alpha_n}$ be the Nehari-type manifold for J_n, the next lemma guarantees that the infimum in (3.41) is well-defined, i.e., $\gamma_n = \gamma_{\alpha_n} < \infty$.

Lemma 3.33 *Under* (3.42), *we have* $\mathscr{N}_n \neq \emptyset$ *for n large. In particular, there eixsts* $t_n > 0$ *such that* $t_n P_n z_n \in \mathscr{N}_n$ *and* $|t_n - 1| = O(\|\nabla J_n(P_n z_n)\|_n)$ *as* $n \to \infty$.

Proof Let's use the same notation $\phi_n = P_n z_n + \chi_n(P_n z_n)$ as in the previous lemma. Then, from (3.42) and Lemma 3.32 (3), it follows that

$$\liminf_{n \to \infty} \langle N_n(\phi_n), \phi_n \rangle_n \geq c_0 \tag{3.44}$$

for some constant $c_0 > 0$. Indeed, up to a subsequence, if $\langle N_n(\phi_n), \phi_n \rangle_n \to 0$ as $n \to \infty$ then

$$\begin{aligned} \delta \|\phi_n\|_n &\leq \|L_n \phi_n\|_n \leq \|\nabla \Phi_n(\phi_n)\|_n + \|N_n(\phi_n)\|_n \\ &\leq \|\nabla \Phi_n(\phi_n)\|_n + K \langle N_n(\phi_n), \phi_n \rangle_n^{\mu} + \mu\delta \|\phi_n\|_n \quad \text{by (H4)} \\ &= o_n(1) + \mu\delta \|\phi_n\|_n. \end{aligned}$$

Since $\mu < 1$, we find $\|\phi_n\|_n \to 0$ as $n \to \infty$. This contradicts to $\Phi_n(\phi_n) \geq \Phi_n(z_n) \geq c_1 > 0$ as assumed in (3.42).

Let us set $\eta_n : [0, \infty) \to \mathbb{R}$ by $\eta_n(t) = \mathcal{K}_n(tP_n z_n)$, where $\mathcal{K}_n = \mathcal{K}_{\alpha_n}$ is defined by (3.38). One easily checks that $\eta_n'(t) = \langle \nabla \mathcal{K}_n(tP_n z_n), P_n z_n \rangle_n$. And hence, by Lemma 3.31 and the Taylor's expansion of η_n around $t = 1$, we get

$$t\eta_n'(t) \leq 2\eta_n(1) - (\kappa - 1)\langle N_n(\phi_n), \phi_n \rangle_n + O(|t-1|) \tag{3.45}$$

where the $O(|t-1|)$ term is independent of n due to the boundedness of $\eta_n'(t)$ (a direct consequence of Lemma 3.32 (1)).

Note that $\eta_n(1) = \langle \nabla J_n(P_n z_n), P_n z_n \rangle_n \to 0$ (by Lemma 3.32 (3)), we can conclude from (3.44) and (3.45) that there exists a small constant $\nu > 0$ such that

$$\eta_n'(t) \leq -\nu \text{ for all } t \in [1-\nu, 1+\nu] \text{ and } n \text{ large enough.}$$

In particular, $\eta_n(1-\nu) \geq \eta_n(1) + \nu^2 > 0$ and $\eta_n(1+\nu) \leq \eta_n(1) - \nu^2 < 0$ for n large. Then, by the Inverse Function Theorem, $t_n = \eta_n^{-1}(0)$ exists and $t_n P_n z_n \in \mathcal{N}_n$ for n large enough. Furthermore, since $|\eta_n'(t)^{-1}| \leq \frac{1}{\nu}$ on $[1-\nu, 1+\nu]$, we consequently get

$$|t_n - 1| = \left|\eta_n^{-1}(0) - \eta_n^{-1}(\mathcal{K}_n(P_n z_n))\right| \leq \frac{1}{\nu}|\mathcal{K}_n(P_n z_n)|.$$

Now the conclusion follows from $\mathcal{K}_n(P_n z_n) = O(\|\nabla J_n(P_n z_n)\|_n)$. □

Combining Lemmas 3.28–3.33, we have the next general result for the abstract problem (3.30). Notice that the proofs of Lemmas 3.28–3.33 and Proposition 3.29 do not need the hypothesis (H6), we state our result as follows.

Theorem 3.34 (Xu [15]) *Under the hypotheses (H1) to (H5). If there exist* $\{\alpha_n\} \subset \mathcal{A}$ *and* $\{z_n\} \subset E$ *verifying* (3.42), *then, for n large, the min-max value* $\gamma_n = \gamma_{\alpha_n} > 0$ *defined in* (3.41) *satisfies*

$$\gamma_n \leq \Phi_n(z_n) + O(\|\nabla \Phi_n(z_n)\|_n^2) \quad \text{as } n \to \infty.$$

Furthermore, if the hypothesis (H6) is satisfied, then γ_n *is a critical value of* Φ_n *for n large.*

Remark 3.35 Clearly, under the hypotheses of Theorem 3.34, there is a natural upper bound of γ_n, that is the constant c_2 in the assumption (3.42). However, this is not enough since sometimes it is crucial to have a more explicit asymptotic characterization for the energies (see for instance the resolution of classical Yamabe problem and its variations, elliptic Brezis-Nirenberg problem, etc.). As we will see later in the applications, the sequence $\{z_n\}$ in Theorem 3.34 plays the role of certain test elements for the functionals Φ_n. Our result shows a refined upper bound estimate for the min-max value γ_n, and makes it computable whenever $\{z_n\}$ is constructible or is known beforehand.

Proof of Theorem 3.34 We set $\phi_n = P_n z_n + \chi_n(P_n z_n)$ as before and $t_n > 0$ as was in Lemma 3.33, we also let $\psi_n = t_n P_n z_n + \chi_n(t_n P_n z_n)$. It follows from Lemma 3.32 and 3.33 that

$$\begin{aligned}\|z_n - \psi_n\|_n &\le \|z_n - \phi_n\|_n + |t_n - 1|\|P_n z_n\|_n + \|\chi_n(P_n z_n) - \chi_n(t_n P_n z_n)\|_n \\ &= O(\|\nabla\Phi_n(z_n)\|_n) + O(\|\nabla J_n(P_n z_n)\|_n)\end{aligned} \tag{3.46}$$

where we have used the inequality

$$\|\chi_n(P_n z_n) - \chi_n(t_n P_n z_n)\|_n \le \||\nabla\chi_n(\theta_n P_n z_n)\||_n \cdot \|P_n z_n\|_n \cdot |t_n - 1|$$

for some θ_n between t_n and 1. Since we have $\|\nabla J_n(P_n z_n)\|_n = \|\nabla\Phi_n(\phi_n)\|_n$ (by Proposition 3.29 (2)), we can see from Lemma 3.32 that

$$\|\nabla J_n(P_n z_n)\|_n \le \|\nabla\Phi_n(z_n)\|_n + O(\|\phi_n - z_n\|_n) = O(\|\nabla\Phi_n(z_n)\|_n).$$

Together with (3.46), this implies $\|z_n - \psi_n\|_n = O(\|\nabla\Phi_n(z_n)\|_n)$.

Now, using Taylor's expansion, we obtain

$$\begin{aligned}\Phi_n(z_n) &= \Phi_n(\psi_n) + \langle\nabla\Phi_n(\psi_n), z_n - \psi_n\rangle_n + O(\|z_n - \psi_n\|_n^2) \\ &= \Phi_n(\psi_n) + \langle\nabla\Phi_n(\psi_n), z_n - \psi_n\rangle_n + O(\|\nabla\Phi_n(z_n)\|_n^2).\end{aligned}$$

Since $\psi_n = t_n P_n z_n + \chi_n(t_n P_n z_n)$ and $t_n P_n z_n \in \mathcal{N}_n$, we have (by (3.34))

$$\begin{aligned}\langle\nabla\Phi_n(\psi_n), z_n - \psi_n\rangle_n &= \langle\nabla\Phi_n(\psi_n), P_n(z_n - \psi_n)\rangle_n \\ &= (1 - t_n)\langle\nabla J_n(t_n P_n z_n), P_n z_n\rangle_n = 0.\end{aligned}$$

Therefore, we have

$$\gamma_n = \inf_{\varphi\in\mathcal{N}_n} J_n(\varphi) \le J_n(t_n P_n z_n) = \Phi_n(\psi_n) = \Phi_n(z_n) + O(\|\nabla\Phi_n(z_n)\|_n^2)$$

as desired.

Finally, for a fixed n large, let us consider a minimizing sequence $\{\varphi_j\} \subset \mathcal{N}_n$ for J_n. By Lemma 3.28 and Proposition 3.29, we see that $\{z_j := \varphi_j + \chi_n(\varphi_j)\}$ is a bounded Palais-Smale sequence for Φ_n and $\liminf_{j\to\infty}\langle N_n(z_j), z_j\rangle_n > 0$ (similar to (3.44)). Thus the conclusion follows from (H6). □

References

1. Ackermann, N.: A nonlinear superposition principle and multibump solution of periodic Schrödinger equations. J. Funct. Anal. **234**, 423–443 (2006)
2. Bartsch, T.: Topological Methods for Variational Problems with Symmetries. Lect. Notes Math., vol. 1560. Springer, Berlin-Heidelberg (1993)
3. Bartsch, T., Ding, Y.H.: Deformation theorems on non-metrizable vector spaces and applications to critical point theory. Math. Nachr. **279**(12), 1267–1288 (2006)
4. Benci, V.: A geometrical index for the group S^1 and some applications to the study of periodic solutions of ordinary differential equations. Commun. Pure Appl. Math. **34**, 393–432 (1981)
5. Buffoni, B., Jeanjean, L., Stuart, C.A.: Existence of a nontrivial solution to a strongly indefinite semilinear equation. Proc. Am. Math. Soc. **119**(1), 179–186 (1993)
6. Clapp, M., Puppe, D.: Critical point theory with symmetries. J. Reine Angew. Math. **418**, 1–29 (1991)
7. de Figueiredo, D.G.: Semilinear elliptic systems: a survey of superlinear problems. Resenhas **2**(4), 373–391 (1996)
8. Ding, Y.H.: Variational Methods for Strongly Indefinite Problems. Interdiscip. Math. Sci., vol. 7. World Scientific Publ., Singapore (2007)
9. Ding, Y.H., Xu, T.: Concentrating patterns of reaction-diffusion systems: a variational approach. Trans. Am. Math. Soc. **369**(1), 97–138 (2017)
10. Ding, Y.H., Xu, T.: Effect of external potentials in a coupled system of multi-component incongruent diffusion. Topol. Methods Nonlinear Anal. **54**, 715–750 (2019)
11. Kryszewski, W., Szulkin, A.: Generalized linking theorem with an application to semilinear Schrödinger equation. Adv. Differ. Equ. **3**, 441–472 (1998)
12. Miguel, R., Hugo, T.: Solutions with multiple spike patterns for an elliptic system. Calc. Var. Partial Differ. Equ. **31**(1), 1–25 (2008)
13. Miguel, R., Yang, J.F.: Spike-layered solutions for an elliptic system with Neumann boundary conditions. Trans. Am. Math. Soc. **357**, 3265–3284 (2005)
14. Rabinowitz, P.H.: Minimax Methods in Critical Point Theory with Applications to Differential Equations. American Mathematical Society, Providence (1986)
15. Xu, T.: A local energy estimate for a class of strongly indefinite functionals. Minimax Theory Appl. **7**(2), 381–408 (2022). Special Issue for The Nehari Manifold: Theory, Applications, and Related Topics

Chapter 4
Semiclassical Standing Waves of Nonlinear Dirac Equations

In this chapter we will deal with the model (2.3) in external fields. We will mainly follow the works [2, 8] dealing with the localized concentrating behavior of solutions of nonlinear Dirac equations. For the sake of completeness, we will also discuss some further existence results in [10].

4.1 Basic Notations and Facts

We will consider the nonlinear Dirac equation like

$$-i\varepsilon\alpha\cdot\nabla u + a\beta u + V(x)u = f(|u|)u, \quad u \in H^1(\mathbb{R}^3, \mathbb{C}^4). \tag{4.1}$$

Here and throughout in the sequel, we fix the constant $a > 0$ and make the following assumption on the potential V

$(V0)$ $\quad V \in C^{0,1}(\mathbb{R}^3) \cap L^\infty(\mathbb{R}^3)$ and $|V|_\infty < a$.

Here we use the notation $|\cdot|_p$ for the various L^p-norms, $1 \le p \le \infty$.

We write $L^q = L^q(\mathbb{R}^3, \mathbb{C}^4)$ for $q \ge 1$ and $H^s = H^s(\mathbb{R}^3, \mathbb{C}^4)$ for $s > 0$. Let $D_a = -i\alpha\cdot\nabla + a\beta$ denote the self-adjoint operator on L^2 with domain $\mathcal{D}(D_a) = H^1$. It is well known that the spectrum of D_a is purely continuous and $\sigma(D_a) = \sigma_c(D_a) = \mathbb{R}\setminus(-a, a)$. Therefore L^2 possesses the orthogonal decomposition

$$L^2 = L^+ \oplus L^-, \quad u = u^+ + u^-, \tag{4.2}$$

so that D_a is positive definite (resp. negative definite) in L^+ (resp. L^-). Now let $E := \mathcal{D}(|D_a|^{1/2})$ be the form domain of D_a endowed with the inner product

$$\langle u, v\rangle = \operatorname{Re}\left(|D_a|^{1/2}u, |D_a|^{1/2}v\right)_2$$

© The Author(s), under exclusive license to Springer Nature Singapore Pte Ltd. 2025
Y. Ding, T. Xu, *Localization Approaches in Strongly Indefinite Problems*, Series in Contemporary Mathematics 6, https://doi.org/10.1007/978-981-97-9539-0_4

and induced norm $\|\cdot\|$; here $(\cdot,\cdot)_2$ denotes the L^2-inner product. This norm is equivalent to the usual $H^{1/2}$-norm, hence E embeds continuously into L^q for all $q \in [2,3]$ and compactly into L^q_{loc} for all $q \in [2,3)$. Clearly E possesses the decomposition

$$E = E^+ \oplus E^- \quad \text{with } E^\pm = E \cap L^\pm, \tag{4.3}$$

orthogonal with respect to both $(\cdot,\cdot)_2$ and $\langle\cdot,\cdot\rangle$. Since $\sigma(D_a) = \mathbb{R} \setminus (-a, a)$, one has

$$a|u|_2^2 \le \|u\|^2 \quad \text{for all } u \in E. \tag{4.4}$$

And remarkably, this decomposition of E induces also a natural decomposition of L^q for every $q \in (1,\infty)$ as proved in [8].

Proposition 4.1 (The L^q-extension of the Spectral Projections) *Setting $E_q^\pm := E^\pm \cap L^q$ for $q \in (1,\infty)$ there holds*

$$L^q = \mathrm{cl}_q\, E_q^+ \oplus \mathrm{cl}_q\, E_q^-$$

with cl_q denoting the closure in L^q. More precisely, for every $q \in (1,\infty)$ there exists $d_q > 0$ such that

$$d_q|u^\pm|_q \le |u|_q \quad \textit{for all } u \in E \cap L^q.$$

Here, in L^q's (for $q \neq 2$), by "$\oplus$" we mean the topologically direct sum. Before proving Proposition 4.1 we would like to introduce the following definition for *Multipliers* (see [16, Chapter 4]) which plays an important role in our arguments.

Definition 4.2 Let m be a bounded measurable function on $\mathbb{R}^n$, we associate a linear operator T_m on $L^2 \cap L^q$ by $(T_m u)\hat{}(\xi) = m(\xi)\hat{u}(\xi)$ where $\hat{u}$ denotes the Fourier transform of u. We say that m is a multiplier for L^q ($1 \le q \le \infty$) if whenever $u \in L^2 \cap L^q$ then $T_m u \in L^q$ (notice it is automatically in L^2), and T_m is bounded, that is,

$$|T_m u|_q \le A \cdot |u|_q, \quad u \in L^2 \cap L^q \text{ (with } A \text{ independent of } u). \tag{4.5}$$

Observe that if (4.5) is satisfied, and $p < \infty$, then T_m has a unique bounded extension to L^q, which again satisfies the same inequality.

Proof of Proposition 4.1 First we remark that in this context, the spatial domain is $\mathbb{R}^3$. Now recall the definitions for the matrices σ_k, $k = 1, 2, 3$ (see Pauli matrices), and let us study $D_a = -i\alpha \cdot \nabla + a\beta$. It is a differential operator with constant coefficients. In the Fourier domain $\xi = (\xi_1, \xi_2, \xi_3)$, it becomes the operator of multiplication by the matrix

$$\hat{D}_a(\xi) = \begin{pmatrix} 0 & \sum_{k=1}^3 \xi_k \sigma_k \\ \sum_{k=1}^3 \xi_k \sigma_k & 0 \end{pmatrix} + \begin{pmatrix} a & 0 \\ 0 & -a \end{pmatrix}.$$

By classical calculus we have that $\hat{D}_a(\xi)$ has two eigenvalues: $\pm\sqrt{a^2 + |\xi|^2}$. Now, denote $P^\pm$ the projections on E with kernel $E^\mp$. We see that in the Fourier domain, $P^\pm$ are multiplication operators by bounded smooth matrix-valued functions of ξ:

$$(P^+u)\hat{}(\xi) = \Big(\frac{1}{2} + \frac{a}{2\sqrt{a^2+|\xi|^2}}\Big) \begin{pmatrix} \mathrm{Id} & \Sigma(\xi) \\ \Sigma(\xi) & A(\xi) \end{pmatrix} \begin{pmatrix} \hat{U} \\ \hat{V} \end{pmatrix} \tag{4.6}$$

$$(P^-u)\hat{}(\xi) = \Big(\frac{1}{2} + \frac{a}{2\sqrt{a^2+|\xi|^2}}\Big) \begin{pmatrix} A(\xi) & -\Sigma(\xi) \\ -\Sigma(\xi) & \mathrm{Id} \end{pmatrix} \begin{pmatrix} \hat{U} \\ \hat{V} \end{pmatrix} \tag{4.7}$$

with Id being the 2×2 identity matrix and

$$A(\xi) = \frac{\sqrt{a^2+|\xi|^2} - a}{a + \sqrt{a^2+|\xi|^2}} \cdot \mathrm{Id}, \quad \Sigma(\xi) = \sum_{k=1}^{3} \frac{\xi_k \sigma_k}{a + \sqrt{a^2+|\xi|^2}}.$$

Here we have used the notation $\hat{u} = (\hat{U}, \hat{V}) \in \mathbb{C}^4$ and $\hat{U} = (\hat{u}_1, \hat{u}_2) \in \mathbb{C}^2$, $\hat{V} = (\hat{u}_3, \hat{u}_4) \in \mathbb{C}^2$.

In order that $P^\pm$ are multipliers for L^q, we need to use the *Marcinkiewicz multiplier theorem* on $\mathbb{R}^3$ (see [16, Chapter 4, Theorem 6']). A direct calculation shows that, for each component, the absolute value of all k-th $(0 < k \le 3)$ order partial derivatives for the multiplication functions are bounded by $B/|\xi|^k$ for some constant $B > 0$. And hence, as an immediate consequence, $P^\pm$ are *multipliers* for L^q for all $q \in (1, \infty)$. This implies that $P^\pm$ are continuous with respect to the L^q-norms. By noting that $P^\pm(E^\mp) = \{0\}$, one easily sees that $P^\pm$ extend to continuous projections on L^q (still denoted by $P^\pm$) with $P^\pm(\mathrm{cl}_q E_q^\mp) = \{0\}$. And this completes the proof. □

The proof of the above result implies even more information about the decomposition of E into $E^+ \oplus E^-$:

Proposition 4.3 *For $u \in H^1$ we have $\partial_k u^\pm = (\partial_k u)^\pm$.*

Proof This proposition follows from the fact that the matrix operations in (4.6) and (4.7) commute with the multiplication by $i\xi_k$ for $k = 1, 2, 3$. □

The proof of our main results will be achieved by variational methods applied to functionals $J : E \to \mathbb{R}$ of the form

$$J(u) = \frac{1}{2}\left(\|u^+\|^2 - \|u^-\|^2\right) + \frac{1}{2}\int_{\mathbb{R}^3} W(x)|u|^2\, dx - \int_{\mathbb{R}^3} G(x, |u|)\, dx. \tag{4.8}$$

The following result is a direct consequence of the infinite-dimensional reduction posed by Proposition 3.13.

Theorem 4.4 *Let $W \in L^\infty$ satisfy $|W|_\infty < a$ and suppose $G : \mathbb{R}^3 \times \mathbb{R}_0^+ \to \mathbb{R}$ has the form $G(x, s) = \int_0^s g(x, t)tdt$ where g is measurable in $x \in \mathbb{R}^3$, of class C^1 in $s \in \mathbb{R}_0^+$ and satisfies*

(i) $0 \le g(x, s)s$ *for all* $x \in \mathbb{R}^3$;
(ii) $g(x, s)s = o(s)$ *as* $s \to 0$ *uniformly in* $x \in \mathbb{R}^3$;
(iii) $0 \le \partial_s\big(g(x, s)s\big) \le Cs$ *for all* $x \in \mathbb{R}^3$, $s > 0$, *some* $C > 0$.

Then the following hold for J as in (4.8).

(a) *There exists a C^1-map $h_J : E^+ \to E^-$ such that for $v \in E^+$ and $w \in E^-$*

$$\nabla J(v + w)[\phi] = 0 \quad \text{for all } \phi \in E^- \qquad \Longleftrightarrow \qquad w = h_J(v)$$

and

$$\|h_J(v)\|^2 \le \frac{2|W|_\infty}{a - |W|_\infty}\|v\|^2 + \frac{2a}{a - |W|_\infty}\int_{\mathbb{R}^3} G(x, |v|)dx.$$

(b) *Setting $J^{red} : E^+ \to \mathbb{R}$, $J^{red}(v) := J(v + h_J(v))$, the sets*

$$\mathcal{M}^+(J) := \{v \in E^+ \setminus \{0\} : \nabla J^{red}(v)[v] = 0\}$$

and

$$\begin{aligned}\mathcal{M}(J) &:= \{v + h_J(v) \in E \setminus \{0\} : v \in \mathcal{M}^+(J)\}\\ &= \{u \in E \setminus \{0\} : \nabla J(u)|_{\mathbb{R}u \oplus E^-} = 0\}\end{aligned}$$

are C^1-submanifolds of E, diffeomorphic to an open subset of the unit sphere $SE^+ = \{v \in E^+ : \|v\| = 1\}$.

(c) *If $(v_n)_n$ is a Palais-Smale sequence for J^{red} then $\{v_n + h_J(v_n)\}_n$ is a Palais-Smale sequence for J.*

(d) *If $|g(x, s)| = O\left(|s|^{p-2}\right)$ as $|s| \to \infty$ for some $p \in (2, 3)$ then h_J is weakly sequentially continuous.*

The diffeomorphisms to an open subset of SE^+ are simply given by $u \mapsto \frac{u^+}{\|u^+\|}$. In the case $W \equiv \nu \in (-a, a)$ the manifold $\mathcal{M}(J)$ is the Nehari-Pankov manifold associated to J. It will be useful that the decomposition $E = E^+ \oplus E^-$ is independent of W and does not necessarily correspond to the decomposition of E into the positive and negative eigenspaces associated to $\nabla^2 J(0) = P^+ - P^- + W(x)$. We call J^{red} the reduced functional, h_J the reduction map, and (J^{red}, h_J) the *reduction couple* of J.

Remark 4.5 In the setting of Theorem 4.4, for each $v \in SE^+$ the map $\varphi_v(t) = J^{red}(tv)$ is C^2 and has at most one critical point $t_v > 0$, which is a nondegenerate maximum. There holds:

$$\mathcal{M}^+(J) = \{t_v v : v \in SE^+,\ \varphi_v'(t_v) = 0\}.$$

If G grows super-quadratically in t as $t \to \infty$ then $J(tu) \to -\infty$ as $t \to \infty$ and $\varphi_v(t)$ has a unique maximum for each $v \in SE^+$. Then $\mathcal{M}(J)$ and $\mathcal{M}^+(J)$ are diffeomorphic to SE^+. It is clear that $\mathcal{M}(J)$ contains all nontrivial critical points of J, and that for $u \in E \setminus \{0\}$ there holds:

$$\nabla J(u) = 0 \qquad \Longleftrightarrow \qquad u^- = h_J(u^+) \text{ and } \nabla J^{red}(u^+) = 0$$

Finally, the infimum of J on $\mathcal{M}(J)$ can be described as follows:

$$\begin{aligned} \gamma(J) := \inf_{u \in \mathcal{M}(J)} J(u) &= \inf_{v \in E^+ \setminus \{0\}} \sup_{u \in \mathbb{R}v \oplus E^-} J(u) \\ &= \inf_{v \in E^+ \setminus \{0\}} \max_{t > 0} J^{red}(tv) = \inf_{v \in \mathcal{M}^+(J)} J^{red}(v) \end{aligned} \tag{4.9}$$

If $\gamma(J)$ is achieved then it is the ground state energy.

The following technical result will be needed later.

Lemma 4.6 *For $v \in E^+ \setminus \{0\}$ the function $H(t) = J^{red}(tv) - \frac{t}{2}\nabla J^{red}(tv)[v]$ is of class C^1 and satisfies $H'(t) > 0$ for all $t > 0$.*

Proof We set $\varphi_v(t) = J^{red}(tv)$ so that $H(t) = \varphi_v(t) - \frac{t}{2}\varphi_v'(t)$. Since

$$H'(t) = \frac{1}{2}\varphi_v'(t) - \frac{t}{2}\varphi_v''(t) = \frac{1}{2t}\left[\varphi_{tv}'(1) - \varphi_{tv}''(1)\right],$$

it is sufficient to check that $\varphi_v'(1) - \varphi_v''(1) > 0$ for all $v \in E^+ \setminus \{0\}$. Setting $K(u) = \int_{\mathbb{R}^3} G(x, |u|)\, dx$, we have by the definition of h_J

$$-\langle h_J(v), \phi\rangle + \mathrm{Re}\int_{\mathbb{R}^3} W(x)(v + h_J(v)) \cdot \overline{\phi}\, dx - \nabla K(v + h_J(v))[\phi] = 0 \tag{4.10}$$

for all $\phi \in E^-$. It follows for $z_v = v + h_J(v)$ and $w_v = \nabla h_J(v)[v] - h_J(v)$ that

$$\varphi_v'(1) = \|v\|^2 + \mathrm{Re}\int_{\mathbb{R}^3} W(x) z_v \cdot \overline{v}\, dx - \nabla K(z_v)[v] = \nabla J(z_v)[z_v + w_v]. \tag{4.11}$$

Since (4.10) is valid for all $v \in E^+$, differentiating yields for all $\phi \in E^-$:

$$0 = -\langle \nabla h_J(v)[v], \phi \rangle + \mathrm{Re} \int_{\mathbb{R}^3} W(x)(v + \nabla h_J(v)[v]) \cdot \overline{\phi}\, dx \\ - \nabla^2 K(v + h_J(v))\big[v + \nabla h_J(v)[v], \phi\big].$$

Choosing $\phi = \nabla h_J(v)[v]$ in the above identity, so that $z_v + w_v = v + \phi$, we get

$$\begin{aligned}
\varphi_v''(1) &= \|v\|^2 + \mathrm{Re} \int_{\mathbb{R}^3} W(x)(v + \nabla h_J(v)[v]) \cdot \overline{v}\, dx - \nabla^2 K(z_v)[z_v + w_v, v] \\
&= \|v\|^2 - \|\phi\|^2 + \int_{\mathbb{R}^3} W(x)|v + \phi|^2\, dx - \nabla^2 K(z_v)[z_v + w_v, v + \phi] \\
&= \nabla^2 J(z_v)[z_v + w_v, z_v + w_v] \\
&= \|v\|^2 - \|h_J(v)\|^2 + \int_{\mathbb{R}^3} W(x)|z_v|^2\, dx - \nabla^2 K(z_v)[z_v, z_v] \\
&\quad + 2\left(-\langle h_J(v), w_v\rangle + \mathrm{Re} \int_{\mathbb{R}^3} W(x) z_v \cdot \overline{w_v}\, dx - \nabla^2 K(z_v)[z_v, w_v]\right) \\
&\quad + \left(-\|w_v\|^2 + \int_{\mathbb{R}^3} W(x)|w_v|^2\, dx - \nabla^2 K(z_v)[w_v, w_v]\right) \\
&= \varphi_v'(1) + \big(\nabla K(z_v)[z_v] - \nabla^2 K(z_v)[z_v, z_v]\big) \\
&\quad + 2\big(\nabla K(z_v)[w_v] - \nabla^2 K(z_v)[z_v, w_v]\big) \\
&\quad - \nabla^2 K(z_v)[w_v, w_v] - \|w_v\|^2 + \int_{\mathbb{R}^3} W(x)|w_v|^2\, dx.
\end{aligned}$$

Finally we obtain:

$$\varphi_v'(1) - \varphi_v''(1) \geq \int_{\mathbb{R}^3} \Big[G'(x, |z_v|)|w_v|^2 \\ + G''(x, |z_v|)|z_v| \Big(|z_v| + \frac{\mathrm{Re}\, z_v \cdot \overline{w_v}}{|z_v|}\Big)^2 \Big] dx > 0$$

□

4.2 The Unperturbed Equation

For $|\nu| < a$ the problem

$$-i\alpha \cdot \nabla u + a\beta u + \nu u = f(|u|)u, \quad u \in E, \tag{4.12}$$

appears as the unperturbed equation of (4.1). We begin with the following model case

$$-i\alpha\cdot\nabla u + a\beta u + \nu u = \lambda|u|^{p-2}u + \kappa|u|u \quad u \in E.$$

And we emphases that Theorem 4.4 applies in particular to (4.12) whose Euler functional will be depending on the parameters $\vec{\mu} = (\kappa, \lambda, \nu, p)$ with $\kappa, \lambda \geq 0$, $|\nu| < a$ and $p \in (2, 3)$:

$$J_{\vec{\mu}}(u) = \frac{1}{2}\big(\|u^+\|^2 - \|u^-\|^2\big) + \frac{\nu}{2}|u|_2^2 - \frac{\lambda}{p}|u|_p^p - \frac{\kappa}{3}|u|_3^3. \tag{4.13}$$

Proposition 4.7 *The infimum $\gamma(J_{\vec{\mu}})$ (given by (4.9)) is attained provided ν satisfies*

$$\Big(\frac{a^2}{a^2-\nu_-^2}\Big)^{\frac{3}{2}}\cdot\kappa^2\cdot\gamma(J_{\vec{\mu}}) < \frac{S^{\frac{3}{2}}}{6}, \tag{4.14}$$

where $\nu_- = \min\{0, \nu\}$ and $S > 0$ stands for the best constant for the embedding $H^1(\mathbb{R}^3, \mathbb{C}^4) \hookrightarrow L^6(\mathbb{R}^3, \mathbb{C}^4)$, i.e.,

$$S := \inf_{0\neq u\in H^1}\frac{|\nabla u|_2^2}{|u|_6^2}.$$

Proof We only give the proof for $\kappa > 0$ since $\kappa = 0$ refers to the subcritical growth and is much easier. Let $(J_{\vec{\mu}}^{red}, h_{\vec{\mu}})$ denote the reduction couple of $J_{\vec{\mu}}$ and let $(v_n)_n$ be a minimizing sequence for $J_{\vec{\mu}}^{red}$ in $\mathcal{M}^+(J_{\vec{\mu}})$. Setting $u_n = v_n + h_{\vec{\mu}}(v_n)$ it is not difficult to check that $(u_n)_n$ is bounded in E, hence it is either vanishing or non-vanishing up to a subsequence (see [14]).

If $(u_n)_n$ has a non-vanishing subsequence then we are done, so let us assume to the contrary that $(u_n)_n$ is vanishing, hence $|u_n|_p \to 0$. We first show that this implies

$$\gamma(J_{\vec{\mu}}) \geq \gamma(J_{\vec{\mu}_0}) \qquad \text{where } \vec{\mu}_0 = (\kappa, 0, \nu, p). \tag{4.15}$$

In order to see this let $t_n > 0$ be defined by $t_n v_n \in \mathcal{M}^+(J_{\vec{\mu}_0})$. Observe that $\|v_n\|$ is bounded away from 0 and the nonlinearity in $J_{\vec{\mu}_0}$ is super-quadratic, so that $(t_n)_n$ is bounded. Theorem 4.4 (d) now implies $|h_{\vec{\mu}_0}(t_n v_n)|_p \to 0$ where $h_{\vec{\mu}_0}$ is the reduction map for $J_{\vec{\mu}_0}$. Now (4.15) follows from

$$\begin{aligned}\gamma(J_{\vec{\mu}_0}) &\leq J_{\vec{\mu}_0}(t_n v_n + h_{\vec{\mu}_0}(t_n v_n))\\ &= J_{\vec{\mu}}(t_n v_n + h_{\vec{\mu}_0}(t_n v_n)) + o_n(1) \leq J_{\vec{\mu}}^{red}(v_n) + o_n(1) = \gamma(J_{\vec{\mu}}) + o_n(1).\end{aligned}$$

Next we show that

$$J_{\vec{\mu}_0}^{red}(v) \ge \frac{1}{6\kappa^2}\left(\frac{\|v\|^2 + \nu|v|_2^2}{|v|_3^2}\right)^3 \qquad \text{for all } v \in \mathcal{M}^+(J_{\vec{\mu}_0}). \tag{4.16}$$

For this we consider the functional

$$I : E \setminus \{0\} \to \mathbb{R}, \qquad u \mapsto \frac{\|u^+\|^2 - \|u^-\|^2 + \nu|u|_2^2}{|u|_3^2}.$$

For any $v \in E^+$ it is easy to see by a direct argument that $\sup_{w\in E^-} I(v+w) > 0$ is achieved by some $w_v \in E^-$. Moreover, for any $c > 0$ the set $\{w \in E^- : I(v+w) \ge c\}$ is strictly convex because

$$w \mapsto \|v\|^2 - \|w\|^2 + \nu|v+w|_2^2 - c|v+w|_3^2$$

is strictly concave on E^-. This also uses $|\nu| < a$. Hence w_v is the unique critical point of $w \mapsto I(v+w)$. On the other hand, for $v \in \mathcal{M}^+(J_{\vec{\mu}_0})$, we have

$$0 = \nabla J_{\vec{\mu}_0}^{red}(v)[v] = \|v\|^2 - \|h_{\vec{\mu}_0}(v)\|^2 + \nu|v + h_{\vec{\mu}_0}(v)|_2^2 - \kappa|v + h_{\vec{\mu}_0}(v)|_3^3, \tag{4.17}$$

hence

$$J_{\vec{\mu}_0}^{red}(v) = J_{\vec{\mu}_0}^{red}(v) - \frac{1}{2}\nabla J_{\vec{\mu}_0}^{red}(v)[v] = \frac{\kappa}{6}|v + h_{\vec{\mu}_0}(v)|_3^3.$$

A direct calculation gives

$$\nabla I\big(v + h_{\vec{\mu}_0}(v)\big)\big|_{E^-} = 0 \quad \text{and} \quad I\big(v + h_{\vec{\mu}_0}(v)\big) > 0$$

which implies $h_{\vec{\mu}_0}(v) = w_v$. Now (4.16) follows, using (4.17) once more:

$$J_{\vec{\mu}_0}^{red}(v) = \frac{\kappa}{6}|v + h_{\vec{\mu}_0}(v)|_3^3 = \frac{1}{6\kappa^2} I^3(v + h_{\vec{\mu}_0}(v)) \ge \frac{1}{6\kappa^2} I^3(v)$$

Finally, the proposition follows from (4.15), (4.16) and

$$\frac{\|v\|^2 + \nu|v|_2^2}{|v|_3^2} \ge \left(\frac{a^2 - \nu_-^2}{a^2}\right)^{\frac{1}{2}} S^{\frac{1}{2}} \qquad \text{for all } v \in \mathcal{M}^+(J_{\vec{\mu}_0}). \tag{4.18}$$

For the proof of (4.18) we pass to the Fourier domain and recall from [8] that

$$\|u\|^2 = \int_{\mathbb{R}^3} (a^2 + |\xi|^2)^{\frac{1}{2}} |\hat{u}|^2 \, d\xi \qquad \text{for all } u \in E.$$

Since $|\nu| < a$ we have

$$(a^2 + t^2)^{\frac{1}{2}} + \nu \geq \left(\frac{a^2 - \nu_-^2}{a^2}\right)^{\frac{1}{2}} |t| \quad \text{for all } t \in \mathbb{R}$$

which implies for $v \in E^+ \setminus \{0\}$:

$$\frac{\|v\|^2 + \nu|v|_2^2}{|v|_3^2} = \frac{\int_{\mathbb{R}^3}[(a^2 + |\xi|^2)^{\frac{1}{2}} + \nu] \cdot |\hat{v}|^2 \, d\xi}{|v|_3^2} \geq \left(\frac{a^2 - \nu_-^2}{a^2}\right)^{\frac{1}{2}} \frac{\int_{\mathbb{R}^3} |\xi||\hat{u}|^2 \, d\xi}{|u|_3^2}$$
$$\geq \left(\frac{a^2 - \nu_-^2}{a^2}\right)^{\frac{1}{2}} S^{\frac{1}{2}}$$

Here the last inequality follows from

$$\frac{\int_{\mathbb{R}^3} |\xi|^2|\hat{u}|^2 d\xi}{|u|_6^2} = \frac{|\widehat{\nabla u}|_2^2}{|u|_6^2} = \frac{|\nabla u|_2^2}{|u|_6^2} \geq S \quad \text{for all } u \in H^1(\mathbb{R}^3, \mathbb{C}^4)$$

and the Calderón-Lions interpolation theorem (see [15]). □

Let us remark that the domain of the quadratic form associated to the Dirac operator D_a is $H^{\frac{1}{2}}(\mathbb{R}^3, \mathbb{C}^4)$. This space embeds into the corresponding L^q-spaces for $2 \leq q \leq 3$, and the embedding is locally compact precisely if $q < 3$. Therefore the nonlinearity $f(|u|)u$ in (4.1) and (4.12) have subcritical growth if $f(s)s \sim s^{p-1}$ with $2 < p < 3$, and it has critical growth if $p = 3$. In the sequel, we denote $F(s) := \int_0^s f(t)t \, dt$ the primitive of the function $f(s)s$, we shall make the assumptions on f as

($f1$) $f \in C^0[0, \infty) \cap C^1(0, \infty)$ satisfies $f(0) = 0$ and $f'(s) > 0$ for $s > 0$.
($f2$) There exist $\lambda > 0$, $p \in (2, 3)$, $\kappa \geq 0$ such that $f(s) \geq \kappa s + \lambda s^{p-2}$ for $s > 0$, and $f'(s) \to \kappa$ as $s \to \infty$.
($f3$) There exists $\theta > 2$ such that $0 < \theta F(s) \leq f(s)s^2 + \frac{\theta-2}{3}\kappa s^3$ for $s > 0$.

These conditions imply that $s \mapsto f(s)s$ is strictly increasing and superlinear. Condition ($f3$) is a weakened Ambrosetti-Rabinowitz condition. If $\kappa > 0$ then the nonlinearity has critical growth.

Now we can consider the energy functional $I_\nu : E \to \mathbb{R}$ associated to (4.12) given by

$$I_\nu(u) = \frac{1}{2}\left(\|u^+\|^2 - \|u^-\|^2\right) + \frac{\nu}{2}|u|_2^2 - \int_{\mathbb{R}^3} F(|u|)dx. \tag{4.19}$$

The hypotheses ($f1$)–($f3$) imply that I_ν satisfies the assumptions of Theorem 4.4 for $|\nu| < a$.

Lemma 4.8 *If* $\nu_0 \in (-a, a)$ *satisfies* (4.14) *then* $\gamma(I_\nu)$ *is achieved for all* $\nu \in (-a, \nu_0]$. *Moreover, the map* $\nu \mapsto \gamma(I_\nu)$ *is continuous and strictly increasing.*

Proof For $\nu \in (-a, \nu_0] \subset (-a, a)$ assumption $(f3)$ implies $I_\nu \le J_{\vec{\mu}_1} \le J_{\vec{\mu}_2}$, where $\vec{\mu}_1 = (\kappa, \lambda, \nu_0, p)$ and $\vec{\mu}_2 = (0, \lambda, \nu_0, p)$. It follows that $\gamma(I_\nu) \le \gamma(J_{\vec{\mu}_1}) \le \gamma(J_{\vec{\mu}_2})$. A similar argument as in the proof of Proposition 4.7 implies the existence of a nontrivial critical point u_ν for I_ν such that u_ν^+ is the minimizer for I_ν^{red} on $\mathcal{M}^+(I_\nu)$.

In order to prove the monotonicity of $\gamma(\nu)$ we consider $-a < \nu_1 < \nu_2 \le \nu_0$. Let $u \in \mathcal{M}(I_{\nu_2})$ be a minimizer for $\gamma(I_{\nu_2})$ and define $s > 0$ by $su^+ \in \mathcal{M}^+(I_{\nu_1})$. Then we have, with $(I_{\nu_1}^{red}, h_{\nu_1})$ denoting the reduction couple for I_{ν_1} and $u_1 := t_1u^+ + h_{\nu_1}(su^+) \in \mathcal{M}(I_{\nu_1})$:

$$\begin{aligned}\gamma(I_{\nu_1}) &\le I_{\nu_1}^{red}(su^+) = I_{\nu_1}(u_1) = I_{\nu_2}(u_1) - \frac{\nu_2 - \nu_1}{2}|u_1|_2^2\\ &\le I_{\nu_2}^{red}(t_1u^+) - \frac{\nu_2 - \nu_1}{2}|u_1|_2^2\\ &\le \max_{t>0} I_{\nu_2}^{red}(tu^+) - \frac{\nu_2 - \nu_1}{2}|u_1|_2^2 = \gamma(I_{\nu_2}) - \frac{\nu_2 - \nu_1}{2}|u_1|_2^2.\end{aligned}$$

Choosing a minimizer $v \in \mathcal{M}(I_{\nu_1})$ for $\gamma(I_{\nu_1})$, defining $t > 0$ by $tv^+ \in \mathcal{M}^+(I_{\nu_2})$, and setting $u_2 := tv^+ + h_{\nu_1}(tv^+) \in \mathcal{M}(I_{\nu_2})$, an analogous argument shows that

$$\gamma(I_{\nu_2}) \le \gamma(I_{\nu_1}) + \frac{\nu_2 - \nu_1}{2}|u_2|_2^2.$$

For the continuity of $\gamma(\nu)$ it remains to prove that s, t are bounded for ν_1, ν_2 in a compact subset of $(-a, \nu_0]$ because then $|\gamma(I_{\nu_2}) - \gamma(I_{\nu_1})| = O(\nu_2 - \nu_1)$. This follows for s from

$$0 < I_{\nu_1}^{red}(su^+) \le \frac{s^2}{2}\left(\|u^+\|^2 + \nu_1|u^+|_2^2\right) - \frac{d_p\lambda}{p}s^p|u^+|_p^p.$$

where $d_p > 0$ is from Proposition 4.1. The bound for t is proved analogously. □

4.3 A Truncation Approach

For a subset $\Lambda \subset \mathbb{R}^3$, let Λ^c denote its complement, and $\Lambda^\varepsilon := \{x \in \mathbb{R}^3 : \varepsilon x \in \Lambda\}$, $\varepsilon > 0$. By the change of variables $x \mapsto \varepsilon x$ and setting $V_\varepsilon(x) = V(\varepsilon x)$, the singularly perturbed problem (4.1) is equivalent to

$$-i\alpha \cdot \nabla u + a\beta u + V_\varepsilon(x)u = f(|u|)u. \qquad (4.20)$$

In the sequel, we will modify the function f similar to [4, 5]. For

$$0 < \delta_0 \le \frac{a - |V|_\infty}{4}, \tag{4.21}$$

we define $\tilde{f} = \tilde{f}_{\delta_0} \in C^1(\mathbb{R}_0^+)$ by $\tilde{f}(0) = 0$ and

$$\frac{d}{ds}\big(\tilde{f}(s)s\big) = \min\big\{f'(s)s + f(s),\ \delta_0\big\}.$$

In case $\kappa = 0$, the choice $\delta_0 = \frac{a-|V|_\infty}{4}$ will be fine. For the critical case $\kappa > 0$ we need to make δ_0 smaller in the course of the proof. Let $\tilde{F}(s) = \int_0^s \tilde{f}(t)t\,dt$ be the primitive of $\tilde{f}(s)s$. By our assumptions on V there exists $R_1 > 0$ so that

$$\nabla V(x) \notin \mathbb{R}x \quad \text{for all } x \in \mathbb{R}^3 \text{ with } |x| = R_1 \text{ and } V(x) = V(0), \tag{4.22}$$

see [3]. We define the cut-off function $\chi : \mathbb{R}^3 \to [0, 1]$ by

$$\chi(x) = \begin{cases} 1, & \text{if } |x| < R_1 \\ \frac{2R_1 - |x|}{R_1}, & \text{if } R_1 \le |x| < 2R_1 \\ 0, & \text{if } |x| \ge 2R_1. \end{cases} \tag{4.23}$$

and consider

$$g(x, s) = \chi(x)f(s) + \big(1 - \chi(x)\big)\tilde{f}(s)$$

as well as

$$G(x, s) = \int_0^s g(x, t)t\,dt = \chi(x)F(s) + \big(1 - \chi(x)\big)\tilde{F}(s).$$

For later use, associated to the above notations, we denote $B_1 = B(0, R_1)$ and $B_2 = B(0, 2R_1)$ the open balls in $\mathbb{R}^3$ of radius R_1 and $2R_1$. The following lemma is easy to prove.

Lemma 4.9 *The function $G(x, s)$ satisfies the conditions (i)–(iii) from Theorem 4.4.*

In what follows, we will mainly focus on the truncated problem

$$-i\alpha \cdot \nabla u + a\beta u + V_\varepsilon(x)u = g_\varepsilon(x, |u|)u, \quad u \in E \tag{4.24}$$

where we write $g_\varepsilon(x,s) = g(\varepsilon x, s)$; we also use the notations χ_ε and G_ε for the dilations of χ and G, respectively. The corresponding energy functional is

$$\Phi_\varepsilon(u) = \frac{1}{2}\big(\|u^+\|^2 - \|u^-\|^2\big) + \frac{1}{2}\int_{\mathbb{R}^3} V_\varepsilon(x)|u|^2\,dx - \int_{\mathbb{R}^3} G_\varepsilon(x,|u|)\,dx.$$

As a direct consequence of Lemma 4.9, we can introduce $(\Phi_\varepsilon^{red}, h_\varepsilon)$ as the reduction couple of Φ_ε.

In order to establish a compactness result for Φ_ε, we first prove a bound for Palais-Smale sequences of Φ_ε that is uniform in ε.

Lemma 4.10 *For $c \in \mathbb{R}$ fixed, $(PS)_c$-sequences of Φ_ε are bounded uniformly in ε.*

Proof Given a $(PS)_c$-sequence $(u_n)_n$ for Φ_ε we have by our conditions on f:

$$\begin{aligned}
&\int_{\mathbb{R}^3} \chi_\varepsilon(x) f(|u_n|)|u_n| \cdot |u_n^+ - u_n^-|dx \\
&\quad\le \Big(\int_{\mathbb{R}^3} \chi_\varepsilon(x)\big(f(|u_n|)|u_n|\big)^{\frac{3}{2}}dx\Big)^{\frac{2}{3}} \cdot |u_n^+ - u_n^-|_3 \\
&\qquad + \delta_0 \int_{\mathbb{R}^3} \chi_\varepsilon(x)|u_n| \cdot |u_n^+ - u_n^-|dx \\
&\quad\le C_\theta\Big(\int_{\mathbb{R}^3} \chi_\varepsilon(x)\big(f(|u_n|)|u_n|^2 - 2F(|u_n|)\big)dx\Big)^{\frac{2}{3}} \|u_n\| + \delta_0|u_n|_2^2,
\end{aligned}$$

where $C_\theta > 0$ only depends on the constant $\theta > 2$ in $(f2)$. It follows from (4.21) that

$$\begin{aligned}
\Big(1 - \frac{|V|_\infty}{a}\Big)\|u_n\|^2 &\le \nabla\Phi_\varepsilon(u_n)[u_n^+ - u_n^-] + \int_{\mathbb{R}^3} g_\varepsilon(x,|u_n|)|u_n| \cdot |u_n^+ - u_n^-|dx \\
&\le C_\theta\Big(2\Phi_\varepsilon(u_n) - \nabla\Phi_\varepsilon(u_n)[u_n]\Big)^{\frac{2}{3}} \|u_n\| + 2\delta_0|u_n|_2^2 + o(\|u_n\|).
\end{aligned}$$

Now the lemma follows using (4.4):

$$\Big(1 - \frac{|V|_\infty + 2\delta_0}{a}\Big)\|u_n\|^2 \le C_\theta\big(2(c+o(1)) + o(\|u_n\|)\big)^{\frac{2}{3}}\|u_n\| + o(\|u_n\|). \qquad (4.25)$$

□

Now we can prove the Palais-Smale condition for Φ_ε. Recall that the nonlinearity G in Φ_ε depends on a constant δ_0; see (4.21).

Proposition 4.11 *If*

$$\kappa^2 \cdot c_0 < \left(\frac{a^2 - |V|_\infty^2}{a^2}\right)^{\frac{3}{2}} \cdot \frac{S^{\frac{3}{2}}}{6},$$

then there exists $\delta_0 > 0$ *such that the truncated functional* Φ_ε *satisfies the* $(PS)_c$*-condition for all* $c \le c_0$*, all* $\varepsilon > 0$.

Proof We choose $\delta_0 > 0$ so that

$$\left(\frac{a^2 - |V|_\infty^2}{a^2}\right)^{\frac{3}{2}} \frac{S^{\frac{3}{2}}}{6} > \left(\frac{a^2 - (|V|_\infty + \delta_0)^2}{a^2}\right)^{\frac{3}{2}} \frac{S^{\frac{3}{2}}}{6} > \kappa^2 \cdot c_0.$$

Let $(u_n)_n$ be a $(PS)_c$-sequence for Φ_ε with $c \le c_0$, any $\varepsilon > 0$. By Lemma 4.10 there exists $u \in E$ such that, along a subsequence, $u_n \rightharpoonup u$ in E and $u_n \to u$ strongly in L^q_{loc} for $q \in [2, 3)$. We want to show that $u_n \to u$ strongly in E.

Set $z_n = u_n - u$ so that $z_n \rightharpoonup 0$ in E and $\|u_n^\pm\|^2 = \|u^\pm\|^2 + \|z_n^\pm\|^2 + o_n(1)$. Note that

$$\lim_{s\to 0} \tilde{f}(s) = \lim_{s\to\infty} \frac{\tilde{f}(s)}{s} = 0 \qquad \text{and} \qquad \lim_{s\to 0} f(s) = \lim_{s\to\infty} \frac{f(s)}{s} - \kappa = 0.$$

By the Brezis-Lieb lemma (see for instance [17, Lemma 1.32]) there holds

$$\int_{\mathbb{R}^3} G_\varepsilon(x, |u_n|) = \int_{\mathbb{R}^3} G_\varepsilon(x, |u|) + \int_{\mathbb{R}^3} (1 - \chi_\varepsilon(x))\tilde{F}(|z_n|) + \frac{\kappa}{3}\int_{\mathbb{R}^3} \chi_\varepsilon(x)|z_n|^3 + o_n(1),$$

and

$$\int_{\mathbb{R}^3} g_\varepsilon(x, |u_n|)|u_n|^2 = \int_{\mathbb{R}^3} g_\varepsilon(x, |u|)|u|^2 + \int_{\mathbb{R}^3} (1 - \chi_\varepsilon(x))\tilde{f}(|z_n|)|z_n|^2 + \kappa\int_{\mathbb{R}^3} \chi_\varepsilon(x)|z_n|^3 + o_n(1).$$

Therefore

$$\Phi_\varepsilon(u_n) = \Phi_\varepsilon(u) + \Phi_\varepsilon(z_n) + o_n(1),$$

and

$$\nabla\Phi_\varepsilon(u_n)[u_n] = \nabla\Phi_\varepsilon(u)[u] + \nabla\Phi_\varepsilon(z_n)[z_n] + o_n(1).$$

Obviously, $\nabla\Phi_\varepsilon(u) = 0$, hence $\nabla\Phi_\varepsilon(z_n)[z_n] = o_n(1)$. We claim that

$$\nabla\Phi_\varepsilon(z_n) \to 0 \text{ as } n \to \infty. \tag{4.26}$$

In fact, consider $\varphi \in E$ with $\|\varphi\| \le 1$ and set $g^1(x,s) = g(x,s) - \kappa\chi(x)s$. We have

$$\begin{aligned}
\nabla\Phi_\varepsilon(u_n)[\varphi] &= \langle u_n^+ - u_n^-, \varphi\rangle + \mathrm{Re}\int_{\mathbb{R}^3} V_\varepsilon(x)u_n \cdot \bar{\varphi} - \mathrm{Re}\int_{\mathbb{R}^3} g_\varepsilon(x,|u_n|)u_n \cdot \bar{\varphi} \\
&= \langle z_n^+, \varphi^+\rangle - \langle z_n^-, \varphi^-\rangle + \langle u^+, \varphi^+\rangle - \langle u^-, \varphi^-\rangle \\
&\quad + \mathrm{Re}\int_{\mathbb{R}^3} V_\varepsilon(x)z_n \cdot \bar{\varphi} + \mathrm{Re}\int_{\mathbb{R}^3} V_\varepsilon(x)u \cdot \bar{\varphi} \\
&\quad - \mathrm{Re}\int_{\mathbb{R}^3} g_\varepsilon^1(x,|z_n|)z_n \cdot \bar{\varphi} - \mathrm{Re}\int_{\mathbb{R}^3} g_\varepsilon^1(x,|u|)u \cdot \bar{\varphi} \\
&\quad - \kappa \cdot \mathrm{Re}\int_{\mathbb{R}^3} \chi_\varepsilon(x)|z_n + u|(z_n + u) \cdot \bar{\varphi} + o_n(\|\varphi\|)
\end{aligned} \tag{4.27}$$

where we used $u_n = z_n + u$ and $\nabla\Phi_\varepsilon(u) = 0$. The estimate for the subcritical part

$$\mathrm{Re}\int_{\mathbb{R}^3} g_\varepsilon^1(x,|u_n|)u_n \cdot \bar{\varphi} - \mathrm{Re}\int_{\mathbb{R}^3} g_\varepsilon^1(x,|z_n|)z_n \cdot \bar{\varphi} - \mathrm{Re}\int_{\mathbb{R}^3} g_\varepsilon^1(x,|u|)u \cdot \bar{\varphi} = o_n(\|\varphi\|)$$

follows from a standard argument in [7, Lemma 7.10]. To estimate the last integral in (4.27), we set $\psi_n := |z_n + u|(z_n + u) - |z_n|z_n - |u|u$ and observe $|\psi_n| \le 2|z_n| \cdot |u|$. By the Egorov theorem there exists $\Theta_\sigma \subset B_2^\varepsilon$ such that $\mathrm{meas}(B_2^\varepsilon \setminus \Theta_\sigma) < \sigma$ and $z_n \to 0$ uniformly on Θ_σ as $n \to \infty$. Thus, by the Hölder inequality, we have

$$\begin{aligned}
\int_{\mathbb{R}^3} \chi_\varepsilon(x)|\psi_n| \cdot |\varphi| &\le \int_{\Theta_\sigma} |\psi_n| \cdot |\varphi| + \int_{B_2^\varepsilon \setminus \Theta_\sigma} |\psi_n| \cdot |\varphi| \\
&\le \int_{\Theta_\sigma} |\psi_n| \cdot |\varphi| + 2\left(\int_{B_2^\varepsilon \setminus \Theta_\sigma} |z_n|^3\right)^{\frac{1}{3}} \cdot \left(\int_{B_2^\varepsilon \setminus \Theta_\sigma} |u|^3\right)^{\frac{1}{3}} \cdot \left(\int_{B_2^\varepsilon \setminus \Theta_\sigma} |\varphi|^3\right)^{\frac{1}{3}}.
\end{aligned}$$

The first integral in the last line converges to 0 as $n \to \infty$ and the remaining integrals go to 0 uniformly in n as $\sigma \to 0$. This shows

$$\int_{\mathbb{R}^3} \chi_\varepsilon(x)|\psi_n| \cdot |\varphi| = o_n(\|\varphi\|) \quad \text{as } n \to \infty$$

and consequently, using again $\nabla\Phi_\varepsilon(u) = 0$,

$$\begin{aligned}\nabla\Phi_\varepsilon(u_n)[\varphi] &= \langle z_n^+, \varphi^+\rangle - \langle z_n^-, \varphi^-\rangle + \langle u^+, \varphi^+\rangle - \langle u^-, \varphi^-\rangle \\ &\quad + \mathrm{Re}\int_{\mathbb{R}^3} V_\varepsilon(x) z_n \cdot \bar{\varphi} + \mathrm{Re}\int_{\mathbb{R}^3} V_\varepsilon(x) u \cdot \bar{\varphi} \\ &\quad - \mathrm{Re}\int_{\mathbb{R}^3} g_\varepsilon^1(x, |z_n|) z_n \cdot \bar{\varphi} - \mathrm{Re}\int_{\mathbb{R}^3} g_\varepsilon^1(x, |u|) u \cdot \bar{\varphi} \\ &\quad - \kappa \cdot \mathrm{Re}\int_{\mathbb{R}^3} \chi_\varepsilon(x)|z_n| z_n \cdot \bar{\varphi} \\ &\quad - \kappa \cdot \mathrm{Re}\int_{\mathbb{R}^3} \chi_\varepsilon(x)|u| u \cdot \bar{\varphi} + o(\|\varphi\|) \\ &= \nabla\Phi_\varepsilon(z_n)[\varphi] + \nabla\Phi_\varepsilon(u)[\varphi] + o_n(\|\varphi\|) \\ &= \nabla\Phi_\varepsilon(z_n)[\varphi] + o_n(\|\varphi\|)\end{aligned}$$

It follows that $\nabla\Phi_\varepsilon(z_n) \to 0$ as $n \to \infty$ as claimed in (4.26). Now $\nabla\Phi_\varepsilon(z_n)[z_n^+ - z_n^-] = o_n(1)$ reads as

$$\begin{aligned}&\|z_n\|^2 + \mathrm{Re}\int_{\mathbb{R}^3} V_\varepsilon(x) z_n \cdot \overline{(z_n^+ - z_n^-)} \\ &= \int_{\mathbb{R}^3} \big(1 - \chi_\varepsilon(x)\big)\tilde{f}(|z_n|) z_n \cdot \overline{(z_n^+ - z_n^-)} \\ &\quad + \kappa \cdot \mathrm{Re}\int_{\mathbb{R}^3} \chi_\varepsilon(x)|z_n| z_n \cdot \overline{(z_n^+ - z_n^-)} + o_n(1).\end{aligned}$$

Then, by using the fact $\tilde{f}(s) \le \delta_0$ and (4.18), we obtain

$$\left(\frac{a^2 - (|V|_\infty + \delta_0)^2}{a^2}\right)^{\frac{1}{2}} S^{\frac{1}{2}} \left(\int_{\mathbb{R}^3} \chi_\varepsilon(x)|z_n|^3 dx\right)^{\frac{2}{3}} \le \kappa \cdot \int_{\mathbb{R}^3} \chi_\varepsilon(x)|z_n|^3 dx + o_n(1).$$

If $b := \lim_{n\to\infty}\int_{\mathbb{R}^3} \chi_\varepsilon(x)|z_n|^3 dx = 0$ then $\|z_n\| = o_n(1)$ and $u_n \to u$ strongly in E, as claimed. Suppose to the contrary that $b > 0$ so that

$$\left(\frac{a^2 - (|V|_\infty + \delta_0)^2}{a^2}\right)^{\frac{3}{2}} S^{\frac{3}{2}} \le \kappa^3 \cdot \int_{\mathbb{R}^3} \chi_\varepsilon(x)|z_n|^3 dx = \kappa^3 \cdot b + o_n(1).$$

In case $\kappa = 0$, this is a contradiction. In case $\kappa > 0$, using

$$\Phi_\varepsilon(u) = \int_{\mathbb{R}^3} \frac{1}{2} g_\varepsilon(x, |u|)|u|^2 - G_\varepsilon(x, |u|) \ge 0,$$

as well as $\Phi_\varepsilon(u_n) \geq \Phi_\varepsilon(z_n) + o_n(1)$ and $\nabla\Phi_\varepsilon(z_n)[z_n] = o_n(1)$, we obtain the contradiction

$$\kappa^2 \cdot c + o_n(1) \geq \frac{\kappa^3}{6} \cdot b + o_n(1) \geq \left(\frac{a^2 - (|V|_\infty + \delta_0)^2}{a^2}\right)^{\frac{3}{2}} \frac{S^{\frac{3}{2}}}{6} + o_n(1).$$

□

We finish this section with a couple of notations that will be of use later. For simplicity, when ν belongs to the range of $V(x)$ that is $\nu \in \{V(x) : x \in \mathbb{R}^3\}$, we denote $\nu_0 = V(0)$ and correspondingly

$$I_{\nu_0} = I_{V(0)}, \quad I_{\nu_0}^{red} = I_{V(0)}^{red}, \quad \gamma(I_{\nu_0}) = \gamma(I_{V(0)}). \tag{4.28}$$

Moreover, given arbitrarily $y \in \mathbb{R}^3$, we can define the functional $\Phi_y : E \to \mathbb{R}$,

$$\Phi_y(u) = \frac{1}{2}\left(\|u^+\|^2 - \|u^-\|^2\right) + \frac{V(y)}{2}|u|_2^2 - \int_{\mathbb{R}^3} G(y, |u|)dx,$$

and (Φ_y^{red}, h_y) the reduction couple associated to Φ_y. Plainly, the critical point of Φ_y are solutions of the problem

$$-i\alpha \cdot \nabla u + a\beta u + V(y)u = g(y, |u|)u.$$

When $y \in B_1$, we have $\Phi_y = I_{V(y)}$ and $h_y = h_{V(y)}$. Let us point out that, by virtue of [9, Lemma 4.3], we can conclude the following splitting type result.

Proposition 4.12 *For $y \in \mathbb{R}^3$, let us define the functional $\Phi_{\varepsilon,y} : E \to \mathbb{R}$,*

$$\Phi_{\varepsilon,y}(u) = \frac{1}{2}\left(\|u^+\|^2 - \|u^-\|^2\right) + \frac{1}{2}\int_{\mathbb{R}^3} V(\varepsilon x + y)|u|^2 dx - \int_{\mathbb{R}^3} G(\varepsilon x + y, |u|)dx,$$

and $(\Phi_{\varepsilon,y}^{red}, h_{\varepsilon,y})$ the associated reduction couple, we have that

(1) *let $\{y_\varepsilon\} \subset \mathbb{R}^3$ be such that $y_\varepsilon \to y$ for some $y \in \mathbb{R}^3$ then, up to a subsequence, $h_{\varepsilon,y_\varepsilon}(w) \to h_y(w)$ as $\varepsilon \to 0$ for each $w \in E^+$;*
(2) *let $\{y_\varepsilon\} \subset \mathbb{R}^3$ be such that $y_\varepsilon \to y$ for some $y \in \mathbb{R}^3$ and let $\{w_\varepsilon\} \subset E^+$ be such that $w_\varepsilon \rightharpoonup w$ for some $w \in E^+$ then, up to a subsequence,*

$$\|h_{\varepsilon,y_\varepsilon}(w_\varepsilon) - h_{\varepsilon,y_\varepsilon}(w_\varepsilon - w) - h_y(w)\| = o_\varepsilon(1)$$

as $\varepsilon \to 0$;

(3) *let* $\{y_\varepsilon\} \subset \mathbb{R}^3$ *be such that* $y_\varepsilon \to y$ *for some* $y \in \mathbb{R}^3$ *and let* $\{w_\varepsilon\} \subset E^+$ *be such that* $w_\varepsilon \rightharpoonup w$ *for some* $w \in E^+$ *then, up to a subsequence,*

$$\Phi^{red}_{\varepsilon,y_\varepsilon}(w_\varepsilon) - \Phi^{red}_{\varepsilon,y_\varepsilon}(w_\varepsilon - w) - \Phi^{red}_{y}(w) = o_\varepsilon(1)$$

and

$$\nabla\Phi^{red}_{\varepsilon,y_\varepsilon}(w_\varepsilon)[\varphi] - \nabla\Phi^{red}_{\varepsilon,y_\varepsilon}(w_\varepsilon - w)[\varphi] - \nabla\Phi_{y}(w)[\varphi] = o_\varepsilon(1)\|\varphi\|$$

uniformly for $\varphi \in E^+$ *as* $\varepsilon \to 0$.

Proof Observe that assertion (1) follows directly from Lemma 3.15 and that assertion (3) can be viewed as an immediate corollary of assertion (2). Hence, to complete the proof, it suffices to show that, as $\varepsilon \to 0$,

$$\begin{cases} y_\varepsilon \to y \text{ in } \mathbb{R}^3 \\ w_\varepsilon \rightharpoonup w \text{ in } E^+ \end{cases} \implies \|h_{\varepsilon,y_\varepsilon}(w_\varepsilon) - h_{\varepsilon,y_\varepsilon}(w_\varepsilon - w) - h_y(w)\| = o_\varepsilon(1). \tag{4.29}$$

To this end, we first claim that

$$\begin{aligned} &y_\varepsilon \to y \text{ in } \mathbb{R}^3 \text{ and } u_\varepsilon \rightharpoonup u \text{ in } E \text{ as } \varepsilon \to 0 \\ &\quad\implies \Phi_{\varepsilon,y_\varepsilon}(u_\varepsilon) - \Phi_{\varepsilon,y_\varepsilon}(u_\varepsilon - u) - \Phi_{\varepsilon,y_\varepsilon}(u) = o_\varepsilon(1) \quad \text{as } \varepsilon \to 0. \end{aligned} \tag{4.30}$$

This can be proved similarly as (4.27) in Proposition 4.11, therefore we omit the details. We only point out here that, for the nonlinear part, it suffices to check

$$\int_{\mathbb{R}^3} \big(G^1(\varepsilon x + y_\varepsilon, |u_\varepsilon|) - G^1(\varepsilon x + y_\varepsilon, |u_\varepsilon - u|) - G^1(\varepsilon x + y_\varepsilon, |u|)\big)dx = o_\varepsilon(1)$$

where $G^1(x,s) = G(x,s) - \frac{\kappa}{3}\chi(x)s^3$. Since G^1 is subcritical, the proof follows from a standard argument in [7, Lemma 7.10].

As a direct consequence of (4.30), we soon conclude that

$$\text{For any sequence } w_\varepsilon \rightharpoonup 0 \text{ in } E^+, \text{ we have that } h_{\varepsilon,y_\varepsilon}(w_\varepsilon) \rightharpoonup 0 \text{ in } E^-. \tag{4.31}$$

Indeed, notice that $h_{\varepsilon,y_\varepsilon}(w_\varepsilon)$ is bounded (see Theorem 4.4), we may assume up to a subsequence that $h_{\varepsilon,y_\varepsilon}(w_\varepsilon) \rightharpoonup u_0 \in E^-$. Then $u_\varepsilon \equiv w_\varepsilon + h_{\varepsilon,y_\varepsilon}(w_\varepsilon) \rightharpoonup u_0$. Now, remark that $\Psi_{\varepsilon,y_\varepsilon} \geq 0$, we conclude from (4.30) that

$$\frac{a - |V|_\infty}{2a}\|u_0\|^2 \leq -\Phi_{\varepsilon,y_\varepsilon}(u_0) = \Phi_{\varepsilon,y_\varepsilon}(u_\varepsilon - u_0) - \Phi_{\varepsilon,y_\varepsilon}(u_\varepsilon) + o_\varepsilon(1) \leq o_\varepsilon(1)$$

as $\varepsilon \to 0$. And hence $u_0 = 0$.

Now we are ready to show (4.29). Let $w_\varepsilon \rightharpoonup w$ in E^+. We may assume $h_{\varepsilon,y_\varepsilon}(w_\varepsilon) \rightharpoonup v$ in E^-. By (4.31), there holds $h_{\varepsilon,y_\varepsilon}(w_\varepsilon - w) \rightharpoonup 0$. Using (4.30) and assertion (1) (i.e. the fact that $h_{\varepsilon,y_\varepsilon}(w) \to h_y(w)$ as $\varepsilon \to 0$), we conclude that

$$\begin{aligned}\Phi_{\varepsilon,y_\varepsilon}\big(w_\varepsilon + h_{\varepsilon,y_\varepsilon}(w_\varepsilon)\big) &= \Phi_{\varepsilon,y_\varepsilon}(w+v) + \Phi_{\varepsilon,y_\varepsilon}\big(w_\varepsilon - w + h_{\varepsilon,y_\varepsilon}(w_\varepsilon) - v\big) + o_\varepsilon(1)\\ &\le \Phi_{\varepsilon,y_\varepsilon}\big(w + h_{\varepsilon,y_\varepsilon}(w)\big)\\ &\quad + \Phi_{\varepsilon,y_\varepsilon}\big(w_\varepsilon - w + h_{\varepsilon,y_\varepsilon}(w_\varepsilon - w)\big) + o_\varepsilon(1)\\ &= \Phi_{\varepsilon,y_\varepsilon}\big(w + h_y(w)\big)\\ &\quad + \Phi_{\varepsilon,y_\varepsilon}\big(w_\varepsilon - w + h_{\varepsilon,y_\varepsilon}(w_\varepsilon - w)\big) + o_\varepsilon(1)\\ &= \Phi_{\varepsilon,y_\varepsilon}\big(w_\varepsilon + h_{\varepsilon,y_\varepsilon}(w_\varepsilon - w) + h_y(w)\big) + o_\varepsilon(1)\end{aligned}$$

as $\varepsilon \to 0$. Now (use the identity (3.11) with obvious change of notations) we can deduce that

$$\frac{a - |V|_\infty}{2a}\|h_{\varepsilon,y_\varepsilon}(w_\varepsilon) - h_{\varepsilon,y_\varepsilon}(w_\varepsilon - w) - h_y(w)\|^2 \le o_\varepsilon(1)$$

and hence (4.29) is proved. □

4.4 Spikes at Isolated Critical Points of Potential *V*

In this section, besides the assumptions $(V0)$ and $(f1)$–$(f3)$ posed previously, we also require one of the following hypotheses:

$(V1)$ V is $\mathcal{C}^1$ in a neighborhood of 0, and 0 is an isolated local maximum or minimum of V.

$(V2)$ V is $\mathcal{C}^2$ in a neighborhood of 0, 0 is an isolated critical point, and there exists a vector space $X \subset \mathbb{R}^3$ such that:

- (a) $V|_X$ has a strict local maximum at 0;
- (b) $V|_{X^\perp}$ has a strict local minimum at 0.

In the case of $(V2)$ we may assume that $\{0\} \ne X \ne \mathbb{R}^3$ so that 0 is a possibly degenerate saddle point of V. The main purpose of this section is to prove the following theorem.

Theorem 4.13 (Bartsch and Xu [2]) *Assume that V satisfies $(V0)$ and one of $(V1)$ or $(V2)$. Suppose that f satisfies $(f1)$, $(f2)$ and $(f3)$ with κ being relatively bounded (in the sense of (4.32)). Then there exists $\varepsilon_0 > 0$ such that (4.1) has a solution u_ε for all $\varepsilon \in (0, \varepsilon_0]$. And these solutions have the following properties.*

(*i*) *$|u_\varepsilon|$ possesses a global maximum point $x_\varepsilon \in \mathbb{R}^3$ such that $x_\varepsilon \to 0$ as $\varepsilon \to 0$, and*

$$|u_\varepsilon(x)| \le C \exp\Big(-\frac{c}{\varepsilon}|x - x_\varepsilon|\Big)$$

with $C, c > 0$ independent of ε.

(*ii*) *The rescaled function $U_\varepsilon(x) = u_\varepsilon(\varepsilon x + x_\varepsilon)$ converges as $\varepsilon \to 0$ uniformly to a least energy solution $U : \mathbb{R}^3 \to \mathbb{C}^4$ of*

$$-i\alpha \cdot \nabla U + a\beta U + V(0)U = f(|U|)U.$$

Remark 4.14 Thus in the subcritical case $\kappa = 0$ equation (4.1) always has solutions with shape as in (*i*) and (*ii*). We do allow critical growth but the factor κ cannot be too large. As we will see later in (4.32), the possible values of κ depends on $|V|_\infty$, $\sup V$, λ and p. Some of its values can be bounded away from 0 by a positive number provided V is bounded away from $-a$ and $\sup V \le 0$. Moreover κ will stay close to 0 when $|V|_\infty$ is close to the constant a. It is an interesting open problem whether the restriction on κ can be removed.

4.4.1 The Min-Max Scheme

In this section, we will prove the existence of solutions to the truncated problem (4.24) and, by virtue of Lemma 4.8, we will restrict ourselves in the barrier $0 \le \kappa < \bar{\kappa}$ where $\bar{\kappa}$ is given by

$$\bar{\kappa} := \Big(\frac{a^2 - |V|_\infty^2}{a^2}\Big)^{\frac{3}{4}} S^{\frac{3}{4}} \big(6\gamma(J_{\vec{\mu}_V})\big)^{-\frac{1}{2}} \qquad \text{with } \vec{\mu}_V := (0, \lambda, \sup V, p). \tag{4.32}$$

We would like to emphasize that such a choice of $\bar{\kappa}$ can be interpreted as we choose $c_0 = \gamma(J_{\vec{\mu}_V})$ in Proposition 4.11. With all these notations, for such choice of κ, we can fix the constant $\delta_0 > 0$ properly small so that the Palais-Smale condition holds automatically in the energy range $\Phi_\varepsilon \le \gamma(J_{\vec{\mu}_V})$.

To begin with, let us mention that, under our hypotheses on V, there always exists a vector space $X \subset \mathbb{R}^3$ such that:

(*a*) $V|_X$ has a strict local maximum at 0;
(*b*) $V|_{X^\perp}$ has a strict local minimum at 0.

In fact, in case $(V1)$, $X = \mathbb{R}^3$ if 0 is local maximum or $X = \{0\}$ if 0 is local minimum, whereas, in case $(V2)$, X is the space spanned by eigenvectors associated to negative eigenvalues of $D^2V(0)$. Let $P_X : \mathbb{R}^3 \to X$ be the orthogonal projection (in the case $X = \{0\}$, P_X is simply the trivial projection).

In the next, solutions of (4.24) will be obtained as critical points of Φ_ε, and a key ingredient for the construction of a min-max scheme is using the reduction couple $(\Phi_\varepsilon^{red}, h_\varepsilon)$. However, due to the lack of information on the exact behavior of the reduction map $h_\varepsilon : E^+ \to E^-$, it seems hopeless to make a "path of least energy spikes" by proper scaling as was employed in [3, 13].

Recalling $\nu_0 = V(0)$, let us focus on functions in the subspace E^+. Denoted by $B_0 := B(0, R_0)$ for some $R_0 < R_1$. Let us choose a minimizer $U \in \mathcal{M}(I_{\nu_0})$ for $\gamma(I_{\nu_0})$ and consider the path $p_\varepsilon : B_0^\varepsilon \to \mathcal{M}^+(\Phi_\varepsilon)$ defined as

$$p_\varepsilon(\xi)(x) = t_{\xi,\varepsilon} U^+(x - \xi), \quad x \in \mathbb{R}^3,$$

where $\mathcal{M}^+(\Phi_\varepsilon) = \{w \in E^+ \setminus \{0\} : \nabla\Phi_\varepsilon^{red}(w)[w] = 0\}$ and $t_{\xi,\varepsilon}$ is the unique $t > 0$ such that

$$t_{\xi,\varepsilon} U^+(\cdot - \xi) \in \mathcal{M}^+(\Phi_\varepsilon).$$

We also define a family of deformations on $\mathcal{M}^+(\Phi_\varepsilon)$

$$\Gamma_\varepsilon \equiv \{\varphi : \mathcal{M}^+(\Phi_\varepsilon) \to \mathcal{M}^+(\Phi_\varepsilon) \text{ homeomorphism} : \\ \varphi(p_\varepsilon(\xi)) = p_\varepsilon(\xi) \text{ if } \xi \in \partial B_0^\varepsilon \cap X\}.$$

Then we define the min-max level

$$\gamma_\varepsilon := \inf_{\varphi\in\Gamma_\varepsilon} \max_{\xi\in\overline{B_0^\varepsilon}\cap X} \Phi_\varepsilon^{red}(\varphi(p_\varepsilon(\xi))). \tag{4.33}$$

We point out here that, in the case $X = \{0\}$, $\gamma_\varepsilon = \gamma(\Phi_\varepsilon) = \inf_{\mathcal{M}^+(\Phi_\varepsilon)} \Phi_\varepsilon^{red}$. A technical point we would like to emphasize, which constitutes a crucial difference with min-max quantity defined in [3], is the fact that the elements $p_\varepsilon(\xi)+h_\varepsilon(p_\varepsilon(\xi))$ do not resemble a least energy solution of I_ν since not much is known about the map $h_\varepsilon : E^+ \to E^-$.

Proposition 4.15 *There exist ε_0, $\delta > 0$ such that for any $\varepsilon \in (0, \varepsilon_0)$*

$$\Phi_\varepsilon^{red}(p_\varepsilon(\cdot))\big|_{\partial B_0^\varepsilon \cap X} \le \gamma(I_{\nu_0}) - \delta.$$

Proof To simplify notation, we use subscript "ξ" to indicate the coordinate translation of a function $u \in E$, that is, $u_\xi(x) = u(x - \xi)$. Then, on a fixed bounded interval $t \in [0, T_0]$ with some T_0 large, we have

$$\Phi_\varepsilon^{red}(tW_\xi) \le \frac{1}{2}\big(\|tW_\xi\|^2 - \|h_\varepsilon(tW_\xi)\|^2\big) + \frac{1}{2}\int_{\mathbb{R}^3} V_\varepsilon(x)|tW_\xi + h_\varepsilon(tW_\xi)|^2 dx \\ - \int_{B_1^\varepsilon} F(|tW_\xi + h_\varepsilon(tW_\xi)|)dx.$$

Let us first remark that there exists $\sigma > 0$ such that $V(\xi) \le \nu_0 - \sigma$ for all $\xi \in \partial B_0^\varepsilon \cap X$. Since $t \in [0, T_0]$ is bounded and $R_0 < R_1$, by (1) in Proposition 4.12, $h_\varepsilon(tW_\xi) = h_\varepsilon(tW)_\xi \to h_{V(\xi)}(tW)$ uniformly in t as $\varepsilon \to 0$. Thus, we deduce

$$\Phi_\varepsilon^{red}(tW_\xi) \le I_{\nu_0-\sigma}^{red}(tW) + o_\varepsilon(1) \quad \forall \xi \in \partial B_0^\varepsilon \cap X.$$

Finally, since $I_{\nu_0-\sigma} < I_{\nu_0}$ strictly on compact subsets, we have that

$$\begin{aligned}\max_{t>0} I_{\nu_0-\sigma}^{red}(tW) &= \max_{t>0} I_{\nu_0-\sigma}(tW + h_{\nu_0-\sigma}(tW)) \\ &< \max_{t>0} I_{\nu_0}(tW + h_{\nu_0-\sigma}(tW)) \\ &\le \max_{t>0} I_{\nu_0}^{red}(tW) = \gamma(I_{\nu_0}),\end{aligned}$$

which completes the proof. □

Proposition 4.16 *We have that*

$$\limsup_{\varepsilon\to 0} \gamma_\varepsilon \le \gamma(I_{\nu_0}).$$

Proof It suffices to show that

$$\limsup_{\varepsilon\to 0} \max_{\xi\in\overline{B_0^\varepsilon}\cap X} \Phi_\varepsilon^{red}(p_\varepsilon(\xi)) \le \gamma(I_{\nu_0}). \tag{4.34}$$

In the following we take a sequence $\varepsilon = \varepsilon_n \to 0$, but we drop the sub-index n for the sake of clarity. For every ε, there exists a maximum point $\xi_\varepsilon \in B_0^\varepsilon \cap X$ such that

$$\max_{\xi_\varepsilon\in\overline{B_0^\varepsilon}\cap X} \Phi_\varepsilon^{red}(p_\varepsilon(\xi)) = \Phi_\varepsilon^{red}(p_\varepsilon(\xi_\varepsilon)).$$

And we see that

$$\begin{aligned}\Phi_\varepsilon^{red}(p_\varepsilon(\xi_\varepsilon)) \le &\frac{1}{2}\left(\|t_\varepsilon W_{\xi_\varepsilon}\|^2 - \|h_\varepsilon(t_\varepsilon W_{\xi_\varepsilon})\|^2\right) \\ &+ \frac{1}{2}\int_{\mathbb{R}^3} V_\varepsilon(x)|t_\varepsilon W_{\xi_\varepsilon} + h_\varepsilon(t_\varepsilon W_{\xi_\varepsilon})|^2 dx \\ &- \int_{B_1^\varepsilon} F(|t_\varepsilon W_{\xi_\varepsilon} + h_\varepsilon(t_\varepsilon W_{\xi_\varepsilon})|)dx,\end{aligned}$$

where $t_\varepsilon = t_{\xi_\varepsilon,\varepsilon}$. Since we have $\{t_\varepsilon\}$ is bounded (up to a subsequence), we can assume that $t_\varepsilon \to t_0$ and $\varepsilon\xi_\varepsilon \to \xi_0 \in \overline{B_0} \cap X$. Then we can conclude that

$$\begin{aligned}\Phi_\varepsilon^{red}(p_\varepsilon(\xi_\varepsilon)) &\le \frac{1}{2}\big(\|t_0W\|^2 - \|h_{V(\xi_0)}(t_0W)\|^2\big)\\ &\quad + \frac{V(\xi_0)}{2}\int_{\mathbb{R}^3} |t_0W + h_{V(\xi_0)}(t_0W)|^2 dx\\ &\quad - \int_{\mathbb{R}^3} F(|t_0W + h_{V(\xi_0)}(t_0W)|)dx + o_\varepsilon(1)\\ &= I_{V(\xi_0)}^{red}(t_0W) + o_\varepsilon(1).\end{aligned}$$

Notice that $V(\xi_0) \le \nu_0$, then

$$I_{V(\xi_0)}^{red}(t_0W) \le \max_{t>0} I_{\nu_0}^{red}(tW) = \gamma(I_{\nu_0}),$$

and hence (4.34) holds. □

In the next, we will show that γ_ε is a critical value of Φ_ε. Motivated by [3, 6], we are going to give an estimate from below on γ_ε and show that $\gamma_\varepsilon \ge \gamma(I_{\nu_0}) + o_\varepsilon(1)$. And in order to do so, we need to compare γ_ε with another auxiliary minimization value. Firstly, set $B_3 = B(0, 3R_1)$ the open ball of radius $3R_1$ and $\zeta : \mathbb{R}^3 \to \mathbb{R}^3$ be a cut-off function

$$\zeta(x) = \begin{cases} x & \text{if } |x| < 3R_1,\\ 3R_1x/|x| & \text{if } |x| \ge 3R_1,\end{cases} \tag{4.35}$$

and let $Q_\varepsilon : \mathbb{R}^3 \to X$ be defined as $Q_\varepsilon(x) = P_X(\zeta(\varepsilon x))$. Then, let us define the barycenter type functional $\mathcal{B}_\varepsilon : E \setminus \{0\} \to \mathbb{R}$,

$$\mathcal{B}_\varepsilon(u) = \frac{\int_{\mathbb{R}^3} Q_\varepsilon(x)|u|^\theta dx}{\int_{\mathbb{R}^3} |u|^\theta dx}, \quad \text{for } u \in E \setminus \{0\}$$

where $\theta \in (2, 3)$ is the constant in $(f2)$. Recall that $(\Phi_\varepsilon^{red}, h_\varepsilon)$ is the reduction couple for Φ_ε and $\mathcal{M}^+(\Phi_\varepsilon) = \big\{w \in E^+ \setminus \{0\} : \nabla\Phi_\varepsilon^{red}(w)[w] = 0\big\}$, let us consider the following subset of functions in $\mathcal{M}^+(\Phi_\varepsilon)$:

$$\widetilde{\mathcal{M}}^+(\Phi_\varepsilon) = \big\{w \in \mathcal{M}^+(\Phi_\varepsilon) : \mathcal{B}_\varepsilon(w) = 0\big\}.$$

We also define the corresponding auxiliary minimization

$$b_\varepsilon \equiv \inf_{w \in \widetilde{\mathcal{M}}^+(\Phi_\varepsilon)} \Phi_\varepsilon^{red}(w). \tag{4.36}$$

When X is trivial, i.e. $X = \{0\}$, we have $\widetilde{\mathcal{M}}^+(\Phi_\varepsilon) = \mathcal{M}^+(\Phi_\varepsilon)$ and then $b_\varepsilon = \gamma_\varepsilon$.
The next lemma shows that b_ε is well-defined in general.

Lemma 4.17 *There exists ε_0, $\varrho > 0$ such that for $\varepsilon \in (0, \varepsilon_0)$,*

$$\gamma_\varepsilon \geq b_\varepsilon \geq \varrho.$$

Technically, the crucial difference with the barycenter quantity defined in [3, 6] is that the integrations in $\mathcal{B}_\varepsilon$ are taken over the whole space $\mathbb{R}^3$. The reason is twofold: firstly, the orthogonal projections associated to the decomposition $E = E^+ \oplus E^-$ are of convolution type with some tempered distributions $\rho^\pm$ (see an abstract result in [12] for operators that commutes with translations), and thus, making the choice of compact-supported functions in $E^\pm$ by simply multiplying smooth cut-off functions would be in our situation hopeless since the convolution with $\rho^\pm$ do not commute with the multiplication in general. Secondly, the barycenter of an element $w \in E^+$ does not exhibit the location of the mass of those $u \in E$ with $u^+ = w$. Therefore, it is not enough if we only consider the barycenter integrations over a bounded domain as was introduced in [3, 6].

Proof of Lemma 4.17 Since $b_\varepsilon \geq \varrho$ follows directly from $(f1)$–$(f3)$ for some $\varrho > 0$, we only need to prove that $\gamma_\varepsilon \geq b_\varepsilon$ for all small ε.

Motivated by [3], let us take an arbitrary $\varphi \in \Gamma_\varepsilon$. We define $\psi_\varepsilon : \overline{B_0} \cap X \to X$ as

$$\psi_\varepsilon(\xi) = \mathcal{B}_\varepsilon\big(\varphi(p_\varepsilon(\xi/\varepsilon))\big).$$

We point out here that, by the definition of Γ_ε, $\varphi(p_\varepsilon(\xi/\varepsilon)) \neq 0$ for all $\xi \in \overline{B_0} \cap X$, and so ψ_ε is well defined.

For $\xi \in \partial B_0 \cap X$, it can be seen from the definition of $\mathcal{B}_\varepsilon$ that

$$\psi_\varepsilon(\xi) = \xi + o_\varepsilon(1) \text{ uniformly in } \xi \in \partial B_0 \cap X, \text{ as } \varepsilon \to 0.$$

Therefore we can choose ε_0 small enough (independent of φ) so that, for all $\varepsilon \in (0, \varepsilon_0)$,

$$\deg(\psi_\varepsilon, B_0 \cap X, 0) = \deg(id, B_0 \cap X, 0) = 1.$$

Then we can conclude that for every ε, there exists $\xi_\varepsilon \in B_0 \cap X$ such that $\psi_\varepsilon(\xi_\varepsilon) = 0$.
Therefore, since $\xi_\varepsilon/\varepsilon \in \overline{B_0^\varepsilon} \cap X$, there follows

$$\max_{\xi \in \overline{B_0^\varepsilon} \cap X} \Phi_\varepsilon^{red}(\varphi(p_\varepsilon(\xi))) \geq \Phi_\varepsilon^{red}(\varphi(p_\varepsilon(\xi_\varepsilon/\varepsilon))) \geq b_\varepsilon,$$

which concludes the proof. □

Proposition 4.18 *We have that*

$$\liminf_{\varepsilon\to 0} b_\varepsilon \geq \gamma(I_{\nu_0}).$$

The proof of this proposition contains a localization argument and it will be presented in the next subsection. Assuming the conclusion for the moment, jointly with Proposition 4.16, we can obtain the following

Proposition 4.19 *There holds*

$$\lim_{\varepsilon\to 0} \gamma_\varepsilon = \gamma(I_{\nu_0}).$$

From Propositions 4.15 and 4.19, we can get $\gamma_\varepsilon > \Phi_\varepsilon^{red}(p_\varepsilon(\cdot))\big|_{\partial B_0^\varepsilon \cap X}$ for all small $\varepsilon > 0$. Recall that we have restricted $\kappa \in [0, \bar{\kappa})$, it follows that

$$\kappa^2 \cdot \gamma(I_{\nu_0}) < \Big(\frac{a^2 - |V|_\infty^2}{a^2}\Big)^{\frac{3}{2}} \frac{S^{\frac{3}{2}}}{6}$$

which guarantees the compactness. Thus, by Proposition 4.11, we easily obtain

Theorem 4.20 *There exists $\varepsilon_0 > 0$ such that for $\varepsilon \in (0, \varepsilon_0)$ there exists a solution z_ε of the problem* (4.24). *Moreover,* $\Phi_\varepsilon^{red}(z_\varepsilon^+) = \Phi_\varepsilon(z_\varepsilon) = \gamma_\varepsilon$.

4.4.2 A Localization Argument

The localization argument will be carried out in several steps. As a first step, we prove the existence of a minimizer u_ε to be auxiliary problem (4.36).

Lemma 4.21 *There exists $\varepsilon_0 > 0$ such that for any $\varepsilon \in (0, \varepsilon_0)$, there exist $u_\varepsilon \in E \setminus \{0\}$ with $\mathcal{B}_\varepsilon(u_\varepsilon^+) = 0$ and $\lambda_\varepsilon \in X$ such that*

$$-i\alpha\cdot\nabla u_\varepsilon + a\beta u_\varepsilon + V_\varepsilon(x)u_\varepsilon = g_\varepsilon(x, |u_\varepsilon|)u_\varepsilon + \big(\lambda_\varepsilon \cdot Q_\varepsilon(x)|u_\varepsilon^+|^{\theta-2}u_\varepsilon^+\big)^+ \tag{4.37}$$

and

$$\Phi_\varepsilon(u_\varepsilon) = b_\varepsilon.$$

Moreover, the sequence $\{u_\varepsilon\}$ is bounded in E.

Proof We sketch the proof as follows. For $\varepsilon > 0$ fixed, by the Ekeland variational principle, there exists a sequence $\{w_n\} \subset \widetilde{\mathcal{M}}^+(\Phi_\varepsilon)$ which is a constrained (PS)-sequence for Φ_ε^{red} at level b_ε, moreover, it can be deduced that there exists $\{\lambda_n\} \subset X$ such that

$$\Phi_\varepsilon^{red}(w_n) \to b_\varepsilon, \quad \text{as } n \to \infty, \tag{4.38}$$

$$\nabla \Phi_\varepsilon^{red}(w_n) - \frac{(\lambda_n \cdot Q_\varepsilon(x)|w_n|^{\theta-2}w_n)^+}{|w_n|_\theta^\theta} \to 0, \quad \text{as } n \to \infty. \tag{4.39}$$

Now, let us set $u_n = w_n + h_\varepsilon(w_n)$. Since $\mathcal{B}_\varepsilon(u_n^+) = \mathcal{B}_\varepsilon(w_n) = 0$, by (4.38) and (4.39), repeating the arguments of Lemma 4.10, we get that $\{u_n\}$ is bounded in E (uniformly with respect to ε) and, therefore, up to a subsequence, it converges weakly to some $u_\varepsilon \in E$. Since we have assumed $0 \le \kappa < \bar{\kappa}$, it follows that $b_\varepsilon \le \gamma_\varepsilon \le \gamma(I_{\nu_0}) + o_\varepsilon(1) \le \gamma(J_{\tilde{\mu}_V})$ for small ε. By Proposition 4.11, $\{u_n\}$ converges strongly in E, i.e. $u_n \to u_\varepsilon$ as $n \to \infty$. Note that $u_\varepsilon \ne 0$, $\liminf_{\varepsilon\to 0} b_\varepsilon > 0$, also the sequence λ_n is bounded, we have u_ε is the desired minimizer and this concludes the proof. □

Lemma 4.22 *We have that $u_\varepsilon^+ \chi_{B_2^\varepsilon}$ is non-vanishing.*

Proof We only consider the case $\kappa > 0$ since it is much easier when $\kappa = 0$. To the contrary, we assume that $u_\varepsilon^+ \chi_{B_2^\varepsilon}$ vanishes. Then we have $u_\varepsilon^+ \chi_{B_2^\varepsilon} \to 0$ in L^q for all $q \in (2, 3)$. At this point we first claim that

$$u_\varepsilon^+ \chi_{B_2^\varepsilon} \not\to 0 \text{ in } L^3. \tag{4.40}$$

Accepting this fact for the moment, let us consider the function

$$t \mapsto \Phi_\varepsilon(tu_\varepsilon^+)$$

and denote $t_\varepsilon > 0$ the unique maximum point which realizes its maximum. Then $\{t_\varepsilon\}$ is bounded. Set $z_\varepsilon = t_\varepsilon u_\varepsilon^+ \in E^+$, we have that $\nabla\Phi_\varepsilon(z_\varepsilon)[z_\varepsilon] = 0$ and hence

$$\begin{aligned}\|z_\varepsilon\|^2 + \int_{\mathbb{R}^3} V_\varepsilon(x)|z_\varepsilon|^2 dx = &\int_{\mathbb{R}^3} (1 - \chi_\varepsilon(x))\tilde{f}(|z_\varepsilon|)|z_\varepsilon|^2 dx \\ &+ \kappa \int_{\mathbb{R}^3} \chi_\varepsilon(x)|z_\varepsilon|^3 dx + o_\varepsilon(1).\end{aligned}$$

Since $u_\varepsilon^+ \chi_{B_2^\varepsilon} \not\to 0$ in L^3, similarly as that was argued in Proposition 4.11, we soon have that

$$\kappa^3 \int_{\mathbb{R}^3} \chi_\varepsilon(x)|z_\varepsilon|^3 dx + o_\varepsilon(1) \ge \Big(\frac{a^2 - (|V|_\infty + \delta_0)^2}{a^2}\Big)^{\frac{3}{2}} S^{\frac{3}{2}}.$$

And hence, thanks to our choice of $\kappa \in (0, \bar{\kappa})$, we get

$$\begin{aligned}\kappa^2\Phi_\varepsilon(z_\varepsilon) &= \kappa^2\Big(\Phi_\varepsilon(z_\varepsilon) - \frac{1}{2}\nabla\Phi_\varepsilon(z_\varepsilon)[z_\varepsilon]\Big)\\ &\geq \Big(\frac{a^2 - (|V|_\infty + \delta_0)^2}{a^2}\Big)^{\frac{3}{2}}\frac{S^{\frac{3}{2}}}{6} + o_\varepsilon(1)\\ &> \kappa^2\gamma(J_{\vec{\mu}_V}).\end{aligned}$$

Therefore, we have that

$$\gamma(J_{\vec{\mu}_V}) < \Phi_\varepsilon(z_\varepsilon) \leq \max_{t>0}\Phi_\varepsilon^{red}(tu_\varepsilon^+) = b_\varepsilon \leq \gamma(I_{\nu_0}) \quad \text{as } \varepsilon \to 0$$

which is impossible due to Lemma 4.8.

Now, it remains to show (4.40) is valid. Indeed, it follows from Lemma 4.21 that, for some $C > 0$,

$$\begin{aligned}b_\varepsilon = \Phi_\varepsilon(u_\varepsilon) &= \max_{t>0}\Phi_\varepsilon^{red}(tu_\varepsilon^+) \geq \max_{t>0}\Phi_\varepsilon(tu_\varepsilon^+)\\ &\geq \max_{t>0}\Big[\frac{t^2}{2}\Big(1 - \frac{|V|_\infty + \delta_0}{a}\Big)\|u_\varepsilon^+\|^2 - C\kappa t^3\int_{B_2^\varepsilon}|u_\varepsilon^+|^3 dx\Big].\end{aligned}$$

Then, if $u_\varepsilon^+\chi_{B_2^\varepsilon} \to 0$ in L^3 as $\varepsilon \to 0$, we can choose $T_0 > 0$ (independent of ε) large enough such that $\Phi_\varepsilon(T_0u_\varepsilon^+) > 2\gamma(J_{\vec{\mu}_V})$ for all small $\varepsilon > 0$, and we soon have that

$$\liminf_{\varepsilon\to 0} b_\varepsilon \geq \liminf_{\varepsilon\to 0}\Phi_\varepsilon(T_0u_\varepsilon^+) > \gamma(J_{\vec{\mu}_V})$$

which is absurd. □

Lemma 4.23 *We have that* $\{\lambda_\varepsilon\} \subset X$ *is bounded.*

Proof Let us assume that $\lambda_\varepsilon \neq 0$, otherwise we are done. In the sequel, let us set $\tilde{\lambda}_\varepsilon = \lambda_\varepsilon/|\lambda_\varepsilon|$. By elliptic regularity arguments we have that $u_\varepsilon \in \cap_{q\geq 2}W^{1,q}(\mathbb{R}^3, \mathbb{C}^4)$ and then, jointly with Proposition 4.3, we are allowed to multiply (4.37) by $\partial_{\tilde{\lambda}_\varepsilon}u_\varepsilon$. Then, we have

$$\begin{aligned}&\mathrm{Re}\int_{\mathbb{R}^3}\Big(-i\alpha\cdot\nabla u_\varepsilon + a\beta u_\varepsilon + V_\varepsilon(x)u_\varepsilon - g_\varepsilon(x, |u_\varepsilon|)u_\varepsilon\Big)\cdot\overline{\partial_{\tilde{\lambda}_\varepsilon}u_\varepsilon}\,dx\\ &= \mathrm{Re}\int_{\mathbb{R}^3}\lambda_\varepsilon\cdot Q_\varepsilon(x)|u_\varepsilon^+|^{\theta-2}u_\varepsilon^+\cdot\overline{\partial_{\tilde{\lambda}_\varepsilon}u_\varepsilon^+}\,dx.\end{aligned} \tag{4.41}$$

Now, let us evaluate each term of the previous equality. We get

$$0 = \mathrm{Re} \int_{\mathbb{R}^3} \partial_{\tilde{\lambda}_\varepsilon} \big[(-i\alpha \cdot \nabla u_\varepsilon) \cdot \overline{u_\varepsilon}\big] dx = 2\,\mathrm{Re} \int_{\mathbb{R}^3} (-i\alpha \cdot \nabla u_\varepsilon) \cdot \overline{\partial_{\tilde{\lambda}_\varepsilon} u_\varepsilon} dx$$

and so

$$\mathrm{Re} \int_{\mathbb{R}^3} (-i\alpha \cdot \nabla u_\varepsilon) \cdot \overline{\partial_{\tilde{\lambda}_\varepsilon} u_\varepsilon} dx = 0 \tag{4.42}$$

Analogously, we have

$$\begin{aligned} 0 &= \int_{\mathbb{R}^3} \partial_{\tilde{\lambda}_\varepsilon} \big[V_\varepsilon(x)|u_\varepsilon|^2\big] dx \\ &= \varepsilon \int_{\mathbb{R}^3} \partial_{\tilde{\lambda}_\varepsilon} V(\varepsilon x)|u_\varepsilon|^2 dx + 2\,\mathrm{Re} \int_{\mathbb{R}^3} V_\varepsilon(x) u_\varepsilon \cdot \overline{\partial_{\tilde{\lambda}_\varepsilon} u_\varepsilon} dx \end{aligned}$$

and so

$$\mathrm{Re} \int_{\mathbb{R}^3} V_\varepsilon(x) u_\varepsilon \cdot \overline{\partial_{\tilde{\lambda}_\varepsilon} u_\varepsilon} dx = -\frac{\varepsilon}{2} \int_{\mathbb{R}^3} \partial_{\tilde{\lambda}_\varepsilon} V(\varepsilon x)|u_\varepsilon|^2 dx = O(\varepsilon). \tag{4.43}$$

It also follows that

$$\mathrm{Re} \int_{\mathbb{R}^3} a\beta u_\varepsilon \cdot \overline{\partial_{\tilde{\lambda}_\varepsilon} u_\varepsilon} dx = 0. \tag{4.44}$$

For the nonlinear part, let us recall the definition of G_ε,

$$\partial_{\tilde{\lambda}_\varepsilon} G_\varepsilon(x, |u_\varepsilon|) = \varepsilon \partial_{\tilde{\lambda}_\varepsilon} \chi(\varepsilon x)\big(F(|u_\varepsilon|) - \tilde{F}(|u_\varepsilon|)\big) + \mathrm{Re}\ g_\varepsilon(x, |u_\varepsilon|) u_\varepsilon \cdot \overline{\partial_{\tilde{\lambda}_\varepsilon} u_\varepsilon},$$

then we have

$$\begin{aligned} 0 &= \int_{\mathbb{R}^3} \partial_{\tilde{\lambda}_\varepsilon} \big[G_\varepsilon(x, |u_\varepsilon|)\big] dx \\ &= \varepsilon \int_{\mathbb{R}^3} \big(F(|u_\varepsilon|) - \tilde{F}(|u_\varepsilon|)\big)\big(\partial_{\tilde{\lambda}_\varepsilon} \chi(\varepsilon x)\big) dx + \mathrm{Re} \int_{\mathbb{R}^3} g_\varepsilon(x, |u_\varepsilon|) u_\varepsilon \cdot \overline{\partial_{\tilde{\lambda}_\varepsilon} u_\varepsilon} dx \end{aligned}$$

and it follows that

$$\mathrm{Re} \int_{\mathbb{R}^3} g_\varepsilon(x, |u_\varepsilon|) u_\varepsilon \cdot \overline{\partial_{\tilde{\lambda}_\varepsilon} u_\varepsilon} dx = O(\varepsilon). \tag{4.45}$$

Finally

$$\begin{aligned} 0 &= \int_{\mathbb{R}^3} \partial_{\bar{\lambda}_\varepsilon}\big[\lambda_\varepsilon \cdot Q_\varepsilon(x)|u_\varepsilon^+|^\theta\big]dx \\ &= \varepsilon|\lambda_\varepsilon| \int_{B_3^\varepsilon} |u_\varepsilon^+|^\theta dx + \varepsilon|\lambda_\varepsilon| \int_{\mathbb{R}^3\setminus B_3^\varepsilon} \frac{3R_1}{\varepsilon|x|}\Big[1 - \frac{(\lambda_\varepsilon\cdot x)^2}{|\lambda_\varepsilon|^2|x|^2}\Big]|u_\varepsilon^+|^\theta dx \\ &\quad + \theta \operatorname{Re} \int_{\mathbb{R}^3} \lambda_\varepsilon \cdot Q_\varepsilon(x)|u_\varepsilon^+|^{\theta-2}u_\varepsilon^+ \cdot \overline{\partial_{\bar{\lambda}_\varepsilon} u_\varepsilon^+} dx \end{aligned}$$

Observe that $0 \le \partial_{\bar{\lambda}_\varepsilon} \lambda_\varepsilon \cdot Q_\varepsilon(x) \le \varepsilon|\lambda_\varepsilon|$ for all $x \in \mathbb{R}^3 \setminus B_3^\varepsilon$; this is the key point of our estimates. And hence

$$\begin{aligned} \operatorname{Re} &\int_{\mathbb{R}^3} \lambda_\varepsilon \cdot Q_\varepsilon(x)|u_\varepsilon^+|^{\theta-2}u_\varepsilon^+ \cdot \overline{\partial_{\bar{\lambda}_\varepsilon} u_\varepsilon^+} dx \\ &= -\frac{\varepsilon|\lambda_\varepsilon|}{\theta} \int_{B_3^\varepsilon} |u_\varepsilon^+|^2 dx - \frac{\varepsilon|\lambda_\varepsilon|}{\theta} \int_{\mathbb{R}^3\setminus B_3^\varepsilon} \frac{3R_1}{\varepsilon|x|}\Big[1 - \frac{(\lambda_\varepsilon\cdot x)^2}{|\lambda_\varepsilon|^2|x|^2}\Big]|u_\varepsilon^+|^2 dx. \end{aligned} \tag{4.46}$$

By (4.41)–(4.46) and Lemma 4.22, we conclude the boundedness of $\lambda_\varepsilon \in X$. □

In what follows, we consider a sequence $\varepsilon_k \to 0$ and assume that $\lambda_{\varepsilon_k} \to \bar{\lambda} \in X$. For simplicity, we still denote ε_k by ε. For a small $\delta > 0$, let us define

$$H_\varepsilon = \big\{x \in \mathbb{R}^3 : \bar{\lambda} \cdot Q_\varepsilon(x) \le \delta\big\}.$$

The next proposition gives a complete description of u_ε as $\varepsilon \to 0$. We recall the notations $B_2 = B(0, 2R_1)$ and $B_3 = B(0, 3R_1)$.

Proposition 4.24 *Passing to a subsequence if necessary, there exist $y_\varepsilon^1 \in H_\varepsilon$, $y_1 \in B_2$ and $u_1 \in E \setminus \{0\}$ with*

$$-i\alpha \cdot \nabla u_1 + a\beta u_1 + V(y_1)u_1 = g(y_1, |u_1|)u_1,$$

such that $\bar{\lambda} \cdot y_1 = 0$ and

$$\varepsilon y_\varepsilon^1 \to y_1, \quad \|u_\varepsilon - u_1(\cdot - y_\varepsilon^1)\| \to 0 \quad \text{as } \varepsilon \to 0.$$

Proof We divide the proof into different steps:

Step 1 $u_\varepsilon^+|_{H_\varepsilon} \not\to 0$ in the L^2-norm and L^θ-norm.

Let us first show that $u_\varepsilon^+ \not\to 0$ in $L^\theta(H_\varepsilon)$. Suppose contrarily that

$$\int_{H_\varepsilon} |u_\varepsilon^+|^\theta dx \to 0, \quad \text{as } \varepsilon \to 0.$$

Since $\mathcal{B}_\varepsilon(u_\varepsilon^+) = 0$ and $\bar{\lambda} \in X$, we have

$$0 = \int_{\mathbb{R}^3} \bar{\lambda} \cdot Q_\varepsilon(x)|u_\varepsilon^+|^\theta dx = \int_{H_\varepsilon} \bar{\lambda} \cdot Q_\varepsilon(x)|u_\varepsilon^+|^\theta dx + \int_{H_\varepsilon^c} \bar{\lambda} \cdot Q_\varepsilon(x)|u_\varepsilon^+|^\theta dx$$

$$\geq \delta \int_{H_\varepsilon^c} |u_\varepsilon^+|^\theta dx + \int_{H_\varepsilon} \bar{\lambda} \cdot Q_\varepsilon(x)|u_\varepsilon^+|^\theta dx.$$

Therefore

$$\delta \int_{H_\varepsilon^c} |u_\varepsilon^+|^\theta dx \leq \left| \int_{H_\varepsilon} \bar{\lambda} \cdot Q_\varepsilon(x)|u_\varepsilon^+|^\theta dx \right| \leq 3R_1|\bar{\lambda}| \int_{H_\varepsilon} |u_\varepsilon^+|^\theta dx$$

and so

$$\int_{H_\varepsilon^c} |u_\varepsilon^+|^\theta dx \to 0, \quad \text{as } \varepsilon \to 0.$$

Then we get $u_\varepsilon^+ \to 0$ in L^θ which is a contradiction with Lemma 4.22. Now, by the boundedness of $\{u_\varepsilon\}$ in E and so in L^3, we can conclude by interpolation: for a suitable $\mu \in (0, 1)$

$$0 < c \leq \|u_\varepsilon^+\|_{L^\theta(H_\varepsilon)} \leq \|u_\varepsilon^+\|^\mu_{L^2(H_\varepsilon)} \|u_\varepsilon^+\|^{1-\mu}_{L^3(H_\varepsilon)} \leq C\|u_\varepsilon^+\|^\mu_{L^2(H_\varepsilon)}.$$

Step 2 Passing to be limit by concentration-compactness.

By Step 1, we can conclude that $\{u_\varepsilon^+|_{H_\varepsilon}\}$ is non-vanishing. And hence, by concentration-compactness arguments (see [14]), there exist $y_\varepsilon^1 \in H_\varepsilon$ and $r > 0$ such that

$$\int_{B(y_\varepsilon^1, r)\cap H_\varepsilon} |u_\varepsilon^+|^2 \geq c > 0.$$

Therefore there exists $u_1 \in E \setminus \{0\}$ such that $v_\varepsilon^1 = u_\varepsilon(\cdot + y_\varepsilon^1) \rightharpoonup u_1$ in E.

Claim 4.1 $\{\varepsilon y_\varepsilon^1\}$ is bounded and, up to a subsequence, $\varepsilon y_\varepsilon^1 \to y_1 \in B_2$ as $\varepsilon \to 0$.

To see this, let us assume that $\varepsilon y_\varepsilon^1 \notin B_2$ and $\text{dist}(\varepsilon y_\varepsilon^1, \partial B_2)/\varepsilon \to \infty$. Observe that v_ε^1 solves the equation

$$\begin{aligned} &-i\alpha \cdot \nabla v_\varepsilon^1 + a\beta v_\varepsilon^1 + V(\varepsilon x + \varepsilon y_\varepsilon^1)v_\varepsilon^1 \\ &= g(\varepsilon x + \varepsilon y_\varepsilon^1, |v_\varepsilon^1|)v_\varepsilon^1 + \left(\lambda_\varepsilon \cdot Q_\varepsilon(x + y_\varepsilon^1)|v_\varepsilon^{1+}|^{\theta-2}v_\varepsilon^{1+}\right)^+, \end{aligned}$$

and if we assume that $V(\varepsilon y_\varepsilon^1) \to \nu_1$ as $\varepsilon \to 0$ (passing to a subsequence), we have that u_1 is a weak solution of

$$-i\alpha \cdot \nabla u + a\beta u + \nu_1 u = \tilde{f}(|u|)u + \left(\bar{\lambda} \cdot \tilde{y}_1 |u^+|^{\theta-2} u^+\right)^+ \tag{4.47}$$

where $\tilde{y}_1 \in B_3$ is given by

$$\tilde{y}_1 = \begin{cases} \lim\limits_{\varepsilon \to 0} \varepsilon y_\varepsilon^1 & \text{if } \varepsilon y_\varepsilon^1 \in B_3, \\ \lim\limits_{\varepsilon \to 0} \dfrac{3R_1 y_\varepsilon^1}{|y_\varepsilon^1|} & \text{if } \varepsilon y_\varepsilon^1 \in B_3^c. \end{cases}$$

Since $y_\varepsilon^1 \in H_\varepsilon$, we have that $\bar{\lambda} \cdot \tilde{y}_1 \le \delta$ and, by the definition of $\tilde{f}$, we easily get that $\bar{\lambda} \cdot \tilde{y}_1 > 0$ (otherwise u_1^+ should be 0). Now we let $\widetilde{\Phi}_1 : E \to \mathbb{R}$ denote the associated energy functional for (4.47), that is

$$\widetilde{\Phi}_1(u) = \frac{1}{2}\left(\|u^+\|^2 - \|u^-\|^2\right) + \frac{\nu_1}{2}|u|_2^2 - \int_{\mathbb{R}^3} \tilde{F}(|u|)dx - \frac{\bar{\lambda} \cdot \tilde{y}_1}{\theta} \int_{\mathbb{R}^3} |u^+|^\theta dx.$$

Remark that, for any $u \in E$ with $u^+ \neq 0$ and arbitrary $v \in E$, there holds that

$$\bar{\lambda} \cdot \tilde{y}_1 \int_{\mathbb{R}^3} |u^+|^{\theta-2}|v^+|^2 dx + (\theta - 2)\bar{\lambda} \cdot \tilde{y}_1 \int_{\mathbb{R}^3} |u^+|^{\theta-2}\left(|u^+| + \frac{\operatorname{Re} u^+ \cdot \overline{v^+}}{|u^+|}\right)^2 dx > 0.$$

As a consequence of [1, Theorem 5.1] (see also [9, Lemma 4.6]), we have that Theorem 4.4 applies to the situation here. So, we can take $(\widetilde{\Phi}_1^{red}, \tilde{h}_1)$ to be the reduction couple for $\widetilde{\Phi}_1$ and let $\tilde{\gamma}_1$ stand for the critical level realized by u_1, we then have

$$\begin{aligned} \tilde{\gamma}_1 = \widetilde{\Phi}_1^{red}(u_1^+) &= \max_{t>0} \widetilde{\Phi}_1^{red}(t u_1^+) \ge \max_{t>0} \widetilde{\Phi}_1(t u_1^+) \\ &\ge \max_{t>0} \frac{t^2}{2}\left(\|u_1^+\|^2 - (|V|_\infty + \delta_0)|u_1|_2^2\right) - \frac{\bar{\lambda} \cdot \tilde{y}_1}{\theta} t^\theta \int_{\mathbb{R}^3} |u_1^+|^\theta dx \\ &\ge \max_{t>0} \frac{t^2}{2}\left(\|u_1^+\|^2 - (|V|_\infty + \delta_0)|u_1|_2^2\right) - \frac{\delta}{\theta} t^\theta \int_{\mathbb{R}^3} |u_1^+|^\theta dx. \end{aligned}$$

Since $\|u_1\| \le \|v_\varepsilon^1\| = \|u_\varepsilon\| < \infty$, we can conclude that $\tilde{\gamma}_1 > 2\gamma(I_{\nu_0})$ provided that δ is fixed small enough. However, by Fatou's lemma, we get

$$\begin{aligned}
\tilde{\gamma}_1 &= \widetilde{\Phi}_1(u_1) - \frac{1}{2}\nabla\widetilde{\Phi}_1(u_1)[u_1] \\
&= \int_{\mathbb{R}^3} \frac{1}{2}\tilde{f}(|u_1|)|u_1|^2 - \tilde{F}(|u_1|)dx + \big(\frac{1}{2} - \frac{1}{\theta}\big)\bar{\lambda} \cdot \tilde{y}_1 |u_1^+|_\theta^\theta \\
&\le \int_{\mathbb{R}^3} \frac{1}{2}\tilde{f}(|u_1|)|u_1|^2 - \tilde{F}(|u_1|)dx + O(\delta) \\
&\le O(\delta) + \liminf_{\varepsilon\to 0} \int_{\mathbb{R}^3} \frac{1}{2}g(\varepsilon x + \varepsilon y_\varepsilon^1, |v_\varepsilon^1|)|v_\varepsilon^1|^2 - G(\varepsilon x + \varepsilon y_\varepsilon^1, |v_\varepsilon^1|)dx \\
&= O(\delta) + \liminf_{\varepsilon\to 0} \Phi_\varepsilon(u_\varepsilon) \le 2\gamma(I_{\nu_0})
\end{aligned}$$

which is impossible. This proves the claim.

Now by Claim 4.1, passing to the limit, we have u_1 is a weak solution of

$$-i\alpha \cdot \nabla u_1 + a\beta u_1 + V(y_1)u_1 = g(y_1, |u_1|)u_1 + \big(\bar{\lambda} \cdot y_1 |u_1^+|^{\theta-2}u_1^+\big)^+,$$

with $\varepsilon y_\varepsilon^1 \to y_1 \in B_2$ such that $\bar{\lambda} \cdot y_1 \le \delta$ and there exists $\bar{c} > 0$ such that

$$\|u_\varepsilon\| \ge \|u_1\| \ge \bar{c} > 0.$$

Let us define $z_{1,\varepsilon} = u_\varepsilon - u_1(\cdot - y_\varepsilon^1)$. We consider two possibilities: either $\|z_{1,\varepsilon}^+\| \to 0$ or not. In the first case the proposition should be proved. In the second case, there are two sub-cases: either $z_{1,\varepsilon}^+|_{H_\varepsilon} \to 0$ in the L^θ-norm or not.

Step 3 Assume that $z_{1,\varepsilon}^+|_{H_\varepsilon} \not\to 0$ in the L^θ-norm.

In this case, we can repeat the previous argument to the sequence $\{z_{1,\varepsilon}\}$ to obtain $y_\varepsilon^2 \in H_\varepsilon$ such that

$$\int_{B(y_\varepsilon^2, r)\cap H_\varepsilon} |z_{1,\varepsilon}^+|^2 \ge c > 0.$$

Therefore there exists $u_2 \in E\backslash\{0\}$ such that $v_\varepsilon^2 = z_{1,\varepsilon}(\cdot + y_\varepsilon^2) \rightharpoonup u_2$ in E. Moreover, $|y_\varepsilon^1 - y_\varepsilon^2| \to \infty$, $\varepsilon y_\varepsilon^2 \to y_2 \in B_2$, $\bar{\lambda} \cdot y_2 \le \delta$ and

$$-i\alpha \cdot \nabla u_2 + a\beta u_2 + V(y_2)u_2 = g(y_2, |u_2|)u_2 + \big(\bar{\lambda} \cdot y_2 |u_2^+|^{\theta-2}u_2^+\big)^+,$$

and $\|u_2\| \ge \bar{c} > 0$. Also, it follows from the weak convergence,

$$\|u_\varepsilon\|^2 \ge \|u_1\|^2 + \|u_2\|^2.$$

Let us set $z_{2,\varepsilon} = u_\varepsilon - u_1(\cdot - y_\varepsilon^1) - u_2(\cdot - y_\varepsilon^2)$. Suppose that $\|z_{2,\varepsilon}^+\| \not\to 0$ and $z_{2,\varepsilon}^+|_{H_\varepsilon} \not\to 0$ in L^θ, then we can argue again as above. And it is all clear that there exists $l \in \mathbb{N}$ such that, after repeating the above argument for l times, we can get that $z_{l,\varepsilon}^+|_{H_\varepsilon} \to 0$ in the L^θ-norm.

Step 4 $\|z_{l,\varepsilon}^+\| \to 0$ as $\varepsilon \to 0$.

To the contrary let us assume that $\|z_{l,\varepsilon}^+\| \not\to 0$. Since $Q_\varepsilon(\cdot)$ is bounded, it follows from a standard argument that

$$\begin{aligned}\mathrm{Re}\int_{\mathbb{R}^3} \lambda_\varepsilon \cdot Q_\varepsilon(x)|u_\varepsilon^+|^{\theta-2}u_\varepsilon^+ \cdot \overline{\varphi^+}dx &= \sum_{j=1}^{l} \bar{\lambda} \cdot y_j \,\mathrm{Re}\int_{\mathbb{R}^3} |u_j^+(\cdot - y_\varepsilon^j)|^{\theta-2}u_j^+(\cdot - y_\varepsilon^j) \cdot \overline{\varphi^+}dx \\ &\quad + \mathrm{Re}\int_{\mathbb{R}^3} \lambda_\varepsilon \cdot Q_\varepsilon(x)|z_{l,\varepsilon}^+|^{\theta-2}z_{l,\varepsilon}^+ \cdot \overline{\varphi^+}dx + o_\varepsilon(1)\|\varphi\|,\end{aligned}$$

uniformly for $\varphi \in E$ as $\varepsilon \to 0$ and, particularly,

$$\begin{aligned}\int_{\mathbb{R}^3} \lambda_\varepsilon \cdot Q_\varepsilon(x)|u_\varepsilon^+|^\theta dx &= \sum_{j=1}^{l} \bar{\lambda} \cdot y_j \int_{\mathbb{R}^3} |u_j^+|^\theta dx \\ &\quad + \int_{\mathbb{R}^3} \lambda_\varepsilon \cdot Q_\varepsilon(x)|z_{l,\varepsilon}^+|^\theta dx + o_\varepsilon(1).\end{aligned} \tag{4.48}$$

Since $\mathcal{B}_\varepsilon(u_\varepsilon^+) = 0$, together with Proposition 4.12, we can deduce from (4.48) that

$$\begin{aligned}o_\varepsilon(1) = \|z_{l,\varepsilon}^+ + h_\varepsilon(z_{l,\varepsilon}^+)\|^2 &+ \mathrm{Re}\int_{\mathbb{R}^3} V_\varepsilon(x)\big(z_{l,\varepsilon}^+ + h_\varepsilon(z_{l,\varepsilon}^+)\big) \cdot \overline{\big(z_{l,\varepsilon}^+ - h_\varepsilon(z_{l,\varepsilon}^+)\big)}dx \\ &- \mathrm{Re}\int_{\mathbb{R}^3} g_\varepsilon\big(x, |z_{l,\varepsilon}^+ + h_\varepsilon(z_{l,\varepsilon}^+)|\big)\big(z_{l,\varepsilon}^+ + h_\varepsilon(z_{l,\varepsilon}^+)\big) \cdot \overline{\big(z_{l,\varepsilon}^+ - h_\varepsilon(z_{l,\varepsilon}^+)\big)}dx \\ &- \int_{\mathbb{R}^3} \lambda_\varepsilon \cdot Q_\varepsilon(x)|z_{l,\varepsilon}^+|^\theta dx.\end{aligned} \tag{4.49}$$

Therefore, by $(f2)$ and Proposition 4.1, we obtain

$$\|z_{l,\varepsilon}^+ + h_\varepsilon(z_{l,\varepsilon}^+)\|^2 \le C|z_{l,\varepsilon}^+ + h_\varepsilon(z_{l,\varepsilon}^+)|_3^3 \le C'\|z_{l,\varepsilon}^+ + h_\varepsilon(z_{l,\varepsilon}^+)\|^3,$$

for some $C, C' > 0$ which implies there exists $c > 0$ such that $\|z_{l,\varepsilon}^+ + h_\varepsilon(z_{l,\varepsilon}^+)\| \ge c$. In what follows, for simplicity of notation, we denote $\bar{z}_{l,\varepsilon} = z_{l,\varepsilon}^+ + h_\varepsilon(z_{l,\varepsilon}^+)$. By (4.49) again, and a similar argument as in the proof of Lemma 4.10, we get that

$$\begin{aligned}\|\bar{z}_{l,\varepsilon}\|^2 &\le C_\theta\Big(\int_{\mathbb{R}^3} \chi_\varepsilon(x)(f(|\bar{z}_{l,\varepsilon}|)|\bar{z}_{l,\varepsilon}|^2 - 2F(|\bar{z}_{l,\varepsilon}|))dx\Big)^{\frac{2}{3}} |\bar{z}_{l,\varepsilon}^+ - \bar{z}_{l,\varepsilon}^-|_3 \\ &\quad + C\int_{\mathbb{R}^3} \lambda_\varepsilon \cdot Q_\varepsilon(x)|\bar{z}_{l,\varepsilon}^+|^\theta dx + o_\varepsilon(1) \\ &\le C'_\theta(2\Phi_\varepsilon^{red}(z_{l,\varepsilon}^+) - \nabla\Phi_\varepsilon^{red}(z_{l,\varepsilon}^+)[z_{l,\varepsilon}^+])^{\frac{2}{3}} \|\bar{z}_{l,\varepsilon}\| \\ &\quad + C\int_{\mathbb{R}^3} \lambda_\varepsilon \cdot Q_\varepsilon(x)|\bar{z}_{l,\varepsilon}^+|^\theta dx + o_\varepsilon(1)\end{aligned}$$

for some $C, C_\theta, C'_\theta > 0$. Remark that $\bar{z}_{l,\varepsilon}^+ = z_{l,\varepsilon}^+ \to 0$ in $L^\theta(H_\varepsilon)$. Then, it follows from $\|\bar{z}_{l,\varepsilon}\| \ge c$ and $(f2)$ that there exists constant $c' > 0$ (independent of R_1) such that

$$\liminf_{\varepsilon\to 0}\Big(\Phi_\varepsilon^{red}(z_{l,\varepsilon}^+) - \frac{1}{2}\nabla\Phi_\varepsilon^{red}(z_{l,\varepsilon}^+)[z_{l,\varepsilon}^+]\Big) \ge c'. \tag{4.50}$$

Next, let us distinguish two possible situations.

- *Case 1.* $\bar{\lambda} \cdot y_j \ge 0$ for all $j = 1, \dots, l$.

  Since $\mathcal{B}_\varepsilon(u_\varepsilon^+) = 0$, we have that

$$0 = \int_{H_\varepsilon} \lambda_\varepsilon \cdot Q_\varepsilon(x)|u_\varepsilon^+|^\theta dx + \int_{H_\varepsilon^c} \lambda_\varepsilon \cdot Q_\varepsilon(x)|u_\varepsilon^+|^\theta dx.$$

By virtue of $z_{l,\varepsilon}^+|_{H_\varepsilon} \to 0$ in the L^θ-norm and $\bar{\lambda} \cdot y_j \ge 0$ for all $j = 1, \dots, l$, we know that

$$\int_{H_\varepsilon} \lambda_\varepsilon \cdot Q_\varepsilon(x)|u_\varepsilon^+|^\theta dx \to \sum_{j=1}^{l} \bar{\lambda} \cdot y_j \int_{\mathbb{R}^3} |u_j^+|^\theta dx \ge 0,$$

whereas $\lambda_\varepsilon \cdot Q_\varepsilon(x) \ge \frac{1}{2}\delta > 0$ in H_ε^c. Thus we have

$$\bar{\lambda} \cdot y_j = 0, \quad \text{for all } j = 1, \dots, l,$$

and so

$$\frac{\delta}{2}\int_{H_\varepsilon^c} |u_\varepsilon^+|^\theta dx \le \int_{H_\varepsilon^c} \lambda_\varepsilon \cdot Q_\varepsilon(x)|u_\varepsilon^+|^\theta dx \to 0, \quad \text{as } \varepsilon \to 0.$$

We also deduce from (4.48) that

$$\int_{H_\varepsilon^c} \lambda_\varepsilon \cdot Q_\varepsilon(x)|z_{l,\varepsilon}^+|^\theta dx \to 0, \quad \text{as } \varepsilon \to 0.$$

With all those information in hand, by Proposition 4.12, we can estimate the energy $\Phi_\varepsilon^{red}(u_\varepsilon^+)$ as

$$\Phi_\varepsilon^{red}(u_\varepsilon^+) = \Phi_\varepsilon^{red}(z_{l,\varepsilon}^+) + \sum_{j=1}^{l} \Phi_{y_j}^{red}(u_j^+) + o_\varepsilon(1).$$

Moreover, we have that

$$\nabla\Phi_\varepsilon^{red}(u_\varepsilon^+)[u_\varepsilon^+] = \nabla\Phi_\varepsilon^{red}(z_{l,\varepsilon}^+)[z_{l,\varepsilon}^+] + \sum_{j=1}^{l} \nabla\Phi_{y_j}^{red}(u_j^+)[u_j^+] + o_\varepsilon(1).$$

Since $\bar{\lambda} \cdot y_j = 0$ for all $j = 1, \ldots, l$, we have u_j^+'s are critical points of $\Phi_{y_j}^{red}$. And so, we get the estimate

$$\liminf_{\varepsilon\to 0} b_\varepsilon = \liminf_{\varepsilon\to 0} \Phi_\varepsilon^{red}(u_\varepsilon^+) = \liminf_{\varepsilon\to 0} \Big(\Phi_\varepsilon^{red}(z_{l,\varepsilon}^+) - \frac{1}{2}\nabla\Phi_\varepsilon^{red}(z_{l,\varepsilon}^+)[z_{l,\varepsilon}^+] \Big) + \sum_{j=1}^{l} \Phi_{y_j}^{red}(u_j^+).$$

Recall that we have denoted $\bar{z}_{l,\varepsilon} = z_{l,\varepsilon}^+ + h_\varepsilon(z_{l,\varepsilon}^+)$, hence, by (4.50), we have

$$\liminf_{\varepsilon\to 0} b_\varepsilon \geq c' + \sum_{j=1}^{l} \Phi_{y_j}^{red}(u_j^+).$$

Since, by Lemma 4.9, we have that $\Phi_{y_j}^{red}(w) \geq I_{V(y_j)}^{red}(w)$, $\forall w \in E^+$, for all $j = 1, \ldots, l$, we can infer that

$$\Phi_{y_j}^{red}(u_j^+) \geq \gamma(I_{V(y_j)}), \quad j = 1, \ldots, l.$$

And therefore

$$\liminf_{\varepsilon\to 0} b_\varepsilon \geq l \cdot \min_{j=1,\ldots,l} \gamma(I_{V(y_j)}) + c'.$$

Remark that $y_j \in B_2 = B(0, 2R_1)$, by shrinking R_1 if necessary, we can conclude from the continuity of the map $\nu \mapsto \gamma(I_\nu)$ that

$$\left|\gamma(I_{V(y_j)}) - \gamma(I_{\nu_0})\right| < \frac{1}{2}c' \quad \text{for all } j = 1, \dots, l,$$

and then we obtain

$$\liminf_{\varepsilon \to 0} b_\varepsilon \geq \gamma(I_{\nu_0}) + \frac{1}{2}c' > \gamma(I_{\nu_0})$$

which contradicts to Proposition 4.16 and Lemma 4.17.

- *Case 2.* There exists $\{j_1, \dots, j_k\} \subset \{1, \dots, l\}$ such that $\bar{\lambda} \cdot y_{j_m} < 0$ for $m = 1, \dots, k$.

  In this case, similar as that in Case 1, we can apply Proposition 4.12 to obtain

$$\Phi_\varepsilon^{red}(u_\varepsilon^+) = \left(\Phi_\varepsilon^{red}(z_{l,\varepsilon}^+) - \frac{1}{2}\nabla\Phi_\varepsilon^{red}(z_{l,\varepsilon}^+)[z_{l,\varepsilon}^+]\right) + \sum_{j=1}^{l}\left(\Phi_{y_j}^{red}(u_j^+) - \frac{1}{2}\nabla\Phi_{y_j}^{red}(u_j^+)[u_j^+]\right) + o_\varepsilon(1).$$

By the definition of $G(x, s)$, we have $\Phi_{y_j}^{red}(u_j^+) - \frac{1}{2}\nabla\Phi_{y_j}^{red}(u_j^+)[u_j^+] \geq 0$ for all $j = 1, \dots, l$. Then, we conclude that

$$\Phi_\varepsilon^{red}(u_\varepsilon^+) \geq \left(\Phi_\varepsilon^{red}(z_{l,\varepsilon}^+) - \frac{1}{2}\nabla\Phi_\varepsilon^{red}(z_{l,\varepsilon}^+)[z_{l,\varepsilon}^+]\right) + \sum_{m=1}^{k}\left(\Phi_{y_{j_m}}^{red}(u_{j_m}^+) - \frac{1}{2}\nabla\Phi_{y_{j_m}}^{red}(u_{j_m}^+)[u_{j_m}^+]\right) + o_\varepsilon(1). \tag{4.51}$$

To evaluate the above inequality, let us denote $\mathcal{M}^+(\Phi_{y_{j_m}}) = \left\{w \in E^+ \setminus \{0\} : \nabla\Phi_{y_{j_m}}^{red}(w)[w] = 0\right\}$, for $m = 1, \dots, k$, and $t_m > 0$ be the unique point such that $t_m u_{j_m}^+ \in \mathcal{M}^+(\Phi_{y_{j_m}})$. Observe that $\bar{\lambda} \cdot y_{j_m} < 0$, by Step 2 and Step 3, we get that

$$\nabla\Phi_{y_{j_m}}^{red}(u_{j_m}^+)[u_{j_m}^+] - \bar{\lambda} \cdot y_{j_m} \int_{\mathbb{R}^3} |u_{j_m}^+|^\theta dx = 0,$$

and hence we have $t_m < 1$. Observe that, by applying Lemma 4.6, we have

$$\Phi_{y_{j_m}}^{red}(u_{j_m}^+) - \frac{1}{2}\nabla\Phi_{y_{j_m}}^{red}(u_{j_m}^+)[u_{j_m}^+] > \Phi_{y_{j_m}}^{red}(t_m u_{j_m}^+) - \frac{1}{2}\nabla\Phi_{y_{j_m}}^{red}(t_m u_{j_m}^+)[t_m u_{j_m}^+].$$

Then, it follows from $t_m u_{j_m}^+ \in \mathcal{M}^+(\Phi_{y_{j_m}})$ that

$$\Phi_{y_{j_m}}^{red}(u_{j_m}^+) - \frac{1}{2}\nabla\Phi_{y_{j_m}}^{red}(u_{j_m}^+)[u_{j_m}^+] > \gamma(I_{V(y_{j_m})}), \quad \text{for all } m = 1, \dots, k.$$

Finally, by (4.50) and (4.51), we obtain the inequality

$$\liminf_{\varepsilon\to 0} b_\varepsilon = \liminf_{\varepsilon\to 0} \Phi_\varepsilon^{red}(u_\varepsilon^+) \geq k \cdot \min_{m=1,\dots,k} \gamma(I_{V(y_{j_m})}) + c'.$$

And therefore, as in Case 1, we conclude easily a contradiction.

Step 5 Complete description of u_ε as $\varepsilon \to 0$.

As was argued in the previous steps, we know that there exists $l \in \mathbb{N}$ and, for any $j = 1, \dots, l$, $y_\varepsilon^j \in H_\varepsilon$, $y_j \in B_2$ and $u_j \in E \setminus \{0\}$ such that

$$\begin{aligned}
&|y_\varepsilon^j - y_\varepsilon^{j'}| \to \infty, \quad \text{if } j \neq j',\\
&\varepsilon y_\varepsilon^j \to y_j,\\
&\Big\| u_\varepsilon^+ - \sum_{j=1}^{l} u_j^+(\cdot - y_\varepsilon^j)\Big\| \to 0,\\
&\nabla\Phi_{y_j}^{red}(u_j^+) - \bar{\lambda}\cdot y_j\big(|u_j^+|^{\theta-2}u_j^+\big)^+ = 0.
\end{aligned}$$

Observe that there strictly holds

$$\Phi_{y_j}^{red}(u_j^+) - \frac{1}{2}\nabla\Phi_{y_j}^{red}(u_j^+)[u_j^+] > \gamma(I_{V(y_j)})$$

provided that $\bar{\lambda}\cdot y_j < 0$. Moreover, Lemma 4.8 implies that $\gamma(I_{V(y_j)}) \geq \gamma(I_{\nu_0}) - \sigma$ for any $y_j \in B_2$, where $\sigma > 0$ can be taken arbitrary small by appropriately shrinking R_1. Therefore, by Proposition 4.16 and Lemma 4.17, we conclude that $l = 1$ and $\bar{\lambda}\cdot y_1 = 0$. And thus we have $\|u_\varepsilon - u_1(\cdot - y_\varepsilon^1)\| \to 0$ as $\varepsilon \to 0$ which complete the proof. □

As an immediate consequence of Proposition 4.24 we have the following corollary, which gives us Proposition 4.18.

Corollary 4.25 $y_1 \in X^\perp$ *and* $\liminf_{\varepsilon\to 0} b_\varepsilon \geq \gamma(I_{\nu_0})$.

Proof Since $\mathcal{B}_\varepsilon(u_\varepsilon^+) = 0$, by Proposition 4.24, we get

$$\begin{aligned}
0 &= \int_{\mathbb{R}^3} Q_\varepsilon(x)|u_\varepsilon^+(x)|^\theta dx\\
&= \int_{\mathbb{R}^3} P_X(\zeta(\varepsilon x + \varepsilon y_\varepsilon^1))|u_\varepsilon^+(x + y_\varepsilon^1)|^\theta dx \to P_X(y_1)\int_{\mathbb{R}^3} |u_1^+|^\theta dx.
\end{aligned}$$

Then $y_1 \in X^{\perp}$, and we soon conclude

$$\liminf_{\varepsilon\to 0} b_\varepsilon \geq \gamma(I_{V(y_1)}) \geq \gamma(I_{\nu_0}).$$

□

In order to study the asymptotic behavior of the solution z_ε obtained in Theorem 4.20. We will show that z_ε is actually a solution of the original problem (4.20), and consequently, we can complete the proof of Theorem 4.13.

Let us recall that z_ε is the critical point of Φ_ε at level γ_ε, that is,

$$-i\alpha\cdot\nabla z_\varepsilon + a\beta z_\varepsilon + V_\varepsilon(x)z_\varepsilon = g_\varepsilon(x,|z_\varepsilon|)z_\varepsilon. \tag{4.52}$$

Moreover, Proposition 4.19 implies that $\Phi_\varepsilon(z_\varepsilon)\to\gamma(I_{\nu_0})$ as $\varepsilon\to 0$.

In what follows, we will give the asymptotic behavior of z_ε as $\varepsilon\to 0$.

Proposition 4.26 *Given a sequence $\varepsilon_j \to 0$, up to a subsequence, there exists $\{y_{\varepsilon_j}\}\subset\mathbb{R}^3$ such that*

$$\varepsilon_j y_{\varepsilon_j}\to 0, \quad \|z_{\varepsilon_j} - Z(\cdot - y_{\varepsilon_j})\|\to 0,$$

where Z is a ground state solution of the limiting equation (4.12).

Proof For the sake of clarity, let us write $\varepsilon = \varepsilon_j$. Our argument here has been used already in the previous section, so we will be sketchy. First of all, analogous to Proposition 4.24, we can conclude that: there exist $\bar{y}_\varepsilon^1 \in \mathbb{R}^3$, $\bar{y}_1\in B_2$ and $z_1\in E\setminus\{0\}$ with

$$-i\alpha\cdot\nabla z_1 + a\beta z_1 + V(\bar{y}_1)z_1 = g(\bar{y}_1,|z_1|)z_1,$$

such that

$$\varepsilon\bar{y}_\varepsilon^1\to\bar{y}_1, \quad \|z_\varepsilon - z_1(\cdot - y_\varepsilon^1)\|\to 0 \quad \text{as } \varepsilon\to 0.$$

So, the only thing that need to be proved is that $\bar{y}_1 = 0$.

By regularity arguments, $\{z_\varepsilon\}\subset\bigcap_{q\geq 2}W^{1,q}(\mathbb{R}^3,\mathbb{C}^4)$. For arbitrary $\xi\in\mathbb{R}^3$, multiplying (4.52) by $\partial_\xi z_\varepsilon$ and integrating, we get

$$-\frac{\varepsilon}{2}\int_{\mathbb{R}^3}\partial_\xi V(\varepsilon x)|z_\varepsilon|^2dx + \varepsilon\int_{\mathbb{R}^3}(F(|z_\varepsilon|) - \tilde{F}(|z_\varepsilon|))\partial_\xi\chi(\varepsilon x)dx = 0. \tag{4.53}$$

And if χ is C^1 around $\bar{y}_1$, we shall divide by ε and pass to the limit to obtain

$$-\frac{\partial_\xi V(\bar{y}_1)}{2}\int_{\mathbb{R}^3}|z_1|^2dx + \partial_\xi\chi(\bar{y}_1)\int_{\mathbb{R}^3}(F(|z_1|) - \tilde{F}(|z_1|))dx = 0. \tag{4.54}$$

At this point, similar as that in [3], we consider three different cases.

- *Case 1.* $\bar{y}_1 \in B_1$.

By (4.54), we get that $\partial_\xi V(\bar{y}_1) = 0$. Since $\xi \in \mathbb{R}^3$ is arbitrary, $\bar{y}_1$ is a critical point of V in B_1, and therefore $\bar{y}_1 = 0$.

- *Case 2.* $\bar{y}_1 \in B_2 \setminus \overline{B_1}$.

In this case, let us first fix $\xi = \frac{1}{|\bar{y}_1|}\bar{y}_1$. By the definition of χ (see (4.23)), we have that $\partial_\xi \chi(\bar{y}_1) = -1/R_1$.

Now, using $(f3)$ and the fact $\tilde{F}(s) \le \frac{\delta_0}{2}s^2$, it follow easily that there exists a constant $c > 0$ (which is independent of the choice of δ_0) such that

$$\int_{\mathbb{R}^3} F(|z_1|)dx \ge c,$$

and so by the boundedness of $z_1 \in E$ (see an argument of Lemma 4.10) we get

$$c' \int_{\mathbb{R}^3} |z_1|^2 dx \le \int_{\mathbb{R}^3} \big(F(|z_1|) - \tilde{F}(|z_1|)\big)dx.$$

Thus, it suffices to take R_1 smaller, if necessary, to get a contradiction with (4.54).

- *Case 3.* $\bar{y}_1 \in \partial B_1$.

In this case, observe that $\chi(\bar{y}_1) = 1$, and so z_1 is a solution of

$$-i\alpha \cdot \nabla z_1 + a\beta z_1 + V(\bar{y}_1)z_1 = f(|z_1|)z_1.$$

Since $I^{red}_{V(\bar{y}_1)}(z_1^+) = I_{V(\bar{y}_1)}(z_1) = \gamma(I_{\nu_0})$, Lemma 4.8 implies that $V(\bar{y}_1) = \nu_0$. Then, by (4.22), there exists $\tau \in \mathbb{R}^3$ tangent to ∂B_1 at $\bar{y}_1$ such that $\partial_\tau V(\bar{y}_1) \neq 0$.

Remark that χ is not C^1 on ∂B_1, let us go back to consider (4.53). Take $\xi = \tau$ and $r < R_1$, we can estimate by the dominated convergence theorem and the strong convergence of $z_\varepsilon(\cdot + \bar{y}_\varepsilon^1)$ that

$$\begin{aligned}
&\Big|\int_{\mathbb{R}^3} \partial_\tau \chi(\varepsilon x)\big[F(|z_\varepsilon|) - \tilde{F}(|z_\varepsilon|)\big]dx\Big| \\
&\le \frac{1}{R_1}\int_{B(0,r/\sqrt{\varepsilon})} \Big[\frac{|x\cdot\tau|}{|x+\bar{y}_\varepsilon^1|} + \frac{|\bar{y}_\varepsilon^1\cdot\tau|}{|x+\bar{y}_\varepsilon^1|}\Big]\big[F(|z_\varepsilon(x+\bar{y}_\varepsilon^1)|) - \tilde{F}(|z_\varepsilon(x+\bar{y}_\varepsilon^1)|)\big]dx \\
&\quad + \frac{1}{R_1}\int_{\mathbb{R}^3\setminus B(0,r/\sqrt{\varepsilon})} \frac{|(x+\bar{y}_\varepsilon^1)\cdot\tau|}{|x+\bar{y}_\varepsilon^1|}\big[F(|z_\varepsilon(x+\bar{y}_\varepsilon^1)|) - \tilde{F}(|z_\varepsilon(x+\bar{y}_\varepsilon^1)|)\big]dx \to 0.
\end{aligned}$$

Dividing by ε and passing to the limit in (4.53), we can conclude

$$\frac{1}{2}\partial_\tau V(\bar{y}_1)\int_{\mathbb{R}^3}|z_1|^2dx = 0,$$

a contradiction. □

Complete Proof of Theorem 4.13 It suffices to show that $|z_\varepsilon(x)| \to 0$ uniformly in $\mathbb{R}^3 \setminus B_1^\varepsilon$ as $\varepsilon \to 0$. In fact, from the regularity argument in [11, Lemma 3.19], we have that there exists $C > 0$ (independent of ε) such that $|z_\varepsilon|_\infty \le C$. Then we can use elliptic estimate to get

$$|z_\varepsilon(x)| \le C_0 \int_{B(x,1)} |z_\varepsilon(y)|dy$$

with $C_0 > 0$ independent of both ε and $x \in \mathbb{R}^3$. And thus, by Proposition 4.26, we have that for any $x \in \mathbb{R}^3 \setminus B_1^\varepsilon$,

$$\begin{aligned}|z_\varepsilon(x)| &\le C_0\Big(\int_{B(x,1)}|z_\varepsilon|^2\Big)^{1/2}\\ &\le C_0\Big(\int_{\mathbb{R}^3}\big|z_\varepsilon - Z(\cdot - y_\varepsilon)\big|^2\Big)^{1/2} + C_0\Big(\int_{B(x,1)}\big|Z(\cdot - y_\varepsilon)\big|^2\Big)^{1/2} \to 0,\end{aligned}$$

as $\varepsilon \to 0$. Finally, by the decay estimates obtained in [8, Lemma 4.2], it is standard to prove that there exists $C, c > 0$ independent of ε such that

$$|z_\varepsilon(x)| \le C\exp\big(-c|x - y_\varepsilon|\big).$$

This concludes the whole proof. □

4.5 Spikes Around Local Minima of Potential V: A Weaker Condition

The main purpose of this section is to prove the following theorem

Theorem 4.27 (Ding and Xu [8]) *Suppose* $(V0)$, $(f1)$ *and*

$(f2')$ (i) *there exist* $p \in (2,3)$, $c_1 > 0$ *such that* $f(s) \le c_1(1 + s^{p-2})$ *for* $s \ge 0$;
  (ii) *there exists* $\theta > 2$ *such that* $0 < \theta F(s) \le f(s)s^2$ *for all* $s > 0$,
$(f3')$ *the function* $s \mapsto f'(s)s + f(s)$ *is nondecreasing*

are satisfied. Assume additionally that there is a bounded domain Λ *in* $\mathbb{R}^3$ *such that*

$$\underline{c} := \min_{\Lambda} V < \min_{\partial\Lambda} V . \tag{4.55}$$

Then for all $\varepsilon > 0$ *small,*

(i) *a solution* $w_\varepsilon \in \cap_{q\geq 2} W^{1,q}(\mathbb{R}^3, \mathbb{C}^4)$ *to (4.1) exists;*
(ii) $|w_\varepsilon|$ *possesses a (global) maximum point* x_ε *in* Λ *such that*

$$\lim_{\varepsilon\to 0} V(x_\varepsilon) = \underline{c} ,$$

and

$$|w_\varepsilon(x)| \leq C \exp\big(-\frac{c}{\varepsilon}|x - x_\varepsilon|\big)$$

for certain constants $C, c > 0$*;*
(iii) *setting* $v_\varepsilon(x) = w_\varepsilon(\varepsilon x + x_\varepsilon)$*, as* $\varepsilon \to 0$*, we have* v_ε *converges in* H^1 *to a least energy solution of*

$$-i\alpha\cdot\nabla v + a\beta v + \underline{c}\, v = f(|v|)v .$$

Comparing with Theorem 4.13, this theorem concerns with a particular case of subcritical growth nonlinearity, and the assumption (4.55) is, to some extent, weaker than $(V1)$ in the previous setting in the sense that we allow . Though it is possible to consider a critical term in the function f, we prefer to use $(f2')$ and $(f3')$ here for a simplified proof.

To frame the problem in the abstract setting, let us assume $0 \in \Lambda$ and $V(0) = \underline{c}$, where Λ is a bounded domain as in the assumptions of Theorem 4.27. And we choose $\xi > 0$ be the value at which $f'(\xi)\xi + f(\xi) = \frac{a-|V|_\infty}{2}$. Let us consider $\tilde{f} \in C^1(0,\infty)$ such that

$$\frac{d}{ds}\big(\tilde{f}(s)s\big) = \begin{cases} f'(s)s + f(s) & \text{if } s < \xi \\ \dfrac{a-|V|_\infty}{2} & \text{if } s > \xi \end{cases} ,$$

and define

$$g(\cdot, s) = \chi_\Lambda f(s) + (1-\chi_\Lambda)\tilde{f}(s) , \tag{4.56}$$

where χ_Λ denotes the characteristic function of Λ. Denoted by $G(x,s) = \int_0^s g(x,t)t\,dt$, we are then led to the modified Euler functional

$$\widetilde{\Phi}_\varepsilon(u) = \frac{1}{2}\big(\|u^+\|^2 - \|u^-\|^2\big) + \frac{1}{2}\int_{\mathbb{R}^3} V_\varepsilon(x)|u|^2 - \Psi_\varepsilon(u) ,$$

where $\Psi_\varepsilon(u) = \int_{\mathbb{R}^3} G(\varepsilon x, |u|)$. Then, we see that $\widetilde{\Phi}_\varepsilon \in C^2(E, \mathbb{R})$. Let us highlight that the functional I_ν defined in (4.19) plays the role of the unperturbed functional.

It is standard to check that g is a Caratheodory function and it satisfies

(g_1) *$g_s(x,s)$ exists everywhere, $g(x,s)s = o(s)$ uniformly in x as $s \to 0$;*
(g_2) *$0 \le g(x,s)s \le f(s)s$ for all x;*
(g_3) *$0 < 2G(x,s) \le g(x,s)s^2 \le \frac{a-|V|_\infty}{2} s^2$ for all $x \notin \Lambda$ and $s > 0$;*
(g_4) *$0 < \theta G(x,s) \le g(x,s)s^2$ for all $x \in \Lambda$ and $s > 0$;*
(g_5) *$\frac{d}{ds}\big(g(x,s)s\big) \ge 0$ for all x and $s > 0$;*
(g_6) *$\widehat{G}(x,s) := \frac{1}{2}g(x,s)s^2 - G(x,s) \to \infty$ as $s \to \infty$ uniformly in x.*

Obviously, for any fixed $u \in E$, we have $V(\varepsilon x)ut \to V(0)u$ and $g(\varepsilon x, u) \to f(u)$ as $\varepsilon \to 0$ and hence $\widetilde{\Phi}$ has the form discussed in Sect. 3.2. The following two lemmas will show that $\widetilde{\Phi}_\varepsilon$ satisfies the remaining assumptions of Theorem 3.21.

Lemma 4.28 *There are $r > 0$ and $\tau > 0$, both independent of ε, such that $\widetilde{\Phi}_\varepsilon|_{B_r^+} \ge 0$ and $\widetilde{\Phi}_\varepsilon|_{S_r^+} \ge \tau$, where*

$$B_r^+ = B_r \cap E^+ = \{u \in E^+ : \|u\| \le r\},$$

$$S_r^+ = \partial B_r^+ = \{u \in E^+ : \|u\| = r\}.$$

Proof Recall that $|u|_p^p \le C_p\|u\|^p$ for all $u \in E$ by Sobolev's embedding theorem. The conclusion follows easily because, for $u \in E^+$

$$\begin{aligned}\widetilde{\Phi}_\varepsilon(u) &= \frac{1}{2}\|u\|^2 - \frac{1}{2}\int V_\varepsilon(x)|u|^2 - \Psi_\varepsilon(u)\\ &\ge \frac{1}{2}\|u\|^2 - \frac{|V|_\infty}{2}|u|_2^2 - \left(\frac{a-|V|_\infty}{4}|u|_2^2 + C|u|_p^p\right)\\ &\ge \frac{a-|V|_\infty}{4a}\|u\|^2 - C'\|u\|^p\end{aligned}$$

with $C, C' > 0$ independent of u and $p > 2$, where we have used the fact

$$G(x,s) \le \frac{a-|V|_\infty}{4}s^2 + Cs^p \quad \text{for all } s \ge 0.$$

□

Lemma 4.29 *For any $e \in E^+ \setminus \{0\}$, we have $I_{V(0)}(u) \to -\infty$ provided $u \in E_e = \mathbb{R}^+e \oplus E^-$ and $\|u\| \to \infty$, where $I_\nu : E \to \mathbb{R}$ is given by (4.19) for $\nu \in (-a, a)$.*

Proof Observe that assumption ($f2'$) implies that for any $\delta > 0$ there is a constant $c_\delta > 0$ such that

$$F(s) \ge c_\delta s^\theta - \delta s^2 \quad \text{for all } s \ge 0.$$

Directly, for $v \in E^-$, $u = te + v \in E_e$, we have that

$$
\begin{aligned}
I_{V(0)}(u) &= \frac{t^2}{2}\|e\|^2 - \frac{\|v\|^2}{2} + \frac{V(0)}{2}\int |te+v|^2 - \int F(|te+v|) \\
&\le \frac{a+|V(0)+2\delta|}{2a}t^2\|e\|^2 - \frac{a-|V(0)+2\delta|}{2a}\|v\|^2 - c_\delta \int |te+v|^\theta .
\end{aligned}
$$

And hence, by Proposition 4.1,

$$
I_{V(0)}(u) \le \frac{a+|V(0)+2\delta|}{2a}t^2\|e\|^2 - \frac{a-|V(0)+2\delta|}{2a}\|v\|^2 - C_{\delta,\theta}t^\theta |e|_\theta^\theta \quad \text{on } E_e .
$$

□

Denoted by

$$
c_\varepsilon = \inf_{e \in E^+\setminus\{0\}} \sup_{u \in E_e} \widetilde{\Phi}_\varepsilon(u)
$$

and $\gamma_{V(0)} = \gamma(I_{V(0)})$. By applying Theorem 3.21 with trivial group action $\mathscr{G} = \{\mathrm{Id}\}$, we conclude that

Lemma 4.30 *$c_\varepsilon \le \gamma_{V(0)} + o(1)$ as $\varepsilon \to 0$.*

Now, similar to the proof of Theorem 4.13, the proof of Theorem 4.27 will be accomplished by the following localized concentration result.

Lemma 4.31 *Assuming (g_1)–(g_6) and, for all ε sufficiently small, let u_ε be a critical point of $\widetilde{\Phi}_\varepsilon$ at critical level c_ε, then $|u_\varepsilon|$ possesses a (global) maximum $x_\varepsilon \in \Lambda_\varepsilon$ such that*

$$
\lim_{\varepsilon \to 0} V(\varepsilon x_\varepsilon) = V(0) = \min_{x \in \Lambda} V(x) .
$$

Moreover, by setting $v_\varepsilon(x) = u_\varepsilon(x + x_\varepsilon)$, we must have $|v_\varepsilon|$ decays uniformly at infinity and $\{v_\varepsilon\}$ converges in H^1 to a ground state solution to

$$
-i\alpha \cdot \nabla v + a\beta v + V(0)v = f(|v|)v .
$$

Proof Let $u_\varepsilon \in E$ be the critical point so that $\widetilde{\Phi}_\varepsilon(u_\varepsilon) = c_\varepsilon$. We have $\{u_\varepsilon\}$ is a bounded set in E.

Step 1 $\{u_\varepsilon\}$ is non-vanishing.

Suppose contrarily that

$$
\sup_{x \in \mathbb{R}^3} \int_{B_R(x)} |u_\varepsilon|^2 \, dx \to 0 \quad \text{as } \varepsilon \to 0
$$

for all $R > 0$. Then, by Lion's concentration principle [14], $|u_\varepsilon|_q \to 0$ for $q \in (2, 3)$. Since $\{u_\varepsilon\}$ is bounded in E, we have $meas\{x \in \mathbb{R} : |u_\varepsilon(x)| \geq r\}$ is uniformly bounded for all $\varepsilon > 0$ provided $r > 0$ is fixed. So we find

$$\int_{\{x\in\mathbb{R}:\,|u_\varepsilon(x)|\geq r\}} |u_\varepsilon|^2\,dx \to 0$$

as $\varepsilon \to 0$. Now, we have

$$c_\varepsilon = \widetilde{\Phi}_\varepsilon(u_\varepsilon) - \frac{1}{2}\widetilde{\Phi}'_\varepsilon(u_\varepsilon)u_\varepsilon = \int \widehat{G}(\varepsilon x, |u_\varepsilon|) = o(1)\,,$$

which contradict to the fact $c_\varepsilon \geq \tau > 0$ (see Lemma 4.28).

Step 2 $\{\chi_{\Lambda_\varepsilon} \cdot u_\varepsilon\}$ is non-vanishing.

Indeed, if $\{\chi_{\Lambda_\varepsilon} \cdot u_\varepsilon\}$ vanishes, by virtue of *Step 1* we have $\{(1 - \chi_{\Lambda_\varepsilon}) \cdot u_\varepsilon\}$ is non-vanishing, that is (together with the fact that each u_ε is decaying at infinity) there exist $x_\varepsilon \in \mathbb{R}^3$ and constants $R > 0$ and $\delta > 0$ such that $B_R(x_\varepsilon) \subset \mathbb{R}^3 \setminus \Lambda_\varepsilon$ and

$$\int_{B_R(x_\varepsilon)} |u_\varepsilon|^2 \geq \delta\,.$$

Set $v_\varepsilon(x) = u_\varepsilon(x + x_\varepsilon)$, then v_ε satifies

$$-i\alpha\cdot\nabla v_\varepsilon + a\beta v_\varepsilon + \hat{V}_\varepsilon(x)v_\varepsilon = g(\varepsilon(x + x_\varepsilon), |v_\varepsilon|)v_\varepsilon\,, \tag{4.57}$$

where $\hat{V}_\varepsilon(x) := V(\varepsilon(x + x_\varepsilon))$. Additionally, $v_\varepsilon \rightharpoonup v$ in E and $v_\varepsilon \to v$ in L^q_{loc} for $q \in [1, 3)$. Now assume without loss of generality that $V(\varepsilon x_\varepsilon) \to V_\infty$, using $\psi \in C_c^\infty(\mathbb{R}^3, \mathbb{C}^4)$ as a test function in (4.57), one gets

$$\begin{aligned} 0 &= \lim_{\varepsilon\to 0}\int \Big(-i\alpha\cdot\nabla v_\varepsilon + a\beta v_\varepsilon + \hat{V}_\varepsilon(x)v_\varepsilon - g(\varepsilon(x + x_\varepsilon), |v_\varepsilon|)v_\varepsilon\Big)\overline{\psi} \\ &= \int \Big(-i\alpha\cdot\nabla v + a\beta v + V_\infty v - \tilde{f}(|v|)v\Big)\overline{\psi}\,. \end{aligned}$$

Hence v satisfies

$$-i\alpha\cdot\nabla v + a\beta v + V_\infty v = \tilde{f}(|v|)v\,. \tag{4.58}$$

However, using the test function $v^+ - v^-$ in (4.58), we have

$$
\begin{aligned}
0 &= \|v\|^2 + V_\infty \int v \cdot \overline{(v^+ - v^-)} - \int \tilde{f}(|v|)v \cdot \overline{(v^+ - v^-)} \\
&\geq \|v\|^2 - \frac{|V|_\infty}{a}\|v\|^2 - \frac{a - |V|_\infty}{2a}\|v\|^2 \\
&= \frac{a - |V|_\infty}{2a}\|v\|^2 .
\end{aligned}
$$

Therefore, we have $v = 0$ a contradiction.

Step 3 Let $x_\varepsilon \in \mathbb{R}^3$ and $R, \delta > 0$ be such that

$$
\int_{B_R(x_\varepsilon)} |\chi_{\Lambda_\varepsilon} \cdot u_\varepsilon|^2 \geq \delta .
$$

Then $\varepsilon x_\varepsilon \to \mathscr{A} := \{x \in \Lambda : V(x) = \underline{c}\}$.

To prove this, clearly, we may first choose $x_\varepsilon \in \Lambda_\varepsilon$, i.e. $\varepsilon x_\varepsilon \in \Lambda$. Suppose that, up to a subsequence if necessary, $\varepsilon x_\varepsilon \to x_0 \in \bar{\Lambda}$ as $\varepsilon \to 0$. Again, set $v_\varepsilon(x) = u_\varepsilon(x + x_\varepsilon)$, we have $v_\varepsilon \rightharpoonup v$ in E and v satisfies

$$
- i\alpha \cdot \nabla v + a\beta v + V(x_0)v = g_\infty(x, |v|)v , \tag{4.59}
$$

where $g_\infty(x, s) = \chi_\infty \cdot f(s) + (1 - \chi_\infty) \cdot \tilde{f}(s)$ and χ_∞ is either a characteristic function of a half-space of $\mathbb{R}^3$ provided

$$
\limsup_{\varepsilon \to 0} \operatorname{dist}(x_\varepsilon, \partial\Lambda_\varepsilon) < +\infty
$$

or $\chi_\infty \equiv 1$ (since Λ is an open set with smooth boundary, this can be see by the fact $\chi_\Lambda(\varepsilon(\cdot + x_\varepsilon))$ converges pointwise a.e. on $\mathbb{R}^3$ to $\chi_\infty(\cdot)$ and $x_\varepsilon \in \Lambda_\varepsilon$). Denote S_∞ to be the associate energy functional to (4.59):

$$
S_\infty(u) := \frac{1}{2}\big(\|u^+\|^2 - \|u^-\|^2\big) + \frac{V(x_0)}{2}|u|_2^2 - \Psi_\infty(u) ,
$$

where

$$
\Psi_\infty(u) := \int G_\infty(x, |u|) \quad \text{and} \quad G_\infty(x, s) = \int_0^s g_\infty(x, \tau)\tau \, d\tau .
$$

By noting that $\Psi_\infty(u) \leq \Psi(u)$ (thanks to (g_2)), we have

$$
S_\infty(u) \geq I_{V(x_0)}(u) = I_{V(0)}(u) + \frac{V(x_0) - V(0)}{2}|u|_2^2 \quad \text{for all } u \in E .
$$

Furthermore, if $s > 0$ we find from $(f1)$, $(f3)$ and the definition of $\tilde{f}$ that $\tilde{f}'(s)s > 0$. Hence, if $u \in E \setminus \{0\}$ and $v \in E$, we have

$$\begin{aligned}&\big(\Psi''_\infty(u)[u,u] - \Psi'_\infty(u)u\big) + 2\big(\Psi''_\infty(u)[u,v] - \Psi'_\infty(u)v\big) + \Psi''_\infty(u)[v,v]\\ &= \int g_\infty(x,|u|)|v|^2 + \int \partial_s g_\infty(x,|u|)|u|\Big(|u| + \frac{\mathrm{Re}\, u\cdot v}{|u|}\Big)^2 > 0.\end{aligned}$$

Now let us define (as before) $h_\infty : E^+ \to E^-$ and $I_\infty : E^+ \to \mathbb{R}$ by

$$S_\infty\big(u + h_\infty(u)\big) = \max_{v\in E^-} S_\infty(u+v),$$

$$S_\infty^{red}(u) = S_\infty\big(u + h_\infty(u)\big).$$

It is standard to see that: if $z \in E^+ \setminus \{0\}$ satisfies $\nabla S_\infty^{red}(z)[z] = 0$, then $\nabla^2 S_\infty^{red}(z)[z,z] < 0$ (see Claim 3.1 in Lemma 3.18). Since we already have $v \neq 0$ is a critical point of S_∞, we then infer v^+ is a critical point of S_∞^{red} and $S_\infty^{red}(v^+) = \max_{t\ge 0} S_\infty^{red}(tv^+)$. Let $\tau > 0$ such that $I_{V(0)}^{red}(\tau v^+) = \max_{t\ge 0} I_{V(0)}^{red}(tv^+)$, we infer

$$\begin{aligned}S_\infty(v) &= S_\infty^{red}(v^+) = \max_{t\ge 0} S_\infty^{red}(tv^+)\\ &\ge S_\infty^{red}(\tau v^+) = S_\infty\big(\tau v^+ + h_\infty(\tau v^+)\big)\\ &\ge S_\infty\big(\tau v^+ + h_{V(0)}(\tau v^+)\big)\\ &\ge I_{V(0)}\big(\tau v^+ + h_{V(0)}(\tau v^+)\big) + \frac{V(x_0) - V_0}{2}\big|\tau v^+ + h_{V(0)}(\tau v^+)\big|_2^2\\ &\ge \gamma_{V(0)} + \frac{V(x_0) - V_0}{2}\big|\tau v^+ + h_{V(0)}(\tau v^+)\big|_2^2.\end{aligned} \tag{4.60}$$

On the other hand, by Fatou's lemma, we deduce

$$\begin{aligned}c_\varepsilon &= \widetilde{\Phi}_\varepsilon(u_\varepsilon) - \frac{1}{2}\widetilde{\Phi}'_\varepsilon(u_\varepsilon)u_\varepsilon\\ &= \int \widehat{G}(\varepsilon x, |u_\varepsilon|) = \int \widehat{G}(\varepsilon(x + x_\varepsilon), |v_\varepsilon|)\\ &\ge \int \widehat{G}_\infty(x, |v|)\\ &= S_\infty(v) - \frac{1}{2}\nabla S_\infty(v)[v] = S_\infty(v),\end{aligned}$$

where $\widehat{G}_\infty(x,s) := \frac{1}{2}g_\infty(x,s)s^2 - G_\infty(x,s)$ for $(x,s) \in \mathbb{R}^3 \times \mathbb{R}^+$. Therefore, together with (4.60), we have $c_\varepsilon \geq \gamma_{V(0)}$ and $c_\varepsilon > \gamma_{V(0)}$ provided $V(x_0) \neq V(0) = \underline{c}$. Therefore, by virtue of Lemma 4.30, we soon have $x_0 \in \mathscr{A}$ and $\chi_\infty \equiv 1$.

Step 4 Let v_ε be defined in *Step 3*, then $v_\varepsilon \to v$ in E.

It sufficient to prove that there is a subsequence $\{v_{\varepsilon_j}\}$ such that $v_{\varepsilon_j} \to v$ in E. Recall that, as the argument shows, v is a ground state solution to

$$- i\alpha \cdot \nabla v + a\beta v + V(0)v = f(|v|)v\,, \tag{4.61}$$

and

$$\lim_{\varepsilon \to 0} \int \widehat{G}(\varepsilon(x + x_\varepsilon), |v_\varepsilon|) = \int \widehat{F}(|v|)\,.$$

Let $\eta : [0,\infty) \to [0,1]$ be a smooth function satisfying $\eta(s) = 1$ if $s \leq 1$, $\eta(s) = 0$ if $s \geq 2$. Define $\tilde{v}_j(x) = \eta(2|x|/j)v(x)$. One has

$$\|\tilde{v}_j - v\| \to 0 \quad \text{and} \quad |\tilde{v}_j - v|_q \to 0 \quad \text{as } j \to \infty \tag{4.62}$$

for $q \in [2,3]$. Set $B_d := \{x \in \mathbb{R}^3 : |x| \leq d\}$ for $d > 0$. We have that there possesses a subsequence $\{v_{\varepsilon_j}\}$ such that, for any $\delta > 0$ there exists $r_\delta > 0$ satisfying

$$\limsup_{j \to \infty} \int_{B_j \setminus B_r} |v_{\varepsilon_j}|^p \leq \delta$$

for all $r \geq r_\delta$ (see an argument of [7, Lemma 5.7]), where $p \in (2,3)$ is the constant in condition $(f2')(i)$. Denote $z_j = v_{\varepsilon_j} - \tilde{v}_j$, we remark that $\{z_j\}$ is bounded in E and

$$\begin{aligned} \lim_{j \to \infty} \bigg| \int G\big(\varepsilon_j(x + x_{\varepsilon_j}), |v_{\varepsilon_j}|\big) - G\big(\varepsilon_j(x + x_{\varepsilon_j}), |z_j|\big) \\ - G\big(\varepsilon_j(x + x_{\varepsilon_j}), |\tilde{v}_j|\big) \bigg| = 0 \end{aligned} \tag{4.63}$$

and

$$\begin{aligned} \lim_{j \to \infty} \bigg| \int \Big[g\big(\varepsilon_j(x + x_{\varepsilon_j}), |v_{\varepsilon_j}|\big)v_{\varepsilon_j} - g\big(\varepsilon_j(x + x_{\varepsilon_j}), |z_j|\big)z_j \\ - g\big(\varepsilon_j(x + x_{\varepsilon_j}), |\tilde{v}_j|\big)\tilde{v}_j \Big]\overline{\varphi} \bigg| = 0 \end{aligned} \tag{4.64}$$

uniformly in $\varphi \in E$ with $\|\varphi\| \le 1$ (see [7, Lemma 7.10]). Using the decay of v and the fact that $\hat{V}_{\varepsilon_j}(x) \to V(0)$, $G(\varepsilon_j(x + x_{\varepsilon_j}), s) \to F(s)$ as $j \to \infty$ uniformly on any bounded set of x, one checks easily the following

$$\mathrm{Re} \int \hat{V}_{\varepsilon_j}(x) v_{\varepsilon_j} \cdot \tilde{v}_j \to \int V(0) \cdot |v|^2, \quad \int G\big(\varepsilon_j(x + x_{\varepsilon_j}), |\tilde{v}_j|\big) \to \int F(|v|).$$

Let us denote $\hat{\Phi}_\varepsilon$ to be the associate energy functional of (4.57), then we have

$$\begin{aligned}
\hat{\Phi}_{\varepsilon_j}(z_j) &= \hat{\Phi}_{\varepsilon_j}(v_{\varepsilon_j}) - S_\infty(v) \\
&\quad + \int G\big(\varepsilon_j(x + x_{\varepsilon_j}), |v_{\varepsilon_j}|\big) - G\big(\varepsilon_j(x + x_{\varepsilon_j}), |z_j|\big) \\
&\quad - G\big(\varepsilon_j(x + x_{\varepsilon_j}), |\tilde{v}_j|\big) + o(1) \\
&= o(1)
\end{aligned}$$

as $j \to \infty$, which implies that $\hat{\Phi}_{\varepsilon_j}(z_j) \to 0$. Similarly,

$$\begin{aligned}
\nabla \hat{\Phi}_{\varepsilon_j}(z_j)[\varphi] &= \mathrm{Re} \int \Big[g\big(\varepsilon_j(x + x_{\varepsilon_j}), |v_{\varepsilon_j}|\big) v_{\varepsilon_j} - g\big(\varepsilon_j(x + x_{\varepsilon_j}), |z_j|\big) z_j \\
&\quad - g\big(\varepsilon_j(x + x_{\varepsilon_j}), |\tilde{v}_j|\big) \tilde{v}_j \Big] \overline{\varphi} + o(1) \\
&= o(1)
\end{aligned}$$

as $j \to \infty$ uniformly in $\|\varphi\| \le 1$, which implies $\nabla \hat{\Phi}_{\varepsilon_j}(z_j) \to 0$. Therefore,

$$o(1) = \hat{\Phi}_{\varepsilon_j}(z_j) - \frac{1}{2} \nabla \hat{\Phi}_{\varepsilon_j}(z_j)[z_j] = \int \widehat{G}\big(\varepsilon_j(x + x_{\varepsilon_j}), |z_j|\big). \tag{4.65}$$

Owning to (g_6) and the regularity result, for any fixed $r > 0$, one has

$$\int \widehat{G}\big(\varepsilon_j(x + x_{\varepsilon_j}), |z_j|\big) \ge C_r \int_{\{x \in \mathbb{R}^3 : |z_j(x)| \ge r\}} |z_j|^2$$

for some constant C_r depends only on r. Hence

$$\int_{\{x \in \mathbb{R}^3 : |z_j(x)| \ge r\}} |z_j|^2 \to 0$$

as $j \to \infty$ for any fixed $r > 0$. Notice $\{|z_j|_\infty\}$ is bounded, as a consequence, we get

$$\begin{aligned}
\Big(1 - \frac{|V|_\infty}{a}\Big)\|z_j\|^2 &\le \|z_j\|^2 + \mathrm{Re}\int \hat{V}_{\varepsilon_j}(x) z_j \cdot \overline{(z_j^+ - z_j^-)} \\
&= \nabla\hat{\Phi}_{\varepsilon_j}(z_j)[z_j^+ - z_j^-] \\
&\quad + \mathrm{Re}\int f\big(\varepsilon_j(x + x_{\varepsilon_j}), |z_j|\big) z_j \cdot \overline{(z_j^+ - z_j^-)} \\
&\le o(1) + \frac{a - |V|_\infty}{2a}\|z_j\|^2 \\
&\quad + C_\infty \int_{\{x \in \mathbb{R}^3 : |z_j(x)| \ge r\}} |z_j| \cdot |z_j^+ - z_j^-| \\
&\le o(1) + \frac{a - |V|_\infty}{2a}\|z_j\|^2 ,
\end{aligned}$$

that is, $\|z_j\| \to 0$ as $j \to \infty$. Together with (4.62) we get $v_{\varepsilon_j} \to v$ in E.

Step 5 $v_\varepsilon(x) \to 0$ as $|x| \to \infty$ uniformly for all small ε.

Assume by contradiction that there exist $\delta > 0$ and $y_\varepsilon \in \mathbb{R}^3$ with $|y_\varepsilon| \to \infty$ such that

$$\delta \le |v_\varepsilon(y_\varepsilon)| \le C_0 \int_{B_1(y_\varepsilon)} |v_\varepsilon(y)|\, dy .$$

Since $v_\varepsilon \to v$ in E, we obtain, as $\varepsilon \to 0$,

$$\begin{aligned}
\delta &\le C_0\Big(\int_{B_1(y_\varepsilon)} |v_\varepsilon|^2\Big)^{1/2} \\
&\le C_0\Big(\int |v_\varepsilon - v|^2\Big)^{1/2} + C_0\Big(\int_{B_1(y_\varepsilon)} |v|^2\Big)^{1/2} \to 0,
\end{aligned}$$

a contradiction.

By virtue of *Step 5* it is clear that one may assume the sequence $\{x_\varepsilon\}$ in *Step 3* to be the maximum points of $|u_\varepsilon|$. Moreover, from the above argument we readily see that, any sequence of such points satisfies $\varepsilon x_\varepsilon$ converging to some point in $\mathscr{A}$ as $\varepsilon \to 0$.

Finally, in order to verify that $v_\varepsilon \to v$ in H^1, we first deduce from (4.57) and (4.61) that

$$H_0(v_\varepsilon - v) = g\big(\varepsilon(x + x_\varepsilon), |v_j|\big)v_j - f(|v|)v - \big(\hat{V}_\varepsilon(x)v_j - V(x_0)v\big) .$$

Using *Step 4* and the L^∞-estimate, it is easy to check that $|H_0(v_\varepsilon - v)|_2 \to 0$ as $\varepsilon \to 0$. Therefore $v_j \to v$ in $H^1(\mathbb{R}^3, \mathbb{C}^4)$, ending the proof. □

4.6 Other Results: Matrix-Type Potentials

In this section we collect some further results on the existence of solutions of (4.1) that exhibit localized concentrating behavior. We will not give the proofs, referring to [10].

We consider such equations in the following form

$$\alpha \cdot (-i\varepsilon\nabla + b(x))w + a\beta w + M(x)w = f(|w|)w \ , \ x \in \mathbb{R}^3 \ , \tag{4.66}$$

where here (and below) we write for notational convenience

$$b(x) = (b_1(x), b_2(x), b_3(x)) \ ,$$

and

$$\alpha \cdot (-i\varepsilon\nabla + b(x)) = \sum_{k=1}^{3} \alpha_k\big(-i\varepsilon\partial_k + b_k(x)\big) \ .$$

It is standard that (4.66) is equivalent to

$$\alpha \cdot \big(-i\nabla + b_\varepsilon(x)\big)u + a\beta u + M_\varepsilon(x)u = f(|u|)u \tag{4.67}$$

where $b_\varepsilon(x) = b(\varepsilon x)$, $M_\varepsilon(x) = M(\varepsilon x)$ and $u(x) = w(\varepsilon x)$.

We set, as before, $F(|u|) = \int_0^{|u|} f(s)s\,ds$ and denote the energy of a solution $w \neq 0$ of (4.66) by

$$\mathcal{E}(w) := \int_{\mathbb{R}^3} \left(\frac{1}{2}\big\langle\big(\alpha \cdot (-i\varepsilon\nabla + b(x)) + a\beta + M(x)\big)w, \ w\big\rangle - G(|w|)\right) dx$$

where here and in the sequel $\langle \cdot, \cdot \rangle$ stands for the hermitian product in $\mathbb{C}^4$, i.e., $\langle u, v\rangle = u \cdot \bar{v}$ for $u, v \in \mathbb{C}^4$. Setting

$$\tau_\varepsilon := \inf\big\{\mathcal{E}(w) : w \neq 0 \ \text{ is a solution of (4.66)}\big\} \ ,$$

a solution w_0 of (4.66) with $\mathcal{E}(w_0) = \tau_\varepsilon$ is called a least energy solution. Let $\mathcal{L}_\varepsilon$ denote the set of all least energy solutions of (4.66).

In what follows, for two symmetric matrices A and B, we write $A \geq (>)B$ if $\langle (A - B)\xi, \xi\rangle \geq (>) 0$ for all $\xi \in \mathbb{C}^4$ with $|\xi| = 1$. For a symmetric matrix valued function $M(x)$, let $\lambda_{min}(x)$ and $\lambda_{max}(x)$ denote, respectively, the minimal and maximal eigenvalues of $M(x)$. Set

$$\Lambda_{min} := \min_{x\in\mathbb{R}^3} \lambda_{max}(x) \quad , \quad \Lambda_{max} := \max_{x\in\mathbb{R}^3} \lambda_{max}(x) \, .$$

Denote

$$\mathscr{M}_{min} := \{x \in \mathbb{R}^3 \ : \ \lambda_{max}(x) = \Lambda_{min}\} \, .$$

The assumptions of our Theorems will imply that Λ_{min} and Λ_{max} exist, and that $\mathscr{M}_{min}$ is not empty.

A matrix $M(x_0)$ is called minimal if $x_0 \in \mathscr{M}_{min}$. Let M_{min} denote any minimal matrix $M(x_0)$, $x_0 \in \mathscr{M}_{min}$. Moreover, we define $\liminf\limits_{|x|\to\infty} M(x)$ to be a matrix whose minimal eigenvalue is $\liminf\limits_{|x|\to\infty} \lambda_{min}(x)$. The following equation

$$-i\alpha\cdot\nabla u + a\beta u + M_{min}u = f(|u|)u. \tag{4.68}$$

will be called *limit equation* (associated with (4.67)).

In order to state our results, we recall the assumptions posed on the nonlinear function f in Theorem 4.27, i.e., $(f1)$ and $(f2')$. A typical example is the power function $f(|u|) = |u|^{p-2}$. Let us first state the results for the equation without magnetic fields $b(x)$:

$$-i\varepsilon\alpha\cdot\nabla w + a\beta w + M(x)w = f(|w|)w. \tag{4.69}$$

The first result concerns an equation with a scalar potential of the form of a potential well: assume that

(V) $M(x) = V(x)\mathrm{Id}$ with $V \in C^1(\mathbb{R}^3, \mathbb{R})$ and

$$-a < V_{min} < \liminf_{|x|\to\infty} V(x), \qquad V_{max} < a.$$

Theorem 4.32 (Ding et al. [10]) *Let (V), $(f1)$ and $(f2')$ be satisfied. Then for all $\varepsilon > 0$ small,*

(i) Equation (4.69) has at least one least energy solution w_ε, with $w_\varepsilon \in W^{1,q}(\mathbb{R}^3, \mathbb{C}^4)$ for all $q \geq 2$;
(ii) The set of least energy solutions $\mathcal{L}_\varepsilon$ is compact in $H^1(\mathbb{R}^3, \mathbb{C}^4)$;
(iii) If additionally ∇V is bounded then there is a maximum point y_ε of $|w_\varepsilon|$ such that $\lim\limits_{\varepsilon\to 0} \mathrm{dist}(y_\varepsilon, \mathscr{M}_{min}) = 0$, and for any subsequence y_{ε_j} of such y_ε, the sequence $u_{\varepsilon_j}(y) := w_{\varepsilon_j}(\varepsilon_j y + y_{\varepsilon_j}) \to u(y)$ in H^1 with u being a least energy solution of the limit equation (4.68). Moreover, $|w_\varepsilon(y)| \leq C\exp(-\frac{c}{\varepsilon}|y - y_\varepsilon|)$ for some $C, c > 0$.

Next, we state a result involving a more general matrix potential. In what follows, for a 4×4 matrix function $P(x)$, we denote

$$\mathrm{Re}[P(x)] := \inf_{|\xi|=1} \mathrm{Re}\left(P(x)\xi \cdot \overline{\xi}\right),$$

and if furthermore $P(x)$ is Hermitian then

$$|P(x)| := \max\{|\lambda| : \lambda \in \sigma(P(x))\}.$$

For the existence of semiclassical solutions, assume the following

(M) M is a C^1 real symmetric matrix function satisfying

$$-a < \Lambda_{min} < \liminf_{|x|\to\infty} \lambda_{min}(x) \quad \text{and} \quad \Lambda_{max} < a.$$

Also, for the concentration phenomena, we assume that

(M_0) ∇M is bounded, $\alpha \cdot M = \kappa M \cdot \alpha$ where $\kappa \in \mathbb{R}$ with $|\kappa| = 1$ and

$$\liminf_{|x|\to\infty} \mathrm{Re}[a^2 + a\beta M(x) - a\kappa M(x)\beta - \kappa M(x)^2 - i\alpha \cdot \nabla M(x)] > 0.$$

Here, denoting $M = (m_{ij})$, we say that ∇M is bounded if ∇m_{ij} is for $i, j = 1, 2, 3, 4$.

Theorem 4.33 (Ding et al. [10]) *Let* (M), $(f1)$ *and* $(f2')$ *be satisfied. Then for all* $\varepsilon > 0$ *small,*

- (i) *Equation* (4.66) *has at least one least energy solution* $w_\varepsilon \in W^{1,q}(\mathbb{R}^3, \mathbb{C}^4)$ *for all* $q \geq 2$*;*
- (ii) $\mathcal{L}_\varepsilon$ *is compact in* $H^1(\mathbb{R}^3, \mathbb{C}^4)$*;*
- (iii) *If additionally* (M_0) *also holds then there is a maximum point* y_ε *of* $|w_\varepsilon|$ *such that* $\lim_{\varepsilon\to 0} \mathrm{dist}(y_\varepsilon, \mathscr{M}_{min}) = 0$, *and for any subsequence* y_{ε_j} *of such* y_ε, *the sequence* $u_{\varepsilon_j}(y) := w_{\varepsilon_j}(\varepsilon_j y + y_{\varepsilon_j}) \to u(y)$ *in* H^1 *with* u *being a least energy solution of the limit equation* (4.68). *Moreover,* $|w_\varepsilon(y)| \leq C \exp(-\frac{c}{\varepsilon}|y - y_\varepsilon|)$ *for some* $C, c > 0$.

Now we consider the case in the presence of a magnetic field.

(B) Assume that one of the following is satisfied:

- (B_0) $b(x) = \nabla\varphi(x)$ for some $\varphi \in C^2(\mathbb{R}^3, \mathbb{R})$;
- (B_1) $b \in C^1$ and there is $\hat{b} \in \mathbb{R}^3$ such that for $\mathcal{P}(x) := \alpha \cdot (b(x) - \hat{b}) + M(x)$, $\tau := |\mathcal{P}|_\infty < a$ and $\mathcal{P}_{min} = \mathcal{P}(x_p)$ for some $x_p \in \mathbb{R}^3$ hold.

Theorem 4.34 (Ding et al. [10]) *Let (M), (B), $(f1)$ and $(f2')$ be satisfied. Then for all $\varepsilon > 0$ small,*

(i) *Equation* (4.66) *has at least one least energy solution $w_\varepsilon \in W^{1,q}(\mathbb{R}^3, \mathbb{C}^4)$ for all $q \geq 2$;*

(ii) *$\mathcal{L}_\varepsilon$ is compact in $H^1(\mathbb{R}^3, \mathbb{C}^4)$;*

(iii) *If additionally ∇b is bounded and (M_0) also holds then there is a maximum point y_ε of $|w_\varepsilon|$ such that $\lim_{\varepsilon\to 0} \operatorname{dist}(y_\varepsilon, \mathscr{M}_{min}) = 0$, and, for any subsequence y_{ε_j} of such y_ε with $y_{\varepsilon_j} \to y_0 \in \mathscr{M}_{min}$, the sequence $u_{\varepsilon_j}(y) := w_{\varepsilon_j}(\varepsilon_j y + y_{\varepsilon_j}) \to u(y) = e^{ib(y_0)\cdot y} v(y)$ in H^1 with v being a least energy solution of equation* (4.68). *Moreover, $|w_\varepsilon(y)| \leq C \exp(-\frac{c}{\varepsilon}|y - y_\varepsilon|)$ for some $C, c > 0$.*

Remark 4.35

(1) The condition (V) obviously implies (M). It satisfies (M_0) with $\kappa = 1$. Moreover, a point $x_0 \in \mathscr{M}_{min}$ if and only if $V(x_0) = V_{min}$; hence $M_{min} = V_{min}\mathrm{Id}$. In addition, $\mathrm{Re}[a^2 + a\beta M(x) - a\kappa M(x)\beta - \kappa M(x)^2 - i\alpha \cdot \nabla M(x)] = a^2 - V(x)^2$.

As another example, let us consider the matrix:

$$M(x) = \begin{pmatrix} v(x)\mathrm{Id}_{2\times 2} & w(x)\mathrm{Id}_{2\times 2} \\ w(x)\mathrm{Id}_{2\times 2} & v(x)\mathrm{Id}_{2\times 2} \end{pmatrix}$$

which possesses the eigenvalues $\lambda_\pm(x) = v(x) \pm |w(x)|$ of multiplicity 2. It satisfies (M) provided that $\max_x |\lambda_+(x)| < a$, $v_{min} < \liminf_{|x|\to\infty} v(x)$ and there is x_v with $v(x_v) = v_{min}$ satisfying $\limsup_{|x|\to\infty} |w(x) - w(x_v)| < \liminf_{|x|\to\infty} v(x) - v_{min}$. Moreover, $\alpha \cdot M(x) = M(x) \cdot \alpha$ and $\mathrm{Re}[a^2 + a\beta M(x) - aM(x)\beta - M(x)^2 - i\alpha \cdot \nabla M(x)] = a^2 - M(x)^2$.

In general, assume $M(x) = L(x) + R(x)$ so that

$$L_{min} < \liminf_{|x|\to\infty} L(x)$$

and there exists x_L satisfying $L(x_L) = L_{min}$ and

$$\limsup_{|x|\to\infty} \big(R(x_L) - R(x)\big) < \liminf_{|x|\to\infty} L(x) - L_{min}.$$

Then $M_{min} < \liminf_{|x|\to\infty} M(x)$.

(2) Concerning (B_1) one has that if there is $x_m \in \mathscr{M}_{min}$ such that

$$\limsup_{|x|\to\infty} |b(x) - b(x_m)| < \liminf_{|x|\to\infty} M(x) - M_{min},$$

then $\mathcal{P}_{min} = \mathcal{P}(x_p)$ for some $x_p \in \mathbb{R}^3$. Indeed, observe that, for any $d = (d_1, d_2, d_3) \in \mathbb{R}^3$, the matrix $\alpha \cdot d = \sum_k d_k \alpha_k$ has eigenvalues $\ell_\pm(x) = \pm|d|$

of multiplicity 2. Writing $\xi = (\xi_1, \xi_2) \in \mathbb{C}^2 \times \mathbb{C}^2$, one sees that

$$\inf_{|\xi|=1} \langle \alpha \cdot d\,\xi, \xi \rangle = \inf_{|\xi|=1} \big(\ell_+(x)|\xi_1|^2 + \ell_-(x)|\xi_2|^2\big) \geq -|d|$$

and, taking $|\xi| = 1$ with $\xi_1 = 0$, respectively $\xi_2 = 0$,

$$\inf_{|\xi|=1} \langle \alpha \cdot d\,\xi, \xi \rangle \leq -|d|,$$

hence

$$\inf_{|\xi|=1} \langle \alpha \cdot d\,\xi, \xi \rangle = -|d|.$$

Letting $d = b(x) - \hat{b}$, by definition

$$\begin{aligned} &\inf_{|\xi|=1} \langle \big(\mathcal{P}(x) - \mathcal{P}(x_m)\big)\xi, \xi \rangle \\ &= \inf_{|\xi|=1} \langle \big(M(x) - M(x_m)\big)\xi, \xi \rangle + \langle \big(\alpha \cdot (b(x) - b(x_m)\big)\xi, \xi \rangle \\ &\geq \inf_{|\xi|=1} \langle \big(M(x) - M(x_m)\big)\xi, \xi \rangle - |b(x) - b(x_m)|, \end{aligned}$$

hence, jointly with (M),

$$\begin{aligned} &\liminf_{|x|\to\infty} \inf_{|\xi|=1} \langle \big(\mathcal{P}(x) - \mathcal{P}(x_m)\big)\xi, \xi \rangle \\ &\geq \liminf_{|x|\to\infty} \inf_{|\xi|=1} \langle \big((M(x) - M_{min}) - |b(x) - b(x_m)|\big)\xi, \xi \rangle > 0 \end{aligned}$$

Take $R > 0$ large so that $\inf_{|\xi|=1} \langle \big(\mathcal{P}(x) - \mathcal{P}(x_m)\big)\xi, \xi \rangle > 0$, for $|x| \geq R$. Since $\mathcal{P}(x)$ is continuous in $B(0, R) = \{x \in \mathbb{R}^3 : |x| \leq R\}$, let $x_p \in B(0, R)$ so that $\min_{x \in B(0,R)} \mathcal{P}(x) = \mathcal{P}(x_p)$. Since $\mathcal{P}(x) \geq \mathcal{P}(x_m) \geq \mathcal{P}(x_p)$ for all $|x| \geq R$, one sees that $\mathcal{P}(x_p) = \mathcal{P}_{min}$.

(3) Assume $M(x) = V(x)\mathrm{Id}$ with $V(x)$ satisfying (V). Observe that the matrix $\mathcal{P}(x)$ has eigenvalues $\lambda_\pm(x) = V(x) \pm |b(x) - \hat{b}|$ of multiplicity 2, and as above

$$\inf_{|\xi|=1} \langle \mathcal{P}(x)\xi, \xi \rangle = V(x) - |b(x) - \hat{b}|.$$

Thus, (B_1) holds if $-a < \min_x \big(V(x) - |b(x) - \hat{b}|\big) \leq \max_x \big(V(x) + |b(x) - \hat{b}|\big) < a$ and

$$\limsup_{|x|\to\infty} |b(x) - \hat{b}| \leq \sup_x |b(x) - \hat{b}|.$$

(4) If $b \equiv 0$, the above remarks show that Theorems 4.32 and 4.33 are consequences of Theorem 4.34.
(5) Observe that, for $\hat{x} \in \mathbb{R}^3$, u solves (4.66) if and only if $v(x) = u(x - \hat{x})$ solves

$$\alpha \cdot (-i\varepsilon\nabla + b(x - \hat{x}))v + a\beta v + M(x - \hat{x})v = f(|v|)v.$$

Thus we may assume without loss of generality that $0 \in \mathscr{M}_{min}$.
(6) In the case with a magnetic field $b(x) = \nabla\varphi(x)$, plainly, w solves (4.66) if and only if $u := e^{\frac{i}{\varepsilon}\varphi} w$ solves (4.69).
(7) Assume (B_1) is satisfied. Note that u solves (4.66) if and only if, for any $\hat{x} \in \mathbb{R}^3$, $v(x) = e^{i(b(\hat{x})+\hat{b})\cdot x} u(x)$ solves

$$\alpha \cdot \big(- i\varepsilon\nabla + (b(x) - (b(\hat{x}) + \hat{b})\big)v + a\beta v + M(x)v = f(|v|)v.$$

Therefore, without loss of generality, we may assume that

$$\hat{b} = 0 \quad \text{and} \quad b(0) = 0\,.$$

Also, for this reason the condition (B_1) does not restrict the "size" of $\|b\|_\infty$. Indeed, one can choose a vector $c = (c_1, c_2, c_3) \in \mathbb{R}^3$ so that $b^c(x) := b(x) - c$ satisfies (B_1) and consider the equation with b replaced by b^c.

References

1. Ackermann, N.: A nonlinear superposition principle and multibump solution of periodic Schrödinger equations. J. Funct. Anal. **234**, 423–443 (2006)
2. Bartsch, T., Xu, T.: Strongly localized semiclassical states for nonlinear Dirac equations. Discrete Contin. Dyn. Syst. **41**(1), 29–60 (2021). Special issue celebrating 25 years of excellence
3. d'Avenia, P., Pomponio, A., Ruiz, D.: Semiclassical states for the nonlinear Schrödinger equation on saddle points of the potential via variational methods. J. Funct. Anal. **262**, 4600–4633 (2012)
4. Del Pino, M., Felmer, P.: Local mountain passes for semilinear elliptic problems in unbounded domains. Calc. Var. Partial Differ. Equ. **4**, 121–137 (1996)
5. Del Pino, M., Felmer, P.: Multi-peak bound states for nonlinear Schrödinger equations. Ann. Inst. Henri Poincare (C) Non Linear Anal. **15**(2), 127–149 (1998). Elsevier Masson
6. Del Pino, M., Felmer, P.: Semi-classical states of nonlinear Schrödinger equations: a variational reduction method. Math. Ann. **324**, 1–32 (2002)
7. Ding, Y.H.: Variational Methods for Strongly Indefinite Problems. Interdiscip. Math. Sci., vol. 7. World Scientific Publ., Singapore (2007)
8. Ding, Y.H., Xu, T.: Localized concentration of semiclassical states for nonlinear Dirac equations. Arch. Ration. Mech. Anal. **216**(2), 415–447 (2015)
9. Ding, Y.H., Xu, T.: Concentrating patterns of reaction-diffusion systems: a variational approach. Trans. Am. Math. Soc. **369**(1), 97–138 (2017)
10. Ding, Y.H., Lee, C., Ruf, B.: On semiclassical states of a nonlinear Dirac equation. Proc. R. Soc. Edinb.: Sect. A Math. **143**(04), 765–790 (2013)

11. Ding, Y.H., Wei, J.C., Xu, T.: Existence and concentration of semi-classical solutions for a nonlinear Maxwell-Dirac system. J. Math. Phys. **54**(06), 061505, 33 pp. (2013)
12. Grafakos, L.: Classical Fourier Analysis, 3rd edn. Graduate Texts in Mathematics, vol. 249, xviii+638 pp. Springer, New York (2014)
13. Jeanjean, L., Tanaka, K.: A remark on least energy solutions in $\mathbb{R}^N$. Proc. Am. Math. Soc. **131**, 2399–2408 (2003)
14. Lions, P.L.: The concentration-compactness principle in the calculus of variations: the locally compact case, Part II. AIP Anal. non linéaire **1**, 223–283 (1984)
15. Reed, M., Simon, B.: Methods of Mathematical Physics, vols. I–IV. Academic Press, Cambridge (1978)
16. Stein, E.M.: Singular Integrals and Differentiability Properties of Functions, vol. 2. Princeton University Press, Princeton (1970)
17. Willem, M.: Minimax Theorems. Birkhäuser, Boston (1996)

Chapter 5
Effect of External Potentials in a Coupled System Reaction-Diffusion

In this chapter we will deal with the Reaction-Diffusion (RD) system

$$\partial_t z = \mathbf{D}\Delta_x z + M(x)z + \big(W(x)|z|^{p-2} + Q(x)|z|^{4/N}\big)\beta z \tag{5.1}$$

where the diffusion coefficients $\mathbf{D}$ is defined in (2.6) and M, W and Q are external potentials with M being one of the following form

$$M(x) = \begin{pmatrix} -\mathrm{Id} - V(x) & 0 \\ 0 & \mathrm{Id} + V(x) \end{pmatrix} \quad \text{or} \quad M(x) = \begin{pmatrix} -\mathrm{Id} & -V(x) \\ V(x) & \mathrm{Id} \end{pmatrix}. \tag{5.2}$$

After a first section devoted to studying the variational structure of (5.1), we consider in Sect. 5.2, the so-called subcritical case in which $W \equiv constant$ and $Q \equiv 0$, where we construct one localized solution of (5.1) in terms of the potential M. The rest of the chapter deals with the number of distinct solutions generated from the geometric properties of the functions V, W and Q.

5.1 Preliminaries

In this section we establish some preliminary results which are needed for the study of (5.1). For a clearer expression let us set

$$\mathcal{J}_0 = \begin{pmatrix} 0 & \mathrm{Id} \\ \mathrm{Id} & 0 \end{pmatrix},$$

© The Author(s), under exclusive license to Springer Nature Singapore Pte Ltd. 2025
Y. Ding, T. Xu, *Localization Approaches in Strongly Indefinite Problems*, Series in Contemporary Mathematics 6, https://doi.org/10.1007/978-981-97-9539-0_5

and by noting that $\mathbf{D} = \varepsilon^2 \mathcal{J}$, we shall consider the scaling $x \to \varepsilon x$ so that (5.1) can be read as

$$\partial_t z = \mathcal{J}\Delta_x z + M_\varepsilon(x)z + g_\varepsilon(x, |z|)\beta z \tag{5.3}$$

with $g : \mathbb{R}^N \times \mathbb{R} \to \mathbb{R}$ being correspondingly defined and $g_\varepsilon(x, s) = g(\varepsilon x, s)$. Remark that $\beta^{-1} = -\beta$ (given by (2.7)), hence if denoted by $\mathscr{L} = -\beta\partial_t + \mathcal{J}_0(-\Delta_x + 1)$ and $\mathcal{V}(x) =$ either $V(x)$ or $V(x)\mathcal{J}_0$, we have (5.3) to be rewritten as

$$\mathscr{L}z + \mathcal{V}_\varepsilon(x)z = g_\varepsilon(s, |z|)z \quad \text{for } z(t, x) \in \mathbb{R}^{2K} \tag{5.4}$$

where obviously we have used $\mathcal{V}_\varepsilon(x) = \mathcal{V}(\varepsilon x)$.

In what follows, we will focus on the equivalent system (5.4), and, throughout the paper we make use of the following notations: for $1 \le q \le \infty$ we set $L^q := L^q(\mathbb{R} \times \mathbb{R}^N, \mathbb{R}^{2K})$, and by $|\cdot|_q$ we denote the usual L^q-norm, and particularly denoted by $(\cdot, \cdot)_2$ the usual L^2-inner product.

Considering the differential operator $\mathscr{L}$ acts on the Hilbert space L^2, it is quite standard to see that $\mathscr{L}$ is a self-adjoint operator with domain

$$\mathcal{D}(\mathscr{L}) = \mathscr{B}^2 := W^{1,2}\big(\mathbb{R}, L^2(\mathbb{R}^N, \mathbb{R}^{2K})\big) \cap L^2\big(\mathbb{R}, W^{2,2}(\mathbb{R}^N, \mathbb{R}^{2K})\big).$$

Let $\sigma(\mathscr{L})$ and $\sigma_e(\mathscr{L})$ be respectively the spectrum and essential spectrum of $\mathscr{L}$, we have $\sigma(\mathscr{L}) = \sigma_e(\mathscr{L}) = \mathbb{R} \setminus (-1, 1)$ (cf. [2, Lemma 8.7]). And as a direct consequence, L^2 possesses the orthogonal decomposition:

$$L^2 = L^+ \oplus L^-, \quad z = z^+ + z^-, \tag{5.5}$$

so that L is positive definite (*resp.* negative definite) in L^+ (*resp.* L^-).

In order to construct the energy functionals whose critical points are the solutions of (5.4), we introduce $E := \mathcal{D}(|\mathscr{L}|^{1/2})$ (which is the form domain of L) be equipped with the inner product

$$\langle z_1, z_2 \rangle = \big(|\mathscr{L}|^{1/2} z_1, |\mathscr{L}|^{1/2} z_2\big)_2$$

and the induced norm $\|z\| = \langle z, z\rangle^{1/2}$, where $|\mathscr{L}|$ and $|\mathscr{L}|^{1/2}$ denote respectively the absolute value of $\mathscr{L}$ and the square root of $|\mathscr{L}|$. Since $\sigma(\mathscr{L}) = \mathbb{R} \setminus (-1, 1)$, one has

$$|z|_2^2 \le \|z\|^2 \quad \text{for all } z \in E. \tag{5.6}$$

As an interpolation space between $\mathscr{B}^2$ and L^2, E (being a Hilbert space) has the decomposition

$$E = E^+ \oplus E^-, \quad \text{where } E^\pm = E \cap L^\pm$$

which is orthogonal with respect to both $(\cdot,\cdot)_2$ and $\langle\cdot,\cdot\rangle$. We write $z = z^+ + z^-$ for $z \in E$ according to this decomposition. The following result is a counterpart of Proposition 4.1:

Proposition 5.1 *Let $E^+ \oplus E^-$ be the decomposition of E according to the positive and negative part of $\sigma(\mathscr{L})$. Then, set $E_q^\pm := E^\pm \cap L^q$ for $q \in (1,\infty)$, there holds*

$$L^q = \mathrm{cl}_q\, E_q^+ \oplus \mathrm{cl}_q\, E_q^-$$

with cl_q denoting the closure in L^q. More precisely, there exists $d_q > 0$ for every $q \in (1,\infty)$ such that

$$d_q|z^\pm|_q \le |z|_q \quad \text{for all } z \in E \cap L^q.$$

Proof First we remark that in this context, the spatial domain is $\mathbb{R}\times\mathbb{R}^N$. Now recall the definitions for the matrices β and $\mathcal{J}_0$, let us study $\mathscr{L} := -\beta\partial_t + \mathcal{J}_0(-\Delta_x + 1)$. It is a differential operator with real constant coefficients. In the Fourier domain $\xi = (\xi_0, \xi_1, \xi_2, \dots, \xi_N)$, it becomes the operator of multiplication by the matrix:

$$\hat{\mathscr{L}}(\xi) = \begin{pmatrix} 0 & \overline{A(\xi)} \\ A(\xi) & 0 \end{pmatrix} \quad \text{with } A(\xi) = \left(i\xi_0 + 1 + \sum_{k=1}^N \xi_k^2\right)\cdot \mathrm{Id}.$$

Here "Id" denotes the $K \times K$ identity matrix.

Denoted by $\lambda(\xi) = \sqrt{\xi_0^2 + \left(1 + \sum_{k=1}^N \xi_k^2\right)^2}$. By classical calculus, we have that $\hat{\mathscr{L}}(\xi)$ has two eigenvalues: $\pm\lambda(\xi)$. Now, denote $P^\pm$ the projections on E with kernel $E^\mp$. We see that in the Fourier domain, $P^\pm$ are multiplication operators by bounded smooth matrix-valued functions of ξ:

$$(P^+z)\hat{}(\xi) = \frac{1}{2\lambda(\xi)}\begin{pmatrix} \lambda(\xi) & \overline{A(\xi)} \\ A(\xi) & \lambda(\xi) \end{pmatrix}\begin{pmatrix} \hat{u} \\ \hat{v} \end{pmatrix},$$

$$(P^-z)\hat{}(\xi) = \frac{1}{2\lambda(\xi)}\begin{pmatrix} \lambda(\xi) & -\overline{A(\xi)} \\ -A(\xi) & \lambda(\xi) \end{pmatrix}\begin{pmatrix} \hat{u} \\ \hat{v} \end{pmatrix}.$$

Here we have used the notation $z = (u, v)$ and $\hat{z} = (\hat{u}, \hat{v})$.

In order that $P^{\pm}$ are multipliers for L^q, we need to use the *Marcinkiewicz multiplier theorem* on $\mathbb{R}\times\mathbb{R}^N$ (see [6, Chapter 4, Theorem 6']). A direct calculation shows that, for each $0 < k \le N+1$, there holds

$$\left|\frac{\partial^k\big(A(\xi)/\lambda(\xi)\big)}{\partial\xi_{i_1}\cdots\partial\xi_{i_k}}\right| \le \frac{B}{\Pi_{j=1}^k|\xi_{i_j}|} \text{ for some constant } B > 0.$$

And hence, as an immediate consequence, $P^{\pm}$ are *multipliers* for L^q for all $q \in (1,\infty)$. This implies that $P^{\pm}$ are continuous with respect to the L^q-norms. By noting that $P^{\pm}(E^{\mp}) = \{0\}$, one easily sees that $P^{\pm}$ extend to continuous projections on L^q (still denoted by $P^{\pm}$) with $P^{\pm}(\mathrm{cl}_q E_q^{\mp}) = \{0\}$. And this completes the proof.

The embedding from E into L^q's can be concluded in the following lemma.

Lemma 5.2 (See e.g. [2]) *E is continuously embedded in L^q for $q \in [2, 2(N+2)/N]$ if $N \ge 2$, and compactly embedded in L^q_{loc} for $q \in [1, 2(N+2)/N)$ if $N \ge 2$.*

Throughout this chapter, we set $2^* = \frac{2(N+2)}{N}$ for the critical exponent in the sense of Lemma 5.2. On E we consider the functionals Φ_ε and Φ_0 defined by

$$\begin{aligned}\Phi_\varepsilon(z) &= \frac{1}{2}\big(\|z^+\|^2 - \|z^-\|^2\big) + \frac{1}{2}\iint \mathcal{V}_\varepsilon(x)z\cdot z\,dtdx - \iint G_\varepsilon(x,|z|)\,dtdx,\\ \Phi_0(z) &= \frac{1}{2}\big(\|z^+\|^2 - \|z^-\|^2\big) + \frac{1}{2}\iint \mathcal{V}(0)z\cdot z\,dtdx - \iint G(0,|z|)\,dtdx,\end{aligned} \tag{5.7}$$

where $G(x,s) := \int_0^s g(x,\tau)\tau d\tau$ (for simplicity, we shall omit the integration set $\mathbb{R}\times\mathbb{R}^N$ in the integrals). By virtue of Lemma 5.2, one sees that Φ_ε is 2-times Frechét differentiable on E and that its critical points correspond to the solutions of (5.4).

5.2 Construction of One Localized Solution

In this section, to get a clear picture of the construction, let us consider the case that the function g in (5.3) (and hence G in Φ_ε and Φ_0) is independent of the spatial variables and has subcritical growth (similar to the case considered in Theorem 4.27), i.e. $g(s) \sim s^{p-2}$ for some $p \in (2, 2^*)$, and the matrix potential M is of the second form in (5.2). For the potential function V, we accordingly make the following assumption which is similar to the one considered in (4.55):

(V) *V is locally Hölder continuous and* $\max|V| < 1$, *furthermore, there exists an open bounded set $\Lambda \subset \mathbb{R}^N$ such that $0 \in \Lambda$ and*

$$V(0) := \min_{\Lambda} V < \min_{\partial\Lambda} V\,.$$

5.2.1 Study of the Functionals

It is to be expected that critical points of Φ_ε can be found by applying Theorem 3.21, and that the asymptotic behavior (which can be characterized by concentration phenomenon) of these critical points will follow subsequently. However, one may find the abstract theorem can not apply directly to Φ_ε due to the lack of $\mathscr{G}$-weak $(C)_c$-condition. In what follows, let us initially give a modification of Φ_ε which guarantees that the assumptions in the abstract theorem are all satisfied.

Choose $s_0 > 0$ be the value at which $g(s_0) + g'(s_0)s_0 = \frac{1-|V|_\infty}{2}$. Let us consider $\tilde{g} \in C[0, \infty) \cap C^1(0, \infty)$ such that

$$\frac{d}{ds}\big(\tilde{g}(s)s\big) = \begin{cases} g(s) + g'(s)s & \text{if } s \le s_0, \\ \dfrac{1-|V|_\infty}{2} & \text{if } s > s_0. \end{cases}$$

Define

$$f(x, s) = \chi_\Lambda(x)g(s) + \big(1 - \chi_\Lambda(x)\big)\tilde{g}(s),$$

where $\Lambda \subset \mathbb{R}^N$ is the bounded domain fixed to satisfy the assumption (V) and χ_Λ denotes its characteristic function. One should keep in mind here that Λ has to be rescaled when we consider the rescaled system (5.4). Write

$$F(x, s) = \int_0^s f(x, \tau)\tau\, d\tau \quad \text{and} \quad \widehat{F}(x, s) = \frac{1}{2} f(x, s)s^2 - F(x, s).$$

Since our model case is that $g(s) \sim s^p$, it implies that $F : [0, \infty) \to [0, \infty)$ is a Caratheodory function satisfying

$(F1)$ $f_s(x, s) > 0$ *exists every where, and* $f(x, s)s = o(s)$ *uniformly in* x *as* $s \to 0$;

$(F2)$ $0 \le f(x, s)s \le g(s)s$ *for all* $x \in \mathbb{R}^N$ *and* $s \ge 0$;

$(F3)$ $0 < 2F(x, s) \le f(x, s)s^2 \le \frac{1-|V|_\infty}{2}s^2$ *for all* $x \notin \Lambda$ *and* $s > 0$;

$(F4)$ $0 < pF(x, s) \le f(x, s)s^2$ *for all* $x \in \Lambda$ *and* $s > 0$;

$(F5)$ $\frac{d}{ds}\big(f(x, s)s\big) \ge 0$ *for all* x *and* $s > 0$;

For simplicity of notation, we let $f_\varepsilon(x, s)$ and $F_\varepsilon(x, s)$ stand for $f(\varepsilon x, s)$ and $F(\varepsilon x, s)$ respectively. Now, let us define the modified functional $J_\varepsilon : E \to \mathbb{R}$ as

$$\begin{aligned} J_\varepsilon(z) &= \frac{1}{2}\big(\|z^+\|^2 - \|z^-\|^2\big) + \frac{1}{2}\int_{\mathbb{R}}\int_{\mathbb{R}^N} \mathcal{V}_\varepsilon(x)z \cdot z\, dxdt - \int_{\mathbb{R}}\int_{\mathbb{R}^N} F_\varepsilon(x, |z|)\, dxdt \\ &= \frac{1}{2}\big(\|z^+\|^2 - \|z^-\|^2\big) + \frac{1}{2}\int_{\mathbb{R}}\int_{\mathbb{R}^N} V_\varepsilon(x)|z|^2\, dxdt - \Psi_\varepsilon(z). \end{aligned}$$

Then, we see that $\Phi_\varepsilon \in C^2(E, \mathbb{R})$ and critical points of Φ_ε correspond to solutions of

$$\mathscr{L}z + V_\varepsilon(x)z = f_\varepsilon(x, |z|)z.$$

Taking the singular limit into account, we find that the limit system is

$$\mathscr{L}z + V_0\, z = g(|z|)z,$$

where $V_0 := V(0)$ and that the associated functional is

$$\begin{aligned}\Phi_0(z) &= \frac{1}{2}\big(\|z^+\|^2 - \|z^-\|^2\big) + \frac{1}{2}\int_{\mathbb{R}}\int_{\mathbb{R}^N} \mathcal{V}(0)z \cdot z\, dxdt - \int_{\mathbb{R}}\int_{\mathbb{R}^N} G(|z|)\, dxdt \\ &= \frac{1}{2}\big(\|z^+\|^2 - \|z^-\|^2\big) + \frac{V_0}{2}\int_{\mathbb{R}}\int_{\mathbb{R}^N} |z|^2\, dxdt - \Psi_0(z).\end{aligned}$$

To apply the Theorem 3.21 stated in Chap. 3, we now analyze the relevant properties of the variational functionals involved. As introduced above we have $E = E^+ \oplus E^-$, in what follows let A_ε denote the self-adjoint operator defined by $z \mapsto |\mathscr{L}|^{-1} V_\varepsilon(\cdot)z$ for $z \in E$. Analogously, A_0 can be defined by $z \mapsto |\mathscr{L}|^{-1} V_0 z$. It is all clear we have Φ_ε and Φ_0 are in the forms that we introduced in Sect. 3.2, so our strategy is to check all the assumptions appeared in Theorem 3.21 are satisfied.

The Group Action

Denote by $\star$ the action of $\mathscr{G} := \mathbb{R}$ on E that arises from translation: for $z \in E$ and $g \in \mathscr{G}$ define $(g \star z)(t, x) = z(t - g, x)$. From the fact V and $\tilde{g}$ are independent of t that we have Φ_ε is $\mathscr{G}$-invariant for all $\varepsilon > 0$. Moreover, if denote by $\bar{\star}$ the action of $\mathscr{G}' := \mathbb{R} \times \mathbb{R}^N$ on E by $(g' \bar{\star} z)(t, x) = z(t - g_1, x - g_2)$ for $g' = (g_1, g_2) \in \mathscr{G}'$, we soon have Φ_0 is invariant under the action of $\mathscr{G}'$.

The Quadratic Part

Recall the definitions of $A_\varepsilon, A_0 \in \mathscr{L}(E)$. By virtue of (5.6), we easily have

$$\begin{aligned}\|A_\varepsilon\| &= \sup\big\{ \langle A_\varepsilon z, z\rangle : u \in E,\ \|z\| = 1\big\} \\ &= \sup\big\{\big(V_\varepsilon(\cdot)z, z\big)_2 : z \in E,\ \|z\| = 1\big\} \\ &\le |V|_\infty \cdot \sup\big\{(z, z)_2 : z \in E,\ \|z\| = 1\big\} \\ &\le |V|_\infty < 1.\end{aligned}$$

This guarantees the condition $(A1)$. To check $(A2)$, that is the convergence of $\{A_\varepsilon\}$ in the strong operator topology, let us remark that $V_\varepsilon(x) \to V_0$ uniformly on bounded sets of x as $\varepsilon \to 0$. Therefore, for each $z \in E$ we deduce that

$$\begin{aligned}
\|A_\varepsilon z - A_0 z\| &= \sup_{\|w\|=1} \langle (A_\varepsilon - A_0)z, w\rangle \\
&= \sup_{\|w\|=1} \big((V_\varepsilon(\cdot) - V_0)z, w\big)_2 \\
&\le \sup_{\|w\|=1} \big|(V_\varepsilon(\cdot) - V_0)z\big|_2 \cdot |w|_2 \\
&\le \big|(V_\varepsilon(\cdot) - V_0)z\big|_2 = o(1)
\end{aligned}$$

as $\varepsilon \to 0$. And thus we obtain $(A2)$.

The Non-quadratic Part

The required properties (see $(N1)$–$(N5)$) of the nonlinear part will be checked based on $(F1)$–$(F5)$. Firstly, by $(F2)$, we observe that

$$G(|z|) = \int_0^{|z|} g(s)s\,ds \ge \int_0^{|z|} f_\varepsilon(x, s)s\,ds = F_\varepsilon(x, |z|),$$

which implies $\Psi_0(z) \ge \Psi_\varepsilon(z) \ge 0$ for all $z \in E$. Note that $\frac{d}{ds}\big(g(s)s\big) \ge 0$ and $\frac{d}{ds}\big(f_\varepsilon(x, s)s\big) \ge 0$ for all $x \in \mathbb{R}^N$, we have $\Psi_\varepsilon''(z)[w, w] \ge 0$ for $z, w \in E$ and a fixed $\varepsilon > 0$.

Since E embeds continuously into L^r for $r \in [2, 2^*]$, it follows that if $z_n \rightharpoonup z$ in E then $\{z_n\}$ is bounded in L^r and converges to z in L^r_{loc}, for $r \in [1, 2^*)$. Moreover, recall we have assumed that $0 \in \Lambda$, we infer that $\chi_\Lambda(\varepsilon x) \to 1$ a.e. on $\mathbb{R}^N$ as $\varepsilon \to 0$. Therefore it is easy to check $(N1)$ and $(N2)$ are satisfied for Ψ_ε. One should keep in mind here that the mapping $|\mathscr{L}|^{-1} : E^* \to E$ is the isomorphism induced from the Riesz representation theorem.

$(N3)$ is much more obvious. Indeed, the modified nonlinearities satisfy

$$\begin{aligned}
|f_\varepsilon(x, s)| &\le \big|\chi_\Lambda(\varepsilon x)g(s)\big| + \big|(1 - \chi_\Lambda(\varepsilon x))\tilde{g}(s)\big| \\
&\le |g(s)| + |\tilde{g}(s)|
\end{aligned}$$

for all $z \in \mathbb{R}^{2K}$. Therefore, by the embedding $E \hookrightarrow L^{2(N+2)/N}$, we the upper bound estimate

$$\big|\nabla^2\Psi_\varepsilon(z)[v, w]\big| \le C_1\|v\| \cdot \|w\| + C_2\|z\|^{4/N} \cdot \|v\| \cdot \|w\|$$

and $(N3)$ is satisfied.

It remains to prove $(N4)$ and $(N5)$. The verification for Ψ_0 is similar to and simpler than that for Ψ_ε, $\varepsilon > 0$, so we only check the latter. First note that $(F3)$ and $(F4)$ implies

$$\widehat{F}(x,s) := \frac{1}{2} f(x,s)s^2 - F(x,s) > 0 \quad \text{for all } x \in \mathbb{R}^N \text{ and } s > 0.$$

An easy calculation shows

$$\widehat{\Psi}_\varepsilon(z) = \int_{\mathbb{R}} \int_{\mathbb{R}^N} \frac{1}{2} f_\varepsilon(x, |z|)|z|^2 - F_\varepsilon(x, |z|)\, dxdt > 0$$

provided $z \neq 0$, and the sequentially lower semi-continuity comes from Fatou's lemma. Next, to see that $(N5)$ holds, we remark that

$$\nabla \Psi_\varepsilon(z)[w] = \int_{\mathbb{R}} \int_{\mathbb{R}^N} f_\varepsilon(x, |z|) z \cdot w\, dxdt$$

and

$$\nabla^2 \Psi_\varepsilon(z)[v, w] = \int_{\mathbb{R}} \int_{\mathbb{R}^N} f_\varepsilon(x, |z|) v \cdot w + \partial_s f_\varepsilon(x, |z|)|z| \frac{z \cdot v}{|z|} \cdot \frac{z \cdot w}{|z|}\, dxdt$$

for any $z, v, w \in E$. Then, we deduce that

$$\begin{aligned} &\big(\nabla^2 \Psi_\varepsilon(z)[z, z] - \nabla \Psi_\varepsilon(z)[z]\big) + 2\big(\nabla^2 \Psi_\varepsilon(z)[z, w] - \nabla \Psi_\varepsilon(z)[w]\big) \\ &\quad + \nabla^2 \Psi_\varepsilon(z)[w, w] = \int_{\mathbb{R}} \int_{\mathbb{R}^N} f_\varepsilon(x, |z|)|w|^2 \\ &\quad + \partial_s f_\varepsilon(x, |z|)|z| \Big(|z| + \frac{z \cdot w}{|z|}\Big)^2 dxdt \end{aligned} \tag{5.8}$$

And by $(F1)$, we soon obtain $(N5)$ from the above formula.

The Geometric Structure and $\mathscr{G}$-weak Compactness

Recall the definition of F, we remark that there exists $C > 0$ such that

$$|G(|z|)| \le \frac{1 - |V|_\infty}{4} |z|^2 + C|z|^{2^*} \tag{5.9}$$

and

$$|F(x, |z|)| \le \frac{1 - |V|_\infty}{4} |z|^2 + C|z|^{2^*} \tag{5.10}$$

for all $(x, z) \in \mathbb{R} \times \mathbb{R}^{2K}$. Hence, we have:

Lemma 5.3 *There are $\rho, \tau > 0$, both independent of ε, such that $J_\varepsilon|_{B_\rho^+} \geq 0$ and $J_\varepsilon|_{S_\rho^+} \geq \tau$ where*

$$B_\rho^+ := B_\rho \cap E^+ = \{z \in E^+ : \|z\| \geq \rho\},$$

$$S_\rho^+ := \partial B_\rho^+ = \{z \in E^+ : \|z\| = \rho\}.$$

Proof Notice that $|z|_{2^*} \leq C\|z\|$ for $z \in E$ by the embedding $E \hookrightarrow L^{2^*}$. The conclusion follows easily because, for $z \in E^+$,

$$\begin{aligned} J_\varepsilon(z) &= \frac{1}{2}\|z\|^2 + \frac{1}{2}\int_{\mathbb{R}}\int_{\mathbb{R}^N} V_\varepsilon(x)|z|^2\,dxdt - \Psi_\varepsilon(z) \\ &\geq \frac{1}{2}\|z\|^2 - \frac{|V|_\infty}{2}|z|_2^2 - \left(\frac{1-|V|_\infty}{4}|z|_2^2 + C|z|_{2^*}^{2^*}\right) \\ &\geq \frac{1-|V|_\infty}{4}\|z\|^2 - C'\|z\|^{2^*} \end{aligned}$$

with $C, C' > 0$ independent of ε.

The preceding lemma shows $(I1)$ is satisfied for the family $\{J_\varepsilon\}_{\varepsilon>0}$. The proof of $(I2)$ will be as follows.

Lemma 5.4 *For $e \in E^+ \setminus \{0\}$, $\Phi_0(z) \to -\infty$ provided $z \in E_e$ and $\|z\| \to \infty$ (recall $E_e := \mathbb{R}^+e \oplus E^-$).*

Proof First remark that for any $\delta > 0$ there exists $c_\delta > 0$ such that

$$G(|z|) \geq c_\delta|z|^p - \delta|z|^2 \quad \text{for all } z \in \mathbb{R}^{2K}.$$

Let $e \in E^+ \setminus \{0\}$, by virtue of Proposition (5.1), we have for $z = se + v \in E_e$

$$\begin{aligned} \Phi_0(z) &= \frac{1}{2}\|se\|^2 - \frac{1}{2}\|v\|^2 + \frac{1}{2}\int_{\mathbb{R}}\int_{\mathbb{R}^N} V_\varepsilon(x)|se+v|^2\,dxdt - \Psi_0(se+v) \\ &\leq \frac{s^2}{2}\|e\|^2 - \frac{1}{2}\|v\|^2 + \frac{|V|_\infty}{2}|se+v|_2^2 + \delta|se+v|_2^2 - c_\delta|se+v|_p^p \\ &\leq \frac{1+|V|_\infty+2\delta}{2}s^2\|e\|^2 - \frac{1-|V|_\infty-2\delta}{2}\|v\|^2 - c_{\delta,p}s^p|e|_p^p. \end{aligned}$$

By noting that $p > 2$, let δ be chosen small enough, we have the assertion proved.

The next lemma will be devoted to show the $\mathcal{G}$-weak compactness of the modified functional J_ε for each $\varepsilon > 0$. We remark that, by virtue of $(F4)$,

$$\widehat{F}(x,s) := \frac{1}{2}f(x,s)s^2 - F(x,s) \geq \frac{p-2}{2p}f(x,s)s^2 \geq \frac{p-2}{2}F(x,s) > 0 \tag{5.11}$$

for all $x \in \Lambda$ and $s > 0$. By setting $\sigma = p/(p-1)$, we have

$$\big(f(x,s)s\big)^{\sigma} \le a_1 f(x,s)s^2 \le a_2\widehat{F}(x,s) \tag{5.12}$$

for all $|z| \ge r_1$ and $x \in \Lambda$, where r_1 is chosen small enough such that

$$|f(x,s)| \le \frac{1-|V|_\infty}{4} \quad \text{for all } s \le r_1,\ x \in \mathbb{R}^N. \tag{5.13}$$

Lemma 5.5 *For each $\varepsilon > 0$, $c \in \mathbb{R}\setminus\{0\}$, J_ε satisfies the $\mathcal{G}$-weak $(C)_c$-condition.*

Proof We begin by proving any $(C)_c$-sequence of J_ε is bounded in E. In fact, let $\{z_n\}$ be a sequence such that

$$J_\varepsilon(z_n) \to c \quad \text{and} \quad (1+\|z_n\|)\nabla J_\varepsilon(z_n) \to 0$$

as $n \to \infty$, the formulation of J_ε implies that there is $C > 0$ such that

$$C \ge J_\varepsilon(z_n) - \frac{1}{2}\nabla J_\varepsilon(z_n)[z_n] = \int_{\mathbb{R}}\int_{\mathbb{R}^N} \widehat{F}_\varepsilon(x,|z_n|)\,dxdt > 0$$

and

$$\begin{aligned} o(1) &= \nabla J_\varepsilon(z_n)[z_n^+ - z_n^-] \\ &= \|z_n\|^2 + \int_{\mathbb{R}}\int_{\mathbb{R}^N} V_\varepsilon(x)z_n\cdot(z_n^+ - z_n^-)\,dxdt \\ &\quad - \int_{\mathbb{R}}\int_{\mathbb{R}^N} f_\varepsilon(x,|z_n|)z_n\cdot(z_n^+ - z_n^-)\,dxdt. \end{aligned}$$

By the definition of F, we soon obtain

$$\begin{aligned} &\|z_n\|^2 - |V|_\infty \int_{\mathbb{R}}\int_{\mathbb{R}^N} |z_n|\cdot|z_n^+ - z_n^-|\,dxdt \\ &\le \int_{\mathbb{R}}\int_{\mathbb{R}^N} f_\varepsilon(x,|z_n|)|z_n|\cdot|z_n^+ - z_n^-|\,dxdt + o(1) \\ &\le \int_{\mathbb{R}}\int_{\Lambda_\varepsilon} f_\varepsilon(x,|z_n|)|z_n|\cdot|z_n^+ - z_n^-|\,dxdt \\ &\quad + \frac{1-|V|_\infty}{2}\int_{\mathbb{R}}\int_{\mathbb{R}^N} |z_n|\cdot|z_n^+ - z_n^-|\,dxdt + o(1), \end{aligned} \tag{5.14}$$

where $\Lambda_\varepsilon := \{x \in \mathbb{R}^N : \varepsilon x \in \Lambda\}$. Thus, from (5.12) and (5.13), we easily check that

$$\begin{aligned}
&\frac{1-|V|_\infty}{4}\|z_n\|^2 \\
&\le \iint_{\{(t,x)\in\mathbb{R}\times\Lambda_\varepsilon:\,|z_n|\ge r_1\}} f_\varepsilon(x,|z_n|)|z_n|\cdot|z_n^+ - z_n^-|\,dxdt + o(1) \\
&\le \left(\iint_{\{(t,x)\in\mathbb{R}\times\Lambda_\varepsilon:\,|z_n|\ge r_1\}} \big(f_\varepsilon(x,|z_n|)|z_n|\big)^\sigma\,dxdt\right)^{1/\sigma} |z_n^+ - z_n^-|_p + o(1).
\end{aligned}$$

It follows from (5.12) and E embeds continuously into L^p, we find

$$\frac{1-|V|_\infty}{4}\|z_n\|^2 \le C_1\|z_n\| + o(1).$$

Then $\{z_n\}$ is bounded in E as desired.

The next step is concerned with the $\mathcal{G}$-weak compactness of the $(C)_c$-sequence $\{z_n\}$, $c \neq 0$. Keep in mind that $\varepsilon > 0$ is now fixed (which implies Λ_ε is a bounded domain in $\mathbb{R}^N$), let us choose $\varphi \in C_c^\infty(\mathbb{R}^N)$ such that $\overline{\Lambda}_\varepsilon \subset \operatorname{supp}\varphi$ and

$$\varphi(x) = \begin{cases} 1 & x \in \overline{\Lambda}_\varepsilon, \\ 0 & x \notin N_1(\overline{\Lambda}_\varepsilon), \end{cases}$$

where $N_1(\overline{\Lambda}_\varepsilon) := \{x \in \mathbb{R}^N : \operatorname{dist}(x, \overline{\Lambda}_\varepsilon) < 1\}$. Denote by $z_n' = \varphi \cdot z_n$, we have $\{z_n'\}$ is a bounded sequence in E since $\{z_n\}$ is bounded. We claim that there exist $\{g_n\} \subset \mathcal{G} := \mathbb{R}$ and $t_0, \delta_0 > 0$ such that

$$\int_{g_n-t_0}^{g_n+t_0}\int_{N_1(\overline{\Lambda}_\varepsilon)} |z_n'|^2\,dxdt \ge \delta_0 \quad \text{for all } n \ge 1. \tag{5.15}$$

Then, from the compact embedding $E \hookrightarrow L^2_{loc}$, we have the new sequence $\{g_n \star z_n\}$ possesses a subsequence converge weakly to a point in $E \setminus \{0\}$ (here we used the inequality $|z_n| \ge |z_n'|$).

To prove (5.15), let us assume by contradiction that

$$\lim_{n\to\infty}\sup_{g\in\mathbb{R}}\int_{g-r}^{g+r}\int_{N_1(\overline{\Lambda}_\varepsilon)} |z_n'|^2\,dxdt = 0 \quad \text{for any } r > 0.$$

Since Λ_ε is bounded, jointly with the definition of z_n', we see that $\{z_n'\}$ is vanishing. By Lion's concentration compactness principle [5], we have $|z_n'|_q \to 0$ for all $q \in (2, 2^*)$. By virtue of (5.13) and the definition of F, we have

$$\begin{aligned}
&\frac{1-|V|_\infty}{4}\|z_n\|^2 \\
&\le \iint_{\{(t,x)\in\mathbb{R}\times\Lambda_\varepsilon:\,|z_n|\ge r_1\}} f_\varepsilon(x,|z_n|)|z_n|\cdot|z_n^+ - z_n^-|\,dxdt + o(1) \\
&\le \iint_{\{(t,x)\in\mathbb{R}\times N_1(\overline{\Lambda}_\varepsilon):\,|z_n'|\ge r_1\}} f_\varepsilon(x,|z_n'|)|z_n'|\cdot|z_n^+ - z_n^-|\,dxdt + o(1).
\end{aligned}$$

Remark that there must exist $C_0 > 0$ such that

$$|f(x,s)| \le C_0 s^{p-2} \quad \text{for all } x \in \mathbb{R}^N \text{ and } s \ge r_1.$$

Then, by Hölder inequality and the fact $|z_n'|_q \to 0$ for all $q \in (2, 2^*)$, we have $z_n \to 0$ in E as $n \to \infty$ which implies $J_\varepsilon(z_n) \to 0$. This contradicts our assumption: $\{z_n\}$ is a $(C)_c$-sequence with $c \ne 0$.

The Autonomous Limit System

What remains is to check the functional Φ_0 admits a critical value

$$c_0 = \inf_{e\in E^+\setminus\{0\}} \sup_{z\in E_e} \Phi_0(z) < \infty$$

where $E_e = \mathbb{R}^+ e \oplus E^-$.

In what follows, let us consider the autonomous system

$$\begin{cases} \partial_t u = \Delta_x u - u - \mu v + H_v(u,v) \\ -\partial_t v = \Delta_x v - v - \mu u + H_u(u,v) \end{cases} \tag{5.16}$$

for $\mu \in (-1,1)$. Remark that $H(\xi) = \int_0^{|\xi|} g(s)s\,ds$ for $\xi \in \mathbb{R}^{2K}$, it is evident that (5.16) can be rewritten as

$$\mathscr{L}z + \mu z = g(|z|)z$$

for $z = (u,v)$. The solutions to (5.16) are critical points of the $\mathscr{G}'$-invariant energy functional

$$\mathscr{T}_\mu := \frac{1}{2}\left(\|z^+\|^2 - \|z^-\|^2\right) + \frac{\mu}{2}\int_{\mathbb{R}}\int_{\mathbb{R}^N} |z|^2\,dxdt - \Psi_0(z)$$

defined for $z = z^+ + z^- \in E = E^+ \oplus E^-$. It is obvious, for the case $\mu = V_0$, we have $\mathscr{T}_{V_0}$ coincide with Φ_0. For notation convenience, let us denote

$$\mathscr{K}_\mu := \{z \in E \setminus \{0\} : \mathscr{T}_\mu'(z) = 0\} \quad \text{and} \quad \gamma_\mu := \inf\{\mathscr{T}_\mu(z) : z \in \mathscr{K}_\mu\}.$$

We state, in the two following propositions, some known results about the existence of solutions of (5.16) that will be used in the sequel.

Proposition 5.6 *If $g(s) = Cs^p$ for some constant $C > 0$. Then the super quadratic nonlinear system* (5.16) *has a nontrivial solution z which lies in $\mathscr{B}^r(\mathbb{R} \times \mathbb{R}^N, \mathbb{R}^{2K})$ for all $r \geq 2$, where*

$$\mathscr{B}^r(\mathbb{R} \times \mathbb{R}^N, \mathbb{R}^{2K}) := W^{1,r}(\mathbb{R}, L^r(\mathbb{R}^N, \mathbb{R}^{2K})) \cap L^r(\mathbb{R}, W^{2,r}(\mathbb{R}^N, \mathbb{R}^{2K})).$$

Proposition 5.6 is obtained just collecting the results in [2] (see e.g. Theorem 8.1 and Theorem 8.6). The proof can be done, by applying the linking theorems associate with the strongly indefinite functionals, see Sect. 3.1 of Chap. 3.

By virtue of Proposition 3.13, let us consider the reduced functional given by

$$\mathscr{T}_\mu^{red} : E^+ \to \mathbb{R}, \quad \mathscr{T}_\mu^{red}(v) = \mathscr{T}_\mu(v + \mathscr{J}_\mu(v))$$

where $\mathscr{J}_\mu : E^+ \to E^-$ is the reduction map. We remark that critical points of $\mathscr{T}_\mu^{red}$ and $\mathscr{T}_\mu$ are in one-to-one correspondence via the injective map $v \mapsto v + \mathscr{J}_\mu(v)$ from E^+ into E. Denoted by

$$\Gamma_\mu = \{\nu \in C([0,1], E^+) : \nu(0) = 0, \ \mathscr{T}_\mu^{red}(\nu(1)) < 0\},$$

and consider the minimax schemes

$$d_\mu^1 = \inf_{\nu \in \Gamma_\mu} \max_{t \in [0,1]} \mathscr{T}_\mu^{red}(\nu(t)) \quad \text{and} \quad d_\mu^2 = \inf_{v \in E^+ \setminus \{0\}} \max_{t \geq 0} \mathscr{T}_\mu^{red}(tv),$$

we have the following useful result, which is a counterpart of Lemma 4.8.

Lemma 5.7 *For the autonomous system* (5.16)*, there holds:*

(1) *$\gamma_\mu > 0$ is attained, and $\gamma_\mu = d_\mu^1 = d_\mu^2$;*
(2) *if $\mu_1 > \mu_2$, then $\gamma_{\mu_1} > \gamma_{\mu_2}$.*

Remarkably, the definition of $\mathscr{J}_\mu$ implies

$$\mathscr{T}_\mu^{red}(te) = \mathscr{T}_\mu(te + \mathscr{J}_\mu(te)) = \max_{w \in E^-} \mathscr{T}_\mu(te + w)$$

and therefore

$$\sup_{t \geq 0} \mathscr{T}_\mu^{red}(te) = \sup_{t \geq 0} \max_{w \in E^-} \mathscr{T}_\mu(te + w) = \sup_{z \in Ee} \mathscr{T}_\mu(z).$$

By taking infimum with respect to $e \in E^+ \setminus \{0\}$, we have

$$\gamma_\mu = \inf_{e \in E^+ \setminus \{0\}} \sup_{z \in Ee} \mathscr{T}_\mu(z).$$

And therefore, for the case $\mu = V_0$, we have Φ_0 admits the critical value

$$c_0 = \inf_{e \in E^+ \setminus \{0\}} \sup_{z \in Ee} \Phi_0(z) < \infty$$

as required.

5.2.2 *Profile of the Localized Solutions*

From the arguments in the preceding subsection, and applying Theorem 3.21, we obtain the following proposition:

Proposition 5.8 *Under the assumption (V), for all $\varepsilon > 0$ small, the modified functional J_ε admits a critical value which can be characterized by*

$$c_\varepsilon = \inf_{e \in E^+ \setminus \{0\}} \sup_{z \in E_e} J_\varepsilon(z).$$

Moreover, c_ε is the ground state for each J_ε and $c_\varepsilon \leq c_0 + o(1)$ as $\varepsilon \to 0$ with

$$c_0 = \inf_{e \in E^+ \setminus \{0\}} \sup_{z \in E_e} \Phi_0(z).$$

It is worth pointing out that the method of proving Proposition 5.8 carries more information on the ranges of c_ε. In fact the proof strongly depended on the assumption that $0 \in \Lambda$, and however, the value of $V(0)$ is irrelevant. Let $\{x_\varepsilon\}$ be a family of points in Λ_ε so chosen that $V_\varepsilon(x_\varepsilon) = V_0$, and consider the associated equation

$$\mathscr{L}z + \hat{V}_\varepsilon(x)z = f_\varepsilon(x + x_\varepsilon, |z|)z \tag{5.17}$$

with energy functional $\hat{J}_\varepsilon : E \to \mathbb{R}$ being written as

$$\begin{aligned}\hat{J}_\varepsilon(z) = \frac{1}{2}\big(\|z^+\|^2 - \|z^-\|^2\big) + \frac{1}{2}\int_{\mathbb{R}}\int_{\mathbb{R}^N} \hat{V}_\varepsilon(x)|z|^2\,dxdt \\ - \int_{\mathbb{R}}\int_{\mathbb{R}^N} F_\varepsilon(x + x_\varepsilon, |z|)\,dxdt\end{aligned}$$

where $\hat{V}_\varepsilon(x) = V\big(\varepsilon(x + x_\varepsilon)\big)$. Noting that let $z_\varepsilon \in E$ be a solution to

$$\mathscr{L}z + V_\varepsilon(x)z = f(x, |z|)z$$

with energy $J_\varepsilon(z_\varepsilon) = c_\varepsilon$ and set $w_\varepsilon(t, x) = z_\varepsilon(t, x + x_\varepsilon)$, it is a simple matter to see that w_ε solves (5.17) with $\hat{J}_\varepsilon(w_\varepsilon) = J_\varepsilon(z_\varepsilon) = c_\varepsilon$.

Lemma 5.9 *Let c_ε be given in Proposition 5.8. There holds*

$$\limsup_{\varepsilon \to 0} c_\varepsilon \le \gamma_{V_0}.$$

Since we have assumed $V_0 = V(0) = \min_\Lambda V$, for ease of notations, let us denote

$$\mathscr{K}_\varepsilon := \big\{z \in E \setminus \{0\} : \nabla J_\varepsilon(z) = 0\big\}, \quad L_\varepsilon := \big\{z \in \mathscr{K}_\varepsilon : J_\varepsilon(z) = c_\varepsilon\big\},$$

and

$$\mathscr{A} := \big\{x \in \Lambda : V(x) = V_0\big\}.$$

Then we have:

Lemma 5.10 *For all $\varepsilon > 0$ small and $z_\varepsilon \in L_\varepsilon$, the time-dependent process $|z_\varepsilon(t, \cdot)|$ possesses a (global) maximum $x_\varepsilon \in \Lambda_\varepsilon$ such that*

$$\lim_{\varepsilon \to 0} V(\varepsilon x_\varepsilon) = V_0.$$

Moreover, by setting $w_\varepsilon(t, x) = z_\varepsilon(t, x + x_\varepsilon)$, we must have $|w_\varepsilon|$ decays uniformly at infinity and $\{w_\varepsilon\}$ converges in $\mathscr{B}^2(\mathbb{R} \times \mathbb{R}^N, \mathbb{R}^{2K})$ to a ground state solution to

$$\mathscr{L}z + V_0 c\, z = g(|z|)z.$$

Proof Since the proof is very similar to that of Lemma 4.31, we shall be sketchy. The proof will be divided into six steps, and the first four steps can be obtained by repeating the arguments of Lemma 4.31 (with minor changes), hence we omit the details (we refer [3] for specific arguments):

Step 1 $\{z_\varepsilon\}$ is non-vanishing.

Step 2 $\{\chi_{\Lambda_\varepsilon} \cdot z_\varepsilon\}$ is non-vanishing, that is: there exist $(t_\varepsilon, x_\varepsilon) \in \mathbb{R} \times \overline{\Lambda}_\varepsilon$ and constants $R, \delta > 0$ such that

$$\int_{t_\varepsilon - R}^{t_\varepsilon + R} \int_{B_R(x_\varepsilon)} |\chi_{\Lambda_\varepsilon} \cdot z_\varepsilon|^2 \ge \delta.$$

Step 3 Let $x_\varepsilon \in \mathbb{R}^N$ and $R, \delta > 0$ be such that

$$\int_{\mathbb{R}} \int_{B_R(x_\varepsilon)} |\chi_{\Lambda_\varepsilon} \cdot z_\varepsilon|^2 \, dxdt \geq \delta.$$

Then $\varepsilon x_\varepsilon \to \mathscr{A}$.

Step 4 Let w_ε be defined by $w_\varepsilon(t, x) = z_\varepsilon(t, x + x_\varepsilon)$, then $w_\varepsilon \to w \neq 0$ in E.

The remaining two steps are as follows.

Step 5 $w_\varepsilon \to w$ in $\mathscr{B}^2(\mathbb{R} \times \mathbb{R}^N, \mathbb{R}^{2K})$ as $\varepsilon \to 0$.

To prove this, we only need to show that $|\mathscr{L}(w_\varepsilon - w)|_2 \to 0$ as $\varepsilon \to 0$ (this can be seen from that $|\mathscr{L}z|_2$ for $z \in \mathscr{B}^2$ defines a equivalent norm on $\mathscr{B}^2$). Notice that we have

$$\mathscr{L}(w_\varepsilon - w) = f_\varepsilon\big(x + x_\varepsilon, |w_\varepsilon|\big)w_\varepsilon - g(|w|)w - \big(\hat{V}_\varepsilon(x)w_\varepsilon - V_0 w\big).$$

Using the result in *Step 4* and the uniform L^∞ estimate, it follows directly that $|\mathscr{L}(w_\varepsilon - w)|_2 \to 0$ as $\varepsilon \to 0$.

Step 6 $w_\varepsilon(t, x) \to 0$ as $|(t, x)| \to \infty$ uniformly for all small ε.

To show this, let us remark that: for $w_\varepsilon = (w_\varepsilon^1, w_\varepsilon^2) : \mathbb{R} \times \mathbb{R}^N \to \mathbb{R}^{2K}$, if denoted by $\widehat{w}_\varepsilon(t, x) = \big(w_\varepsilon^1(t, x), w_\varepsilon^2(-t, x)\big)$, it is clear that $\widehat{w}_\varepsilon$ satisfies a heat equation of the form

$$\partial_t \widehat{w}_\varepsilon - \Delta_x \widehat{w}_\varepsilon + \widehat{w}_\varepsilon = \widehat{f}_\varepsilon(t, x) \quad \text{in } \mathbb{R} \times \mathbb{R}^N.$$

By virtue of the well-known regularity results, we have $\widehat{f}_\varepsilon \in L^q$ for all $q \geq 2$. According to *Step 5* and the interpolation theory, we have $w_\varepsilon \to w$ in $\mathscr{B}^r(\mathbb{R} \times \mathbb{R}^N, \mathbb{R}^{2K})$ for all $q \geq 2$. So, an easy calculation shows $\widehat{f}_\varepsilon \to \widehat{f}_0$ in L^q for some f_0 and all $q \geq 2$. Then an trivial application of the interior estimates shows that $|\widehat{w}_\varepsilon(t, x)| \to 0$ as $|(t, x)| \to \infty$, which yields the uniformly decay property of $\{w_\varepsilon\}$ as required.

Now, by collecting all the results proved in *Step 1-6*, we have the lemma proved.

We are ready to state our main results for the RD-system (5.3).

Theorem 5.11 (Ding and Xu [3]) *Assuming* (V), *then for* $\varepsilon > 0$ *small problem* (5.3) *has a solution* $\tilde{z}_\varepsilon = (u_\varepsilon, v_\varepsilon) \in \mathscr{B}^r(\mathbb{R} \times \mathbb{R}^N, \mathbb{R}^{2K})$ *for all* $r \geq 2$ *such that*

(*i*) *there exists a family of points* $\{y_\varepsilon\}$ *in* Λ *with* $\lim_{\varepsilon \to 0} V(y_\varepsilon) = \underline{c} := \min_\Lambda V$ *such that*

$$\liminf_{\varepsilon \to 0} \varepsilon^{-N} \int_{\mathbb{R}} \int_{B_{\varepsilon\rho}(y_\varepsilon)} |\tilde{z}_\varepsilon|^2 \, dxdt > 0 \quad \textit{for every } \rho > 0$$

and

$$\lim_{\substack{R\to\infty\\ \varepsilon\to 0}} \|\tilde{z}_\varepsilon(t,\cdot)\|_{L^\infty(\mathbb{R}^N\setminus B_{\varepsilon R}(y_\varepsilon))} = 0 \quad \textit{for every } t \in \mathbb{R};$$

(*ii*) *the transition sequence* $w_\varepsilon(t,x) = \tilde{z}_\varepsilon(t, \varepsilon x + y_\varepsilon)$ *converges in* $\mathscr{B}^2(\mathbb{R} \times \mathbb{R}^N, \mathbb{R}^{2K})$, *as* $\varepsilon \to 0$, *to a ground state solution of*

$$\begin{cases} \partial_t u = \Delta_x u - u - \underline{c}\, v + \partial_v H(u,v)\,, \\ \partial_t v = -\Delta_x v + v + \underline{c}\, u - \partial_u H(u,v)\,, \end{cases}$$

where $H(z) = G(|z|)$ *for* $z = (u,v) \in \mathbb{R}^{2K}$.

Proof We follow the notation used in Lemma 5.10 and define

$$\tilde{z}_\varepsilon(t,x) = z_\varepsilon(t, x/\varepsilon) \quad \text{and} \quad y_\varepsilon = \varepsilon x_\varepsilon.$$

Then $\tilde{z}_\varepsilon = (\tilde{u}, \tilde{v})$ is a solution of

$$\begin{cases} \partial_t u = \varepsilon^2 \Delta_x u - u - V(x)v + f(x,|z|)v \\ -\partial_t v = \varepsilon^2 \Delta_x v - v - V(x)u + f(x,|z|)u \\ z = (u,v) \in \mathscr{B}^2(\mathbb{R} \times \mathbb{R}^N, \mathbb{R}^{2K}) \end{cases}$$

for ε small. Since y_ε is the maximum point, and due to the fact

$$\lim_{\substack{R\to\infty\\ \varepsilon\to 0}} \|z_\varepsilon(t,\cdot)\|_{L^\infty(\mathbb{R}^N\setminus B_R(x_\varepsilon))} = 0,$$

we have

$$\lim_{\substack{R\to\infty\\ \varepsilon\to 0}} \|\tilde{z}_\varepsilon(t,\cdot)\|_{L^\infty(\mathbb{R}^N\setminus B_{\varepsilon R}(y_\varepsilon))} = 0. \tag{5.18}$$

Noting that $y_\varepsilon \to \mathscr{A}$ as $\varepsilon \to 0$, the assumption

$$\min_{\Lambda} V < \min_{\partial\Lambda} V$$

and (5.18) implies: for $\varepsilon > 0$ sufficiently small, there holds $|\tilde{z}_\varepsilon(t,x)| < s_0$ provided $x \notin \Lambda$ (here $s_0 > 0$ is so chosen that $g(s_0) + g'(s_0)s_0 = \frac{1-|V|_\infty}{2}$). Therefore, by the definition of F, we have $F(x, |\tilde{z}_\varepsilon|) = H(\tilde{z}_\varepsilon)$ when $\varepsilon > 0$ is small enough. Note that we have actually proved that $\tilde{z}_\varepsilon$ is a solution to (5.3) for all small ε, and consequently the proof of the theorem is thereby completed by combing (5.18) with Lemma 5.10.

5.3 More Distinct Solutions

In this section we will study the problem (5.1), i.e.,

$$\partial_t z = \mathbf{D}\Delta_x z + M(x)z + \big(W(x)|z|^{p-2} + Q(x)|z|^{4/N}\big)\beta z$$

where the diffusion coefficients $\mathbf{D}$ is defined in (2.6) and the matrix potential M in the form of (5.2) involving a further function $V : \mathbb{R}^N \to \mathbb{R}$.

Throughout the section, let us denote

$$\nu_\infty := \liminf_{|x|\to\infty} V(x), \quad \lambda_\infty := \limsup_{|x|\to\infty} W(x), \quad \kappa_\infty := \limsup_{|x|\to\infty} Q(x)$$

and make the following assumptions on the external potentials M, W and Q:

$(A1)$ $\|V\|_{L^\infty} < 1$, $\inf_{x\in\mathbb{R}^N} W(x) > 0$ and $Q(x) \ge 0$ for all x.

$(A2)$ There exists a constant $\hat{\kappa} = \hat{\kappa}(V, W, p) > 0$ and $x_0 \in \mathbb{R}^N$ such that $Q(x_0) = \max_{x\in\mathbb{R}^N} Q(x) < \hat{\kappa}$ and

$$\nu_\infty \ge \nu_0 := V(x_0), \quad \lambda_\infty \le \lambda_0 := W(x_0), \quad \kappa_\infty \le Q(x_0)$$

with one of the first two inequalities being strict.

It is worth mentioning that here we will not make the sharpest assumptions in order to avoid technicalities, the main purpose being to highlight the ideas of the approach we use. For (further and) more general results as well as for more details, we still refer to [4].

5.3.1 *Necessary Condition for Existence of Spikes*

In what follows, we consider weak solutions to the equation

$$\mathscr{L}z + \omega z = \lambda|z|^{p-2}z + \kappa|z|^{4/N}z \qquad \text{on } \mathbb{R}\times\mathbb{R}^N \tag{5.19}$$

belonging to the class $\mathscr{B}^2$ with ω being a $2K \times 2K$ symmetric constant matrix with its eigenvalues $\sigma(\omega) \subset (-1, 1)$, $\lambda > 0$ and $\kappa \ge 0$. As before, let us introduce the associated functional

$$\Phi(z) = \frac{1}{2}\big(\|z^+\|^2 - \|z^-\|^2\big) + \frac{1}{2}\iint \omega z\cdot z\, dt dx - \frac{\lambda}{p}|z|_p^p - \frac{\kappa}{2_*}|z|_{2_*}^{2_*} \quad \text{on } E = E^+ \oplus E^-.$$

And denoted by (Φ^{red}, ℓ) the reduction couple for Φ and set $\mathscr{N} = \{u \in E^+ \setminus \{0\} : \nabla\Phi^{red}(u)[u] = 0\}$. Then, by the same arguments used in [1, 3], we have $\mathscr{N}$ is a smooth manifold of codimension 1 in E^+, and $\mathscr{N}$ is diffeomorphic to S_1^+ by a C^1 diffeomorphism. Particularly, the function $t \mapsto \Phi^{red}(tu)$ attains its unique critical point $t = t(u) > 0$ for each $u \in E^+ \setminus \{0\}$, and $t : S_1^+ \to \mathbb{R}$ is a C^1 function. If denoted by

$$\gamma(\omega, \lambda, \kappa) \equiv \inf_{e \in E^+ \setminus \{0\}} \sup_{u \in \mathbb{R}e \oplus E^-} \Phi(u),$$

it can be also seen that $\gamma(\omega, \lambda, \kappa) = \inf_{\mathscr{N}} \Phi^{red} > 0$. We will write $\gamma(\omega, \lambda)$ for simplicity in the case when $\kappa \equiv 0$.

The main result of this subsection is the following:

Proposition 5.12 *Let $\nu(\omega) = \min\{\nu : \nu \in \sigma(\omega)\}$, and set $\nu_* = \min\{\nu(\omega), 0\}$. Then $\gamma(\omega, \lambda, \kappa)$ is attained provided that*

$$(1 + \nu_*)^{-\frac{N+2}{2}} \cdot \kappa^{\frac{N}{2}} \cdot \gamma(\omega, \lambda, \kappa) < \frac{S^{\frac{N+2}{2}}}{N+2} \tag{5.20}$$

where S denotes the best constant for the embedding $E \hookrightarrow L^{2^*}$.

This proposition yields information about the competing effect of ω, λ against κ. In the case κ vanishes, (5.20) is satisfied automatically, and hence $\gamma(\omega, \lambda)$ is always attained. Before proving Proposition 5.12, we begin with some preliminary materials. Let us first consider the following functional

$$\mathcal{F}_\omega : E \setminus \{0\} \to \mathbb{R}, \quad z \mapsto \frac{\|z^+\|^2 - \|z^-\|^2 + \iint \omega z \cdot z \, dt dx}{|z|_{2^*}^2},$$

and the minimax scheme

$$T_\omega = \inf_{z \in E^+ \setminus \{0\}} \sup_{w \in E^-} \mathcal{F}_\omega(z + w).$$

Denoted by $\mathscr{F} : L^2 \to L^2$ the Fourier transform, then in Fourier domain $\xi = (\xi_0, \xi_1, \ldots, \xi_N)$ we have $\|z\|^2 = \iint (\xi_0^2 + (1 + \sum_{k=1}^N \xi_k^2)^2)^{\frac{1}{2}} |\mathscr{F}z(\xi)|^2 d\xi$. Remark that S denotes the best constant for the embedding $E \hookrightarrow L^{2^*}$, i.e. $S|z|_{2^*}^2 \le \|z\|_2^2$, and that $\nu(\omega)$ is the smallest eigenvalue of ω, we soon infer that for any $z \in E^+ \setminus \{0\}$

$$\begin{aligned}
\sup_{w \in E^-} \mathcal{F}_\omega(z + w) &\ge \mathcal{F}_\omega(z) = \frac{\|z\|^2 + \iint \omega z \cdot z \, dt dx}{|z|_{2^*}^2} \\
&\ge \frac{\iint [(\xi_0^2 + (1 + \sum_{k=1}^N \xi_k^2)^2)^{\frac{1}{2}} + \nu(\omega)] |\mathscr{F}z(\xi)|^2 d\xi}{|z|_{2^*}^2}.
\end{aligned}$$

Taking into account that

$$\inf_{|\xi|>0} \frac{\left(\xi_0^2 + \left(1+\sum_{k=1}^N \xi_k^2\right)^2\right)^{\frac{1}{2}} + \nu(\omega)}{\left(\xi_0^2 + \left(1+\sum_{k=1}^N \xi_k^2\right)^2\right)^{\frac{1}{2}}} = \begin{cases} 1 & \text{if } \nu(\omega) \geq 0, \\ 1+\nu(\omega) & \text{if } \nu(\omega) < 0, \end{cases}$$

we have

$$T_\omega \geq (1+\nu_*)S \quad \text{with} \quad \nu_* = \min\{\nu(\omega), 0\}. \tag{5.21}$$

Next, let us consider the equation

$$\mathscr{L}z + \omega z = |z|^{4/N} z \quad \text{on } \mathbb{R} \times \mathbb{R}^N \tag{5.22}$$

and the corresponding functional

$$\hat{\Phi}(z) = \frac{1}{2}\left(\|z^+\|^2 - \|z^-\|^2\right) + \frac{1}{2}\iint \omega z \cdot z\, dt dx - \frac{1}{2^*}|z|_{2^*}^{2^*} \quad \text{on } E = E^+ \oplus E^-.$$

Denoted by $(\hat{\Phi}^{red}, \hat{\ell})$ the reduction couple for $\hat{\Phi}$ and set $\hat{\mathscr{N}} = \{u \in E^+ \setminus \{0\} : \nabla\hat{\Phi}^{red}(u)[u] = 0\}$. We have $\hat{\mathscr{N}}$ is a smooth manifold of codimension 1 in E^+, and $\hat{\mathscr{N}}$ is diffeomorphic to S_1^+ by a C^1 diffeomorphism. Particularly, the function $t \mapsto \hat{\Phi}^{red}(tu)$ attains its unique critical point $\hat{t} = \hat{t}(u) > 0$ for each $u \in E^+ \setminus \{0\}$. It would be also standard to see that $\hat{\gamma}_\omega := \inf_{\hat{\mathscr{N}}} \hat{\Phi}^{red} > 0$, and in particular, we have

Lemma 5.13 $T_\omega = \left((N+2)\hat{\gamma}_\omega\right)^{\frac{2}{N+2}}$.

Proof We sketch the proof as follows: Let $z \in E^+ \setminus \{0\}$ be fixed, and set $\pi_z(\cdot) = \mathcal{F}_\omega(z + \cdot)$, then elementary calculation shows that any critical point $w \in E^-$ for π_z satisfies $\nabla^2\pi_z(w)[\eta, \eta] < 0$ for all $\eta \in E^-$. Hence, π_z has a unique critical point in E^- which realize its maximum (if there exists).

For any $z \in \hat{\mathscr{N}}$, we have $\|z\|^2 - \|\hat{\ell}(z)\|^2 + \iint \omega \cdot (z + \hat{\ell}(z)) \cdot (z + \hat{\ell}(z)) dt dx - |z + \hat{\ell}(z)|_{2^*}^{2^*} = 0$, and hence $\pi_z(\hat{\ell}(z)) = |z + \hat{\ell}(z)|_{2^*}^{2^*-2}$. Moreover, it is standard to check that $\nabla\pi_z(\hat{\ell}(z))[w] = 0$ for all $w \in E^-$. Thus, we have $|z + \hat{\ell}(z)|_{2^*}^{2^*-2} = \max_{w\in E^-} \pi_z(w)$, i.e. $\hat{\ell}(z)$ is the unique critical point for π_z.

Now, using the fact $\mathcal{F}_\omega(z) = \mathcal{F}_\omega(tz)$ for all $z \in E$ and $t > 0$, we can conclude

$$\begin{aligned} T_\omega &= \inf_{z\in S_1^+} \sup_{w\in E^-} \mathcal{F}_\omega(z+w) = \inf_{z\in\hat{\mathscr{N}}} \sup_{w\in E^-} \mathcal{F}_\omega(z+w) \\ &= \inf_{z\in\hat{\mathscr{N}}} |z + \hat{\ell}(z)|_{2^*}^{2^*-2} = \inf_{z\in\hat{\mathscr{N}}} \left((N+2)\hat{\Phi}^{red}(z)\right)^{\frac{2}{N+2}} \\ &= \left((N+2)\hat{\gamma}_\omega\right)^{\frac{2}{N+2}} \end{aligned}$$

as is desired. For further reference, we mention that the uniqueness of $\hat{\ell}(z)$ as critical point for π_z also implies that $\hat{\ell}(tz) = t\hat{\ell}(z)$ for all $z \in E^+ \setminus \{0\}$.

Now, we give the proof of the proposition.

Proof of Proposition 5.12 We only give the proof when $\kappa > 0$ since it is much easier for the case $\kappa = 0$.

Let $\{z_n\} \subset \mathscr{N}$ be a minimizing sequence for Φ^{red}. It is no difficult to check that $\{w_n = z_n + \ell(z_n)\}$ is bounded in E. Then by Lion's result (see [5]) it follows that $\{w_n\}$ is either vanishing or non-vanishing.

If $\{w_n\}$ is non-vanishing then we are done, so let us assume contrarily that $\{w_n\}$ is vanishing. Then $|w_n|_s \to 0$ for all $s \in (2, 2^*)$. And thus we have

$$\kappa^{\frac{N}{2}}\Phi(w_n) = \hat{\Phi}\big(\kappa^{\frac{N}{4}}w_n\big) + o_n(1) \le \hat{\Phi}(\hat{w}_n) + o_n(1) \le \kappa^{\frac{N}{2}}\Phi(w_n) + o_n(1)$$

where we used the notation $\hat{w}_n := \hat{t}_n u_n + \hat{\ell}(\hat{t}_n u_n)$ with $\hat{t}_n = \hat{t}(u_n)$ being bounded and such that $\hat{t}_n u_n \in \hat{\mathscr{N}}$ (the last inequality comes from the fact that $\hat{\ell}(tz) = t\hat{\ell}(z)$ guarantees $\{\hat{w}_n\}$ is vanishing).

By the above observation, and $\Phi(w_n) = \Phi^{red}(z_n) = \gamma(\omega, \lambda, \kappa) + o_n(1)$, we easily deduce from Lemma 5.13 and (5.21) that

$$\kappa^{\frac{N}{2}} \cdot \gamma(\omega, \lambda, \kappa) + o_n(1) = \hat{\Phi}(\hat{w}_n) \ge \hat{\gamma}_\omega = \frac{1}{N+2}T_\omega^{\frac{N+2}{2}} \ge (1+\nu_*)^{\frac{N+2}{2}}\frac{S^{\frac{N+2}{2}}}{N+2},$$

which contradict to (5.20). Therefore we have $\{w_n\}$ is non-vanishing, and this ends the proof.

Next, let us go back to our model problems by denoting ω_ν the constant matrix $\nu \cdot \mathrm{id}$ or $\nu\mathcal{J}_0$, and consider

$$\mathscr{L}z + \omega_\nu z = \lambda|z|^{p-2}z + \kappa|z|^{4/N}z \qquad \text{on } \mathbb{R}\times\mathbb{R}^N \tag{5.23}$$

with the associated functional

$$\Phi_{\nu\lambda\kappa}(z) = \frac{1}{2}\big(\|z^+\|^2 - \|z^-\|^2\big) + \frac{1}{2}\iint \omega_\nu z \cdot z\,dtdx - \frac{\lambda}{p}|z|_p^p - \frac{\kappa}{2^*}|z|_{2^*}^{2^*}$$

and minimal energy $\gamma(\omega_\nu, \lambda, \kappa)$. We then end this section by concluding

Lemma 5.14 *Let* $\nu_1, \nu_2 \in (-1, 1)$, $\lambda_1, \lambda_2 > 0$ *and* $\kappa_1, \kappa_2 \ge 0$. *If* $\min\{\nu_2-\nu_1, \lambda_1-\lambda_2, \kappa_1-\kappa_2\} \ge 0$ *and either* $(\omega_{\nu_1}, \lambda_1, \kappa_1)$ *or* $(\omega_{\nu_2}, \lambda_2, \kappa_2)$ *satisfies* (5.20), *then* $\gamma(\omega_{\nu_1}, \lambda_1, \kappa_1) \le \gamma(\omega_{\nu_2}, \lambda_2, \kappa_2)$. *And if in additional* $\max\{\nu_2-\nu_1, \lambda_1-\lambda_2, \kappa_1-\kappa_2\} > 0$, *then* $\gamma(\omega_{\nu_1}, \lambda_1, \kappa_1) < \gamma(\omega_{\nu_2}, \lambda_2, \kappa_2)$.

Proof We only prove the second statement.

Case 1 $\omega_\nu = \nu \cdot \mathrm{id}$.

As mentioned before, let us introduce the reduction couple (Φ_j^{red}, ℓ_j) for $\Phi_{\nu_j\lambda_j\kappa_j}$, $j = 1, 2$, and

$$m_j : S^+ \to \mathcal{N}_j, \quad j = 1, 2$$

to be the C^1 diffeomorphism between $S^+ := \{z \in E^+ : \|z\| = 1\}$ and $\mathcal{N}_j := \{z \in E^+ \setminus \{0\} : \nabla\Phi_j^{red}(z)[z] = 0\}$.

Let $z \in S^+$ be arbitrary, there holds

$$\begin{aligned}
\gamma(\omega_{\nu_1}, \lambda_1, \kappa_1) &\le \Phi_1^{red}\big(m_1(z)\big) \\
&\le \Phi_2^{red}\big(m_1(z)\big) + \frac{\nu_1 - \nu_2}{2}\big|m_1(z) + \ell_1(m_1(z))\big|_2^2 \\
&\quad - \frac{\lambda_1 - \lambda_2}{p}\big|m_1(z) + \ell_1(m_1(z))\big|_p^p - \frac{\kappa_1 - \kappa_2}{2^*}\big|m_1(z) + \ell_1(m_1(z))\big|_{2^*}^{2^*} \\
&\le \Phi_2^{red}\big(m_2(z)\big) + \frac{\nu_1 - \nu_2}{2}\big|m_1(z) + \ell_1(m_1(z))\big|_2^2 \\
&\quad - \frac{\lambda_1 - \lambda_2}{p}\big|m_1(z) + \ell_1(m_1(z))\big|_p^p - \frac{\kappa_1 - \kappa_2}{2^*}\big|m_1(z) + \ell_1(m_1(z))\big|_{2^*}^{2^*}.
\end{aligned} \tag{5.24}$$

If $(\omega_{\nu_2}, \lambda_2, \kappa_2)$ satisfies (5.20), it is all clear that $\gamma(\omega_{\nu_2}, \lambda_2, \kappa_2)$ is achieved. Now, we can fix $z \in S^+$ be such that $\Phi_2^{red}\big(m_2(z)\big) = \gamma(\omega_{\nu_2}, \lambda_2, \kappa_2)$, by (5.24) we have

$$\begin{aligned}
\gamma(\omega_{\nu_1}, \lambda_1, \kappa_1) \le \gamma(\omega_{\nu_2}, \lambda_2, \kappa_2) &+ \frac{\nu_1 - \nu_2}{2}\big|m_1(z) + \ell_1(m_1(z))\big|_2^2 \\
&- \frac{\lambda_1 - \lambda_2}{p}\big|m_1(z) + \ell_1(m_1(z))\big|_p^p \\
&- \frac{\kappa_1 - \kappa_2}{2^*}\big|m_1(z) + \ell_1(m_1(z))\big|_{2^*}^{2^*}.
\end{aligned}$$

Therefore, by $\max\{\nu_2 - \nu_1, \lambda_1 - \lambda_2, \kappa_1 - \kappa_2\} > 0$, we have $\gamma(\omega_{\nu_1}, \lambda_1, \kappa_1) < \gamma(\omega_{\nu_2}, \lambda_2, \kappa_2)$.

Else if $(\omega_{\nu_1}, \lambda_1, \kappa_1)$ satisfies (5.20), let us consider a sequence $\{z_n\} \subset S^+$ be such that

$$\Phi_2^{red}\big(m_2(z_n)\big) \to \gamma(\omega_{\nu_2}, \lambda_2, \kappa_2).$$

Then the sequence $\{m_2(z_n)\}$ is bounded. If $\{m_2(z_n)\}$ is vanishing, as was argued in Proposition 5.12, we have

$$\gamma(\omega_{\nu_1}, \lambda_1, \kappa_1) < \frac{\big((1 + \nu_{1*})S\big)^{\frac{N+2}{2}}}{\kappa_1^{\frac{N}{2}}(N+2)} \le \frac{\big((1 + \nu_{2*})S\big)^{\frac{N+2}{2}}}{\kappa_2^{\frac{N}{2}}(N+2)} \le \gamma(\omega_{\nu_2}, \lambda_2, \kappa_2),$$

where $\nu_{j*} = \min\{\nu_j, 0\}$ for $j = 1, 2$. And hence the proof is done.

Case 2 $\omega_\nu = \nu \mathcal{J}_0$.

In this situation, let us consider the scaling transform $t \mapsto (1+\nu)t$, $x \mapsto \sqrt{1+\nu}\,x$. Then system (5.23) is equivalent to

$$\mathscr{L}z = \frac{\lambda}{1+\nu}|z|^{p-2}z + \frac{\kappa}{1+\nu}|z|^{4/N}z$$

with the minimal energy denoted by $\bar{\gamma}(\omega_\nu, \lambda, \kappa)$. Remark that, by the scaling transformation, we always have

$$\gamma(\omega_\nu, \lambda, \kappa) = (1+\nu)^{\frac{N+4}{2}}\bar{\gamma}(\omega_\nu, \lambda, \kappa). \tag{5.25}$$

And then, at this stage, we can go back to *Case 1*, and the monotonicity of $\bar{\gamma}$ will directly imply our desired conclusion for $\gamma(\omega_\nu, \lambda, \kappa)$ which completes the proof.

Remark 5.15 Lemma 5.14 provides a criterion for checking (5.20). Indeed, since $\kappa \geq 0$, we always have $\gamma(\omega_\nu, \lambda, \kappa) \leq \gamma(\omega_\nu, \lambda)$. Hence, we can set an upper bound for κ as

$$\hat{\kappa}(\nu, \lambda) = \frac{\big((1 - |V|_\infty)S\big)^{\frac{N+2}{N}}}{\big((N+2)\gamma(\omega_\nu, \lambda)\big)^{\frac{2}{N}}}$$

such that $(\omega_\nu, \lambda, \kappa)$ satisfies (5.20) provided that $0 \leq \kappa < \hat{\kappa}(\nu, \lambda)$. Furthermore, we can assert form the relations (5.24) and (5.25) that $\hat{\kappa}(\nu, \lambda)$ is decreasing in ν and increasing in λ. And hence for fixed $\nu \in (-1, 1)$ and $\lambda > 0$, let $\kappa \in [0, \hat{\kappa}(\nu, \lambda))$, then the triple $(\omega_{\nu'}, \lambda', \kappa)$ will satisfy (5.20) provided that $\nu' \leq \nu$ and $\lambda' \geq \lambda$.

5.3.2 *The Compactness*

In order to apply the abstract multiplicity result stated in Theorem 3.21, we need Φ_ε (see (5.7)) to satisfy a compactness condition. However, due to the non-compactness of the Sobolev embedding $E \hookrightarrow L^{2^*}_{loc}$, it is not difficult to see that such a condition cannot be fulfilled in general. Nevertheless, recalling Φ_ε admits a group action of the time-axis translation and denoting such action by $\mathscr{G} := \mathbb{R}$, we can recover the compactness in the sense of $\mathscr{G}$-$(PS)_c$-condition holds below some energy threshold, which related to the minimal energy "at infinity". For ease of notation, let us set $\bar{\kappa} = \max_{x\in\mathbb{R}^N} Q(x)$. Then, inspired by the priori bound $\hat{\kappa}(\nu, \lambda)$ for the factor κ in Remark 5.15, our compactness result can be stated as follows.

Proposition 5.16 *Assume* $(A1)$ *and* $(A2)$. *Suppose that* $\bar{\kappa} < \hat{\kappa}(\nu_\infty, \lambda_\infty)$. *Then, for any* $\varepsilon > 0$, *the functional* Φ_ε *satisfies* $\mathscr{G}$-$(PS)_c$-*condition in the sublevel* $\{z \in E : \Phi_\varepsilon(z) < \bar{\gamma}_\infty\}$, *where* $\bar{\gamma}_\infty := \gamma(\omega_{\nu_\infty}, \lambda_\infty, \bar{\kappa})$.

Let us mention here that for the case Q vanishes identically we can infer that $(\omega_{\nu_\infty}, \lambda_\infty)$ satisfies (5.20) automatically, and hence Φ_ε will satisfy the $\mathscr{G}$-$(PS)_c$-condition with no more additional assumptions except $c < \gamma_\infty(\omega_{\nu_\infty}, \lambda_\infty)$. Before proving Proposition 5.16, let us remark that the following lemma will be the key ingredient in the proofs.

Proof of Proposition 5.16 To begin with, let $c < \bar{\gamma}_\infty$ and let $\{z_n\}$ be a Palais-Smale sequence for Φ_ε at level c, namely

$$\Phi_\varepsilon(z_n) = c + o_n(1), \quad \nabla\Phi_\varepsilon(z_n) = o_n(1).$$

It follows easily that $\{z_n\}$ is bounded (independent of ε). Remark that, for any $\{g_n\} \subset \mathscr{G}$, the time-axis translated sequence $\{g_n z_n\}$ is also a $(PS)_c$-sequence. Hence, up to a subsequence, it has a associated weak limit $z \in E$ and we have to prove that $g_n z_n \to z$ in E for some $\{g_n\} \subset \mathscr{G}$.

Without loss of generality, for arbitrary $\{g_n\} \subset \mathscr{G}$, let us denote $\bar{z}_n := g_n z_n$ and z to be its weak limit (here and subsequently, one should always keep in mind that such weak limit will change correspondingly when $\{g_n\}$ is replaced by other elements). For simplicity, we set $\zeta_n = \bar{z}_n - z$, then $\zeta_n \rightharpoonup 0$ in E and $\|\bar{z}_n^\pm\|^2 = \|z^\pm\|^2 + \|\zeta_n^\pm\|^2 + o_n(1)$. By using the Brézis-Lieb type result (see for example [7]), we have

$$\iint W_\varepsilon(x)|\bar{z}_n|^p\,dtdx = \iint W_\varepsilon(x)|z|^p\,dtdx + \iint W_\varepsilon(x)|\zeta_n|^p\,dtdx,$$

and

$$\iint Q_\varepsilon(x)|\bar{z}_n|^{2^*}\,dtdx = \iint Q_\varepsilon(x)|z|^{2^*}\,dtdx + \iint Q_\varepsilon(x)|\zeta_n|^{2^*}\,dtdx.$$

Thus,

$$\Phi_\varepsilon(\bar{z}_n) = \Phi_\varepsilon(z) + \Phi_\varepsilon(\zeta_n) + o_n(1), \tag{5.26}$$

and

$$\nabla\Phi_\varepsilon(\bar{z}_n)[\bar{z}_n] = \nabla\Phi_\varepsilon(z)[z] + \nabla\Phi_\varepsilon(\zeta_n)[\zeta_n] + o_n(1).$$

Obviously, $\nabla\Phi_\varepsilon(z) = 0$. Therefore, $\nabla\Phi_\varepsilon(\zeta_n)[\zeta_n] = o_n(1)$.

Claim 5.1 $\nabla\Phi_\varepsilon(\zeta_n) \to 0$ as $n \to \infty$.

In fact, let $\varphi \in E$ with $\|\varphi\| \le 1$ be arbitrary. We have

$$\begin{aligned}
\nabla\Phi_\varepsilon(\bar{z}_n)[\varphi] &= \langle \bar{z}_n^+, \varphi^+\rangle - \langle \bar{z}_n^-, \varphi^-\rangle + \iint \mathcal{V}_\varepsilon(x)\bar{z}_n \cdot \varphi\, dtdx - \Psi_\varepsilon'(\bar{z}_n)[\varphi] \\
&= \langle \zeta_n^+ + z^+, \varphi^+\rangle - \langle \zeta_n^- + z^-, \varphi^-\rangle + \iint \mathcal{V}_\varepsilon(x)(\zeta_n + z)\cdot\varphi\, dtdx \\
&\quad - \Psi_\varepsilon'(\zeta_n + z)[\varphi] \\
&= \langle \zeta_n^+, \varphi^+\rangle - \langle \zeta_n^-, \varphi^-\rangle + \langle z^+, \varphi^+\rangle - \langle z^-, \varphi^-\rangle \\
&\quad + \iint \mathcal{V}_\varepsilon(x)\zeta_n \cdot \varphi\, dtdx + \iint \mathcal{V}_\varepsilon(x) z\cdot\varphi\, dtdx \\
&\quad - \iint W_\varepsilon(x)|\zeta_n|^{p-2}\zeta_n\cdot\varphi\, dtdx - \iint W_\varepsilon(x)|z|^{p-2}z\cdot\varphi\, dtdx \\
&\quad - \iint Q_\varepsilon(x)|\zeta_n + z|^{\frac{4}{N}}(\zeta_n + z)\cdot\varphi\, dtdx + o_n(\|\varphi\|).
\end{aligned} \tag{5.27}$$

Here the estimate for the subcritical part

$$\iint W_\varepsilon(x)|\bar{z}_n|^{p-2}\bar{z}_n\cdot\varphi - \iint W_\varepsilon(x)|\zeta_n|^{p-2}\zeta_n\cdot\varphi - \iint W_\varepsilon(x)|z|^{p-2}z\cdot\varphi = o_n(\|\varphi\|)$$

follows from a standard argument. To estimate the last integral in (5.27), we set $\psi_n := |\zeta_n + z|^{\frac{4}{N}}(\zeta_n + z) - |\zeta_n|^{\frac{4}{N}}\zeta_n - |z|^{\frac{4}{N}}z$. It is not difficult to see that exists a constant $C > 0$ independent of n such that $|\psi_n| \le C|\zeta_n|^{\frac{4}{N}}\cdot|z| + C|\zeta_n|\cdot|z|^{\frac{4}{N}}$. Then using the Egorov theorem on any bounded domains, we have

$$\iint |\psi_n|\cdot|\varphi|\, dtdx = o_n(\|\varphi\|) \quad \text{as } n \to \infty.$$

And hence,

$$\begin{aligned}
\nabla\Phi_\varepsilon(\bar{z}_n)[\varphi] &= \nabla\Phi_\varepsilon(\zeta_n)[\varphi] + \nabla\Phi_\varepsilon(z)[\varphi] + o_n(\|\varphi\|) \\
&= \nabla\Phi_\varepsilon(\zeta_n)[\varphi] + o_n(\|\varphi\|)
\end{aligned}$$

where we have used the fact $\nabla\Phi_\varepsilon(z) = 0$. The above estimation shows $\nabla\Phi_\varepsilon(\zeta_n) \to 0$ as $n \to \infty$ as was claimed.

Now to show that $\|\zeta_n\| \to 0$ as $n \to \infty$, the concentration compactness argument will be used.

Claim 5.2 There exists a sequence $\{g_n\} \subset \mathscr{G}$ such that $\{\zeta_n\}$ is vanishing.

Assuming Claim 5.2 for the moment, we have $\zeta_n \to 0$ in L^q for $q \in (2, 2^*)$. And it follows from Claim 5.1 that

$$\|\zeta_n\|^2+\iint \mathcal{V}_\varepsilon(x)\zeta_n\cdot(\zeta_n^+-\zeta_n^-)\,dtdx = \iint Q_\varepsilon(x)|\zeta_n|^{\frac{4}{N}}\zeta_n\cdot(\zeta_n^+-\zeta_n^-)\,dtdx+o_n(1).$$

By noting that $|V|_\infty < 1$, we soon obtain

$$(1-|V|_\infty)\|\zeta_n\|^2 \le \Big(\iint Q_\varepsilon(x)|\zeta_n|^{2^*}dtdx\Big)^{\frac{N+4}{2(N+2)}} \Big(\iint Q_\varepsilon(x)|\zeta_n^+-\zeta_n^-|^{2^*}dtdx\Big)^{\frac{1}{2^*}} + o_n(1).$$

It should be point out that, since $Q(x) \le \bar{\kappa} := \max Q$, there holds

$$\Big(\iint Q_\varepsilon(x)|z|^{2^*}dtdx\Big)^{\frac{2}{2^*}} \le \bar{\kappa}^{\frac{2}{2^*}}\cdot S^{-1}\cdot\|z\|^2, \quad \forall z \in E$$

where S is the best constant for the embedding $E \hookrightarrow L^{2^*}$. Then, we deduce

$$(1-|V|_\infty)\cdot S\cdot\bar{\kappa}^{-\frac{2}{2^*}} \le \Big(\iint Q_\varepsilon(x)|\zeta_n|^{2^*}dtdx\Big)^{\frac{4}{2(N+2)}} + o_n(1) \tag{5.28}$$

provided that $\zeta_n \not\to 0$ in L^{2^*}. On the other hand, by (5.26) and Claim 5.1, we have

$$\begin{aligned}\gamma(\omega_{\nu_\infty},\lambda_\infty,\bar{\kappa}) = \bar{\gamma}_\infty &> \Phi_\varepsilon(\zeta_n) - \frac{1}{2}\Phi_\varepsilon'(\zeta_n)[\zeta_n] \\ &= \frac{1}{N+2}\iint Q_\varepsilon(x)|\zeta_n|^{2^*}dtdx + o_n(1)\end{aligned}$$

as $n \to \infty$. Thus, by (5.28), we can get

$$\gamma(\omega_{\nu_\infty},\lambda_\infty) \ge \gamma(\omega_{\nu_\infty},\lambda_\infty,\bar{\kappa}) > \frac{\big((1-|V|_\infty)S\big)^{\frac{N+2}{2}}}{N+2}\cdot\bar{\kappa}^{-\frac{N}{2}}$$

which contradicts to the fact $\bar{\kappa} < \hat{\kappa}(\nu_\infty,\lambda_\infty)$. Therefore, we have $\zeta_n \to 0$ in L^{2^*} and the compactness for $\{\bar{z}_n\}$ follows.

Now it remains to prove Claim 5.2. Suppose, contrary to our claim, that the $\mathscr{G}$-dependent sequence $\{\zeta_n\}$ is non-vanishing for any $\{g_n\} \subset \mathscr{G}$. Then there exists $(t_n, x_n) \in \mathbb{R}\times\mathbb{R}^N$ and $R, \delta > 0$ such that

$$\int_{t_n-R}^{t_n+R}\int_{B_R(x_n)} |\zeta_n|^2\,dtdx \ge \delta.$$

Without loss of generality we can assuming $\{t_n\}$ is bounded, and hence we have $|x_n| \to \infty$ as $n \to \infty$ since $\zeta_n \to 0$ in L^2_{loc}. At this point, let us assume $V_\varepsilon(x_n) \to \tilde{\nu}$, $W_\varepsilon(x_n) \to \tilde{\lambda}$ and $Q_\varepsilon(x_n) \to \tilde{\kappa}$ as $n \to \infty$. Consider the translated functions $\tilde{z}_n(t, x) = \zeta_n(t, x + x_n)$, it follows that $\tilde{z}_n \rightharpoonup z_0$ in E where z_0 is a non-trivial solution to

$$\mathscr{L}z + \omega_{\tilde{\nu}} z = \tilde{\lambda}|z|^{p-2}z + \tilde{\kappa}|z|^{\frac{4}{N}}z.$$

And then, we can argue similarly as in (5.26) to obtain

$$\gamma(\omega_{\nu_\infty}, \lambda_\infty, \bar{\kappa}) > c \geq \Phi_\varepsilon(\zeta_n) \geq \Phi_{\tilde{\nu}\tilde{\lambda}\tilde{\kappa}}(\tilde{z}_n) + o_n(1) \geq \gamma(\omega_{\tilde{\nu}}, \tilde{\lambda}, \tilde{\kappa}) + o_n(1).$$

Since we have $\tilde{\nu} \geq \nu_\infty$, $\tilde{\lambda} \leq \lambda_\infty$ and $\tilde{\kappa} \leq \bar{\kappa}$, the above inequality becomes absurd. This completes the proof.

5.3.3 A Lower Bound on the Number of Solutions

We are now in position to prove the main result of this section.

Theorem 5.17 (Ding and Xu [4]) *Assume the matrix M is in one of the form in (5.2), $p \in (2, 2^*)$ and V, W, Q are bounded, Hölder continuous functions on $\mathbb{R}^N$ satisfying (A1) and (A2). Then RD system (5.1) possesses at least θ distinct solutions z_ε^k, $k = 1, \ldots, \theta$ for small $\varepsilon > 0$, where θ is the largest integer such that*

$$\theta < \left(\frac{1 + \nu_\infty}{1 + \nu_0}\right)^{\frac{4 - N(p-2)}{2(p-2)}} \left(\frac{\lambda_0}{\lambda_\infty}\right)^{\frac{2}{p-2}}. \tag{5.29}$$

Moreover, among these solutions, z_ε^1 lies in the ground state energy level and has the following properties:

(1) *$|z_\varepsilon^1(t, \cdot)|$ has exactly one global maximum point at some x_ε with $\lim_{\varepsilon \to 0} \operatorname{dist}(x_\varepsilon, \mathscr{C}) = 0$ and*

$$\lim_{\substack{R \to \infty \\ \varepsilon \to 0}} \|\tilde{z}_\varepsilon(t, \cdot)\|_{L^\infty(\mathbb{R}^N \setminus B_{\varepsilon R}(x_\varepsilon))} = 0 \quad \textit{uniformly for } t \in \mathbb{R};$$

(2) *the rescaled function $w_\varepsilon(t, x) = z_\varepsilon(t, \varepsilon x + x_\varepsilon)$ converges as $\varepsilon \to 0$ uniformly to a ground state solution $z : \mathbb{R} \times \mathbb{R}^N \to \mathbb{R}^{2K}$ of*

$$\partial_t z = \mathcal{J}\Delta_x z + M(y_0)z + \big(W(y_0)|z|^{p-2} + Q(y_0)|z|^{\frac{4}{N}}\big)\beta z.$$

We emphasis here that, according to the threshold found in Proposition 5.16, we only need to find critical levels below the threshold $\bar{\gamma}_\infty := \gamma(\omega_{\nu_\infty}, \lambda_\infty, \bar{\kappa})$. Since V, W, Q are functions on $\mathbb{R}^N$, according to Remark 5.15, we can precisely fix Q in a range as

$$0 \le Q(x) < \hat{\kappa}\big(|V|_\infty, \inf W\big), \quad \forall x \in \mathbb{R}^N. \tag{5.30}$$

Note that $\hat{\kappa}\big(|V|_\infty, \inf W\big) \le \hat{\kappa}(\nu_\infty, \lambda_\infty)$, we have $(\omega_{V(y)}, W(y), Q(y))$ satisfies (5.20) for all $y \in \mathbb{R}^N$.

1. The Ground State Energy Solution
Let us first mention that

$$c_\varepsilon = \inf_{e \in E^+ \setminus \{0\}} \sup_{z \in E_e} \Phi_\varepsilon(z)$$

characterizes the ground state critical level for Φ_ε and, particularly, $c_0 = \gamma(\omega_{V(0)}, W(0), Q(0))$. Remark that it is also not difficult to see that c_0 is attained and hence we have $c_\varepsilon \le c_0 + o_\varepsilon(1)$. By noting that $\gamma(\omega_{V(0)}, W(0), Q(0)) < \gamma(\omega_{\nu_\infty}, \lambda_\infty, \bar{\kappa})$ due to Lemma 5.14, we thus conclude that Φ_ε will satisfy the $\mathscr{G}$-(PS)-condition at level c_ε for all small $\varepsilon > 0$ thanks to Proposition 5.16. And therefore, we shall have the existence of a critical point for Φ_ε which lies in the minimal energy level associated to Eq. (5.4).

Let us remark here that: let z_ε be a ground state energy solution to (5.4), taking into account that the energy estimate $0 < c_\varepsilon \le c_0 + o_\varepsilon(1)$, it is possible to prove that the family $\{z_\varepsilon\}$ has a concentration behavior. We will divide this proof into three steps.

Step I For $y \in \mathbb{R}^N$ being fixed arbitrarily, let us set $V_\varepsilon^y(x) = V_\varepsilon(x + y/\varepsilon)$, $W_\varepsilon^y(x) = W_\varepsilon(x + y/\varepsilon)$ and $Q_\varepsilon^y(x) = Q_\varepsilon(x + y/\varepsilon)$. Then, we have $V_\varepsilon^y(0) = V(y)$, $W_\varepsilon^y(0) = W(y)$ and $Q_\varepsilon^y(0) = Q(y)$. Now, let us consider the transformed equation

$$\mathscr{L}z + \mathcal{V}_\varepsilon^y(x)z = W_\varepsilon^y(x)|z|^{p-2}z + Q_\varepsilon^y(x)|z|^{\frac{4}{N}}z \tag{5.31}$$

where $\mathcal{V}_\varepsilon^y(x) =$ either $V_\varepsilon^y(x)$ or $V_\varepsilon^y(x)\mathcal{J}_0$. Denoted by Φ_ε^y the associated energy functional to (5.31) and Φ_0^y the energy functional defined correspondingly by $V(y)$, $W(y)$ and $Q(y)$, let us set

$$c_\varepsilon(y) = \inf_{e \in E^+ \setminus \{0\}} \sup_{z \in \mathbb{R}e \oplus E^-} \Phi_\varepsilon^y(z) \quad \text{and} \quad c_0(y) = \inf_{e \in E^+ \setminus \{0\}} \sup_{z \in \mathbb{R}e \oplus E^-} \Phi_0^y(z).$$

As was mentioned before, we can immediately conclude the fact that $c_0(y) = \gamma(\omega_{V(y)}, W(y), Q(y))$ is always attained. Then, repeated application of Theorem 3.21 enables us to have $c_\varepsilon(y) \le c_0(y) + o_\varepsilon(1)$ for each $y \in \mathbb{R}^N$.

Since a trivial verification would show that

$$c_\varepsilon(y) = \inf_{e\in E^+\setminus\{0\}} \sup_{z\in\mathbb{R}e\oplus E^-} \Phi_\varepsilon^y(z) \equiv \inf_{e\in E^+\setminus\{0\}} \sup_{z\in\mathbb{R}e\oplus E^-} \Phi_\varepsilon(z) = c_\varepsilon \quad \forall y \in \mathbb{R}^N,$$

we thus have actually have that

$$c_\varepsilon \le \min_{y\in\mathbb{R}^N} c_0(y) + o_\varepsilon(1) = \min_{y\in\mathbb{R}^N} \gamma(\omega_{V(y)}, W(y), Q(y)) + o_\varepsilon(1).$$

Step II Due to the fact that the family $\{z_\varepsilon\}$ is bounded and non-vanishing, let us choose $(t_\varepsilon, x_\varepsilon) \in \mathbb{R} \times \mathbb{R}^N$ and $R, \delta > 0$ be such that

$$\int_{t_\varepsilon-R}^{t_\varepsilon+R} \int_{B_R(x_\varepsilon/\varepsilon)} |z_\varepsilon|^2 \, dtdx \ge \delta.$$

Set $w_\varepsilon(t, x) = z_\varepsilon(t + t_\varepsilon, x + x_\varepsilon/\varepsilon)$, we then have w_ε become a minimal energy solution to the equation

$$\mathscr{L}z + V_\varepsilon^{x_\varepsilon}(x)z = W_\varepsilon^{x_\varepsilon}(x)|z|^{p-2}z + Q_\varepsilon^{x_\varepsilon}(x)|z|^{\frac{4}{N}}z. \tag{5.32}$$

Suppose now $V(x_\varepsilon) \to V_\infty$, $W(x_\varepsilon) \to W_\infty$ and $Q(x_\varepsilon) \to Q_\infty$ as $\varepsilon \to 0$. By the Hölder continuity, we then have

$$V_\varepsilon^{x_\varepsilon}(x) \to V_\infty, \quad W_\varepsilon^{x_\varepsilon}(x) \to W_\infty, \quad Q_\varepsilon^{x_\varepsilon}(x) \to Q_\infty \quad \text{as } \varepsilon \to 0$$

uniformly on bounded sets of x. Remark that, if denoted by w the weak limit of w_ε, we soon have $w \ne 0$. And hence, a multiplication by compactly supported functions in (5.32) will generally imply us that w solves

$$\mathscr{L}z + \omega_{V_\infty} z = W_\infty |z|^{p-2}z + Q_\infty |z|^{\frac{4}{N}} z$$

with its critical level being estimated by

$$\begin{aligned}
\Phi_{V_\infty W_\infty Q_\infty}(w) &= \Big(\frac{1}{2} - \frac{1}{p}\Big) W_\infty |w|_p^p + \Big(\frac{1}{2} - \frac{1}{2^*}\Big) Q_\infty |w|_{2^*}^{2^*} \\
&\ge \gamma(\omega_{V_\infty}, W_\infty, Q_\infty) \\
&\ge \min_{y\in\mathbb{R}^N} \gamma(\omega_{V(y)}, W(y), Q(y)).
\end{aligned}$$

On the other hand, by Fatou's lemma and Step I, we have

$$
\begin{aligned}
&\Big(\frac{1}{2}-\frac{1}{p}\Big)W_\infty|w|_p^p+\Big(\frac{1}{2}-\frac{1}{2^*}\Big)Q_\infty|w|_{2^*}^{2^*}\\
&\le \liminf_{\varepsilon\to 0}\Big(\frac{1}{2}-\frac{1}{p}\Big)\iint W_\varepsilon^{x_\varepsilon}(x)|w_\varepsilon|^p\,dtdx+\Big(\frac{1}{2}-\frac{1}{2^*}\Big)\iint Q_\varepsilon^{x_\varepsilon}(x)|w_\varepsilon|^{2^*}\,dtdx\\
&=\liminf_{\varepsilon\to 0}c_\varepsilon(x_\varepsilon)=\liminf_{\varepsilon\to 0}c_\varepsilon\le\limsup_{\varepsilon\to 0}c_\varepsilon\le\min_{y\in\mathbb{R}^N}\gamma(\omega_{V(y)},W(y),Q(y)).
\end{aligned}
$$

Therefore, we can conclude

$$
\lim_{\varepsilon\to 0}c_\varepsilon=\lim_{\varepsilon\to 0}c_0(x_\varepsilon)=\min_{y\in\mathbb{R}^N}\gamma(\omega_{V(y)},W(y),Q(y)). \tag{5.33}
$$

Remark that by the Hölder continuity of V, W and Q, jointly with a similar estimate like (5.24), we deduce that $\gamma(\omega_{V(\cdot)},W(\cdot),Q(\cdot))$ is also Hölder continuous. Hence $\gamma(\omega_{\nu_\infty},\lambda_\infty,\bar{\kappa})>\gamma(\omega_{V(0)},W(0),Q(0))=c_0$ implies $\{x_\varepsilon\}$ is bounded in $\mathbb{R}^N$. And due to

$$
c_0(y)=\gamma(\omega_{V(y)},W(y),Q(y))>\min_{y\in\mathbb{R}^N}\gamma(\omega_{V(y)},W(y),Q(y))
$$

$$
\text{provided that } \operatorname{dist}(y,\mathscr{C})>0,
$$

thus we conclude $\lim_{\varepsilon\to 0}\operatorname{dist}(x_\varepsilon,\mathscr{C})=0$ from (5.33), where

$$
\mathscr{C}=\big\{y\in\mathbb{R}^N:\ c_0(y)=\min_{\xi\in\mathbb{R}^N}c_0(\xi)\big\}.
$$

Actually, by (5.33) and the Brézis-Lieb type result, we also get $|w_\varepsilon-w|_{2^*}\to 0$ and then $|w_\varepsilon^\pm-w^\pm|_{2^*}\to 0$ by Proposition 5.1. And using $w_\varepsilon^\pm-w^\pm$ as test functions in (5.32), we finally get $w_\varepsilon\to w$ in E as $\varepsilon\to 0$.

Step III According to Step II, let us assume without loss of generality that $x_\varepsilon\to y_0\in\mathscr{C}$ as $\varepsilon\to 0$. Then we have $V(x_\varepsilon)\to V(y_0)$, $W(x_\varepsilon)\to W(y_0)$ and $Q(x_\varepsilon)\to Q(y_0)$ as $\varepsilon\to 0$ and w solves the limit equation

$$
\mathscr{L}z+\mathcal{V}(y_0)z=W(y_0)|z|^{p-2}z+Q(y_0)|z|^{\frac{4}{N}}z,\ \text{ where } \mathcal{V}(y_0)=\omega_{V(y_0)}. \tag{5.34}
$$

Hence, by (5.32), we have

$$
\begin{aligned}
\mathscr{L}(w_\varepsilon-w)&=W_\varepsilon^{x_\varepsilon}(x)|w_\varepsilon|^{p-2}w_\varepsilon-W(y_0)|w|^{p-2}w+Q_\varepsilon^{x_\varepsilon}(x)|w_\varepsilon|^{\frac{4}{N}}w_\varepsilon\\
&\quad-Q(y_0)|w|^{\frac{4}{N}}w-\big(\mathcal{V}_\varepsilon^{x_\varepsilon}(x)w_\varepsilon-\mathcal{V}(y_0)w\big).
\end{aligned}
$$

Using the fact $w_\varepsilon \to w$ in E and the uniform L^∞ estimate, it is easy to check that $|L(w_\varepsilon - w)|_2 \to 0$ as $\varepsilon \to 0$. And therefore, we have $w_\varepsilon \to w$ in $\mathscr{B}^2$ as $\varepsilon \to 0$.

Next, let us remark that: for $w_\varepsilon = (w_\varepsilon^1, w_\varepsilon^2) : \mathbb{R} \times \mathbb{R}^N \to \mathbb{R}^{2K}$ solves (5.34), if denoted by $\widehat{w}_\varepsilon(t,x) = \big(w_\varepsilon^1(t,x), w_\varepsilon^2(-t,x)\big)$, it is clear that $\widehat{w}_\varepsilon$ satisfies a equation of the form

$$\partial_t \widehat{w}_\varepsilon - \Delta_x \widehat{w}_\varepsilon + \widehat{w}_\varepsilon = \widehat{f_\varepsilon}(t,x) \quad \text{in } \mathbb{R} \times \mathbb{R}^N.$$

And we have $\widehat{f_\varepsilon} \in L^q$ for all $q \ge 2$. According to the interpolation theory, we infer that $w_\varepsilon \to w$ in $\mathscr{B}^r$ for all $q \ge 2$. So, we get $\widehat{f_\varepsilon} \to \widehat{f_0}$ in L^q for some f_0 and all $q \ge 2$. Then an trivial application of [3, Corollary A.4] would show that $|\widehat{w}_\varepsilon(t,x)| \to 0$ uniformly as $|(t,x)| \to \infty$, which yields the uniformly decay property of $\{w_\varepsilon\}$.

By collecting all the results proved in Step I - Step III, we actually proved that $\tilde{z}_\varepsilon(t,x) = z_\varepsilon(t, x/\varepsilon)$ is a minimal energy solution to the multi-component incongruent RD system

$$\partial_t z = \mathbf{D}\Delta_x z + M(x)z + \big(W(x)|z|^{p-2} + Q(x)|z|^{4/N}\big)\beta z$$

for all small $\varepsilon > 0$, and $|\tilde{z}_\varepsilon(t,\cdot)|$ has a maximum point x_ε which converge to a suitable $y_0 \in \mathscr{C}$. Moreover, we have

$$\lim_{\substack{R\to\infty \\ \varepsilon\to 0}} \|\tilde{z}_\varepsilon(t,\cdot)\|_{L^\infty(\mathbb{R}^N \setminus B_{\varepsilon R}(x_\varepsilon))} = 0.$$

After a translation in time-axis if necessary, the rescaled functions $w_\varepsilon(t,x) = z_\varepsilon(t, x + x_\varepsilon/\varepsilon)$ will converge in $\mathscr{B}^2$ (and hence in $\mathscr{B}^r$, $r \ge 2$) to a minimal energy solution of the limit system

$$\partial_t z = \mathcal{J}\Delta_x z + M(y_0)z + \big(W(y_0)|z|^{p-2} + Q(y_0)|z|^{4/N}\big)\beta z,$$

and this finishes the whole characterization of the asymptotic behavior of the minimal energy solution to our original problem.

2. Existence of Further Solutions

By virtue of Theorem 3.24 in Chap. 3, in order to show the existence of other critical points, it is sufficient to construct a finite dimensional subspace $X_0 \subset E^+$ for Φ_ε such that

$$d := \sup \Phi_\varepsilon(X_0 \oplus E^-) < \bar{\gamma}_\infty = \gamma(\omega_{\nu_\infty}, \lambda_\infty, \bar{\kappa}). \tag{5.35}$$

Thus, since the $\mathscr{G}$-$(PS)_c$-condition is satisfied in the sublevel $\{z \in E : \Phi_\varepsilon(z) < \bar{\gamma}_\infty\}$, the change of pseudo-index (3.28) will imply the existence of at least $n := \dim X_0$ distinct critical points in the energy range $(0, d\,]$.

For simplify notation, we set $\nu_0 = V(0)$, $\lambda_0 = W(0)$ and $\kappa_0 = Q(0)$. Since we have assumed $Q(0) = \sup_{x\in\mathbb{R}^N} Q(x)$, we get $\kappa_0 = \bar{\kappa}$. In what follows, let us consider

$$\mathscr{L}z + \omega_{\nu_\infty} z = \lambda_\infty |z|^{p-2} z + \bar{\kappa}|z|^{\frac{4}{N}} z$$

and

$$\mathscr{L}z + \omega_{\nu_0} z = \lambda_0 |z|^{p-2} z + \kappa_0 |z|^{\frac{4}{N}} z$$

with the minimal energy $\gamma(\omega_{\nu_\infty}, \lambda_\infty, \bar{\kappa})$ and $\gamma(\omega_{\nu_0}, \lambda_0, \kappa_0)$ respectively. And for a clear expression, let us rewrite the above equations as

$$-\beta\partial_t z + \mathcal{J}_0(-\Delta_x)z + (\mathcal{J}_0 + \omega_{\nu_\infty})z = \lambda_\infty |z|^{p-2} z + \bar{\kappa}|z|^{\frac{4}{N}} z \tag{5.36}$$

and

$$-\beta\partial_t z + \mathcal{J}_0(-\Delta_x)z + (\mathcal{J}_0 + \omega_{\nu_0})z = \lambda_0 |z|^{p-2} z + \kappa_0 |z|^{\frac{4}{N}} z. \tag{5.37}$$

Considering the scaling transform $w(t,x) = bz(at, \sqrt{a}x)$ with $a, b > 0$, we have (5.36) is equivalent to

$$-\beta\partial_t w + \mathcal{J}_0(-\Delta_x)w + a\cdot(\mathcal{J}_0 + \omega_{\nu_\infty})w = \frac{a\lambda_\infty}{b^{p-2}}|w|^{p-2}w + \frac{a\bar{\kappa}}{b^{\frac{4}{N}}}|w|^{\frac{4}{N}} w \tag{5.38}$$

Now, let us take $a = \frac{1+\nu_0}{1+\nu_\infty}$, $b = a^{\frac{N}{4}}$ and denote $\tilde{\gamma}(\omega_{\nu_\infty}, \lambda_\infty, \bar{\kappa})$ the minimal energy for (5.38), then we have

$$\tilde{\gamma}(\omega_{\nu_\infty}, \lambda_\infty, \bar{\kappa}) = \frac{b^2}{a^{\frac{N}{2}}}\gamma(\omega_{\nu_\infty}, \lambda_\infty, \bar{\kappa}) = \gamma(\omega_{\nu_\infty}, \lambda_\infty, \bar{\kappa}).$$

Moreover, by the choice of a and b, we also get

$$a\cdot(\mathcal{J}_0 + \omega_{\nu_\infty}) \geq (\mathcal{J}_0 + \omega_{\nu_0}) \quad \text{for the quadratic part,} \tag{5.39}$$

and

$$\frac{a\lambda_\infty}{b^{p-2}} < \lambda_0, \quad \frac{a\bar{\kappa}}{b^{\frac{4}{N}}} = \bar{\kappa} \quad \text{for the nonlinear coefficients.} \tag{5.40}$$

Here in (5.39), for two matrices A and B, by $A \geq B$ we mean $A - B$ is positive definite. Denoted by $\lambda' = \frac{a\lambda_\infty}{b^{p-2}}$, we are then led to the stage to consider the equation

$$-\beta\partial_t w + \mathcal{J}_0(-\Delta_x)w + (\mathcal{J}_0 + \omega_{\nu_0})w = \lambda' |w|^{p-2} w + \bar{\kappa}|w|^{\frac{4}{N}} w \tag{5.41}$$

and its minimal energy $\gamma(\omega_{\nu_0}, \lambda', \bar{\kappa})$. Comparing (5.37) and (5.41), remark that we have the relations $\kappa_0 = \bar{\kappa}$ and

$$\gamma(\omega_{\nu_0}, \lambda, \bar{\kappa}) = \lambda^{-\frac{2}{p-2}} \cdot \gamma\Big(\omega_{\nu_0}, 1, \frac{\bar{\kappa}}{\lambda^{4/N(p-2)}}\Big), \quad \forall \lambda > 0$$

jointly with (5.39) and (5.40) we soon deduce that

$$\begin{aligned}\frac{\gamma(\omega_{\nu_\infty}, \lambda_\infty, \bar{\kappa})}{\gamma(\omega_{\nu_0}, \lambda_0, \kappa_0)} &= \frac{\tilde{\gamma}(\omega_{\nu_\infty}, \lambda_\infty, \bar{\kappa})}{\gamma(\omega_{\nu_0}, \lambda_0, \kappa_0)} \geq \frac{\gamma(\omega_{\nu_0}, \lambda', \bar{\kappa})}{\gamma(\omega_{\nu_0}, \lambda_0, \bar{\kappa})} \\ &= \Big(\frac{\lambda_0}{\lambda'}\Big)^{\frac{2}{p-2}} \cdot \frac{\gamma\big(\omega_{\nu_0}, 1, \frac{\bar{\kappa}}{\lambda'^{4/N(p-2)}}\big)}{\gamma\big(\omega_{\nu_0}, 1, \frac{\bar{\kappa}}{\lambda_0^{4/N(p-2)}}\big)}.\end{aligned}$$

Let us emphasis here that the above estimate become an equality when ω_ν is in the form of $\omega_\nu = \nu \mathcal{J}_0$. Now, let us set $n, n' \in \mathbb{N}$ to be the largest integers satisfying

$$n < \Big(\frac{\lambda_0}{\lambda'}\Big)^{\frac{2}{p-2}} \quad \text{and} \quad n' < \Big(\frac{\lambda_0}{\lambda'}\Big)^{\frac{2}{p-2}} \cdot \frac{\gamma\big(\omega_{\nu_0}, 1, \frac{\bar{\kappa}}{\lambda'^{4/N(p-2)}}\big)}{\gamma\big(\omega_{\nu_0}, 1, \frac{\bar{\kappa}}{\lambda_0^{4/N(p-2)}}\big)},$$

then we can infer from $\lambda' < \lambda_0$ that $n \geq 1$. We should remark that n and n' are not equal in general because $\gamma\big(\omega_{\nu_0}, 1, \frac{\bar{\kappa}}{\lambda'^{4/N(p-2)}}\big) < \gamma\big(\omega_{\nu_0}, 1, \frac{\bar{\kappa}}{\lambda_0^{4/N(p-2)}}\big)$, which implies $n \geq n'$. Nevertheless, we can still conclude from the continuity of γ in the third variable that: in addition to (5.30), if we consider the function Q in a properly smaller range then n and n' will be equal as an invariant constant with respect to the factor $\bar{\kappa}$. At this moment, there holds

$$\gamma(\omega_{\nu_\infty}, \lambda_\infty, \bar{\kappa}) > n \cdot \gamma(\omega_{\nu_0}, \lambda_0, \kappa_0). \tag{5.42}$$

Next, recalling the notations introduced in the previous subsection, let us take $w \in E$ be the minimal energy solution of

$$\mathscr{L}z + V(y_0)z = W(y_0)|z|^{p-2}z + Q(y_0)|z|^{\frac{4}{N}}z$$

such that w is also the limit of the rescaled sequence w_ε defined in Step II. For $r > 0$ let us choose a cut-off function $\eta_r : \mathbb{R}^N \to [0, 1]$ such that $\eta_r(x) \equiv 1$ for $|x| \leq r$ and $\eta_r(x) \equiv 0$ for $|x| \geq r + 1$. Set $w_r = \eta_r \cdot w$, then we have $\|w_r - w\| \to 0$ as $r \to \infty$, particularly, $\{w_r\}$ is a (PS)-sequence for $\Phi_0^{y_0}$ at the minimal energy level $c_0(y_0) = \gamma(\omega_{V(y_0)}, W(y_0), Q(y_0))$.

Now, let us take some $x_r \in \mathbb{R}^N$ with $|x_r| = 2(r + 1)$ and set $x_{rj} = (j - 1)x_r$ for $j = 1, \ldots, n$, where $n \in \mathbb{N}$ is fixed in (5.42). Define $w_{rj}(t, x) = w_r(t, x - x_{rj})$, then it is all clear that for r large enough $\{w_{rj}^+\}_{j=1}^n$ is linearly independent. Indeed, for constants c_j such that $\varphi^+ = \sum_{j=1}^n c_j w_{rj}^+ = 0$ where $\varphi := \sum_{j=1}^n c_j w_{rj}$, we can deduce that

$$\begin{aligned}
0 &\geq -\|\varphi^-\|^2 + \iint \mathcal{V}(y_0)\varphi^- \cdot \varphi^- \, dtdx \geq \|\varphi^+\|^2 - \|\varphi^-\|^2 - |V|_\infty |\varphi|_2^2 \\
&= \sum_{j=1}^n c_j^2 \cdot \left(\|w_{rj}^+\|^2 - \|w_{rj}^-\|^2 - |V|_\infty |w_{rj}|_2^2\right) \\
&= \left(\|w_r^+\|^2 - \|w_r^-\|^2 - |V|_\infty |w_r|_2^2\right) \sum_{j=1}^n c_j^2
\end{aligned}$$

which implies $c_j = 0$ for all j thanks to the fact $\|w_r^+\|^2 - \|w_r^-\|^2 - |V|_\infty |w_r|_2^2$ is strictly positive for large r. Having disposed of this preliminary step, we can now set

$$X_j^r = \mathbb{R} w_{rj}^+ = \operatorname{span}\{w_{rj}^+\} \quad \text{and} \quad X^r = \oplus_{j=1}^n X_j^r.$$

Then let $x_\varepsilon \in \mathbb{R}^N$ be the maximum point of w_ε found in the previous subsection, by using the Brézis-Lieb type result again, we can obtain

$$\begin{aligned}
\max_{z \in X^r \oplus E^-} \Phi_\varepsilon^{x_\varepsilon}(z) &\leq \max_{z \in X^r \oplus E^-} \Phi_0^{y_0}(z) + o_\varepsilon(1) \\
&\leq \sum_{j=1}^n \max_{z \in X_j^r \oplus E^-} \Phi_0^{y_0}(z) + o_r(1) + o_\varepsilon(1) \\
&= \sum_{j=1}^n \max_{z \in X_1^r \oplus E^-} \Phi_0^{y_0}(z) + o_r(1) + o_\varepsilon(1) \\
&= n \cdot \gamma(\omega_{V(y_0)}, W(y_0), Q(y_0)) + o_r(1) + o_\varepsilon(1)
\end{aligned} \tag{5.43}$$

where the first inequality can be derived from the Arzelà-Ascoli theorem (for related details, we refer the readers to [3, Corollary 4.4]). Therefore, by setting $w_{rj\varepsilon}(t, x) = w_{rj}(t, x - x_\varepsilon/\varepsilon)$,

$$X_{j\varepsilon}^r = \mathbb{R} w_{rj\varepsilon}^+ = \operatorname{span}\{w_{rj\varepsilon}^+\} \quad \text{and} \quad X_\varepsilon^r = \oplus_{j=1}^n X_{j\varepsilon}^r,$$

we can conclude from (5.42) and (5.43) that, for $r_0 > 0$ fixed large enough, there exists $\varepsilon_0 > 0$ such that $X_0 := X_{\varepsilon_0}^{r_0} \subset E^+$ satisfies (5.35) for all $\varepsilon \in (0, \varepsilon_0]$. And then the proof is hereby completed.

References

1. Ackermann, N.: A nonlinear superposition principle and multibump solution of periodic Schrödinger equations. J. Funct. Anal. **234**, 423–443 (2006)
2. Ding, Y.H.: Variational Methods for Strongly Indefinite Problems. Interdiscip. Math. Sci., vol. 7. World Scientific Publ., Singapore (2007)
3. Ding, Y.H., Xu, T.: Concentrating patterns of reaction-diffusion systems: a variational approach. Trans. Am. Math. Soc. **369**(1), 97–138 (2017)
4. Ding, Y.H., Xu, T.: Effect of external potentials in a coupled system of multi-component incongruent diffusion. Topol. Methods Nonlinear Anal. **54**, 715–750 (2019)
5. Lions, P.L.: The concentration-compactness principle in the calculus of variations: the locally compact case, Part II. AIP Anal. non linéaire **1**, 223–283 (1984)
6. Stein, E.M.: Singular Integrals and Differentiability Properties of Functions, vol. 2. Princeton University Press, Princeton (1970)
7. Willem, M.: Minimax Theorems. Birkhäuser, Boston (1996)

Chapter 6
The Spinorial Brezis-Nirenberg Problem

Let (M, g) be an m-dimensional compact manifold. We assume that M is spin, and we fix a spin structure σ on M. We denote by $\mathbb{S}(M) = Spin(TM) \times_\rho \mathbb{S}_m$ the spinor bundle on M with hermitian metric $(\cdot, \cdot)$ and compatible spin connection $\nabla^{\mathbb{S}}$. The Clifford multiplication

$$TM \otimes \mathbb{S}(M) \to \mathbb{S}(M)$$

is denoted by $X \otimes \psi \mapsto X \cdot \psi$. Let $D = D_g$ be the (Atiyah-Singer) Dirac operator defined on $\Gamma(\mathbb{S}(M))$, i.e. $D = \sum_{k=1}^{m} e_k \cdot \nabla^{\mathbb{S}}_{e_k}$ for a local orthonormal frame $\{e_1, \dots, e_m\}$ of TM.

In this chapter, we want to find solutions $\psi \in \Gamma(\mathbb{S}(M))$ of the nonlinear Dirac equation

$$D\psi = \lambda\psi + |\psi|^{\frac{2}{m-1}}\psi. \tag{6.1}$$

The exponent $2^* := \frac{2m}{m-1}$ is critical in the sense of the Sobolev embedding $H^{\frac{1}{2}}(M, \mathbb{S}(M)) \subset L^q(M, \mathbb{S}(M))$ for $1 \le q \le 2^*$, which is compact precisely if $q < 2^*$.

6.1 Selected Facts on Spinors

6.1.1 Spin Structure and the (Atiyah-Singer) Dirac Operator

Let (M, g) be an m-dimensional Riemannian manifold with a chosen orientation. Let $P_{SO}(M)$ be the set of positively oriented orthonormal frames on (M, g). This is a $SO(m)$-principal bundle over M. A *spin structure* on M is a pair $\sigma = (P_{Spin}(M), \vartheta)$ where $P_{Spin}(M)$ is a $Spin(m)$-principal bundle over M and

© The Author(s), under exclusive license to Springer Nature Singapore Pte Ltd. 2025

Y. Ding, T. Xu, *Localization Approaches in Strongly Indefinite Problems*, Series in Contemporary Mathematics 6, https://doi.org/10.1007/978-981-97-9539-0_6

$\vartheta : P_{Spin}(M) \to P_{SO}(M)$ is a map such that the diagram

$$\begin{array}{ccccc} P_{Spin}(M) \times Spin(m) & \longrightarrow & P_{Spin}(M) & & \\ \Big\downarrow \vartheta \times \Theta & & \Big\downarrow \vartheta & \searrow & \\ & & & & M \\ & & & \nearrow & \\ P_{SO}(M) \times SO(m) & \longrightarrow & P_{SO}(M) & & \end{array}$$

commutes, where $\Theta : Spin(m) \to SO(m)$ is the nontrivial double covering of $SO(m)$. There is a topological condition for the existence of a spin structure, namely, the vanishing of the second Stiefel-Whitney class $\omega_2(M) \in H^2(M, \mathbb{Z}_2)$. Furthermore, if a spin structure exists, it need not be unique. For these results we refer to [12, 15].

In order to introduce the spinor bundle, we recall that the Clifford algebra $Cl(\mathbb{R}^m)$ is the associative $\mathbb{R}$-algebra with unit, generated by $\mathbb{R}^m$ satisfying the relation $x \cdot y - y \cdot x = -2(x, y)$ for $x, y \in \mathbb{R}^m$ (here $(\cdot, \cdot)$ is the Euclidean scalar product on $\mathbb{R}^m$). It turns out that $Cl(\mathbb{R}^m)$ has a smallest representation $\rho : Spin(m) \subset Cl(\mathbb{R}^m) \to End(\mathbb{S}_m)$ of dimension $\dim_{\mathbb{C}}(\mathbb{S}_m) = 2^{[\frac{m}{2}]}$ such that $\mathbb{C}l(\mathbb{R}^m) := Cl(\mathbb{R}^m) \otimes \mathbb{C} \cong End_{\mathbb{C}}(\mathbb{S}_m)$ as $\mathbb{C}$-algebra. The spinor bundle is then defined as the associated vector bundle

$$\mathbb{S}(M) := P_{Spin}(M) \times_\rho \mathbb{S}_m.$$

Note that the spinor bundle carries a natural Clifford multiplication, a natural hermitian metric and a metric connection induced from the Levi-Civita connection on TM (see [12, 15]), this bundle satisfies the axioms of Dirac bundle in the sense that

(i) for any $x \in M$, $X, Y \in T_x M$ and $\psi \in \mathbb{S}_x(M)$

$$X \cdot Y \cdot \psi + Y \cdot X \cdot \psi + 2g(X, Y)\psi = 0;$$

(ii) for any $X \in T_x M$ and $\psi_1, \psi_2 \in \mathbb{S}_x(M)$,

$$(X \cdot \psi_1, \psi_2) = -(\psi_1, X \cdot \psi_2),$$

where $(\cdot, \cdot)$ is the hermitian metric on $\mathbb{S}(M)$;

(iii) for any $X, Y \in \Gamma(TM)$ and $\psi \in \Gamma(\mathbb{S}(M))$,

$$\nabla_X^{\mathbb{S}}(Y \cdot \psi) = (\nabla_X Y) \cdot \psi + Y \cdot \nabla_X^{\mathbb{S}} \psi,$$

where $\nabla^{\mathbb{S}}$ is the metric connection on $\mathbb{S}(M)$.

The Dirac operator is then defined on the spinor bundle $\mathbb{S}(M)$ as the composition

$$D : \Gamma(\mathbb{S}(M)) \xrightarrow{\nabla^{\mathbb{S}}} \Gamma(T^*M \otimes \mathbb{S}(M)) \longrightarrow \Gamma(TM \otimes \mathbb{S}(M)) \xrightarrow{\mathfrak{m}} \Gamma(\mathbb{S}(M))$$

where $\mathfrak{m}$ stands for the Clifford multiplication $\mathfrak{m} : X \otimes \psi \mapsto X \cdot \psi$.

6.1.2 The Dirac Spectrum and $H^{1/2}$-spinors

Let $\operatorname{spec}(D)$ denote the spectrum of the Dirac operator D. It is well-known that D is essentially self-adjoint in $L^2(M, \mathbb{S}(M))$ and has compact resolvents (see [12, 13, 15]). Moreover, $\operatorname{spec}(D) = \{\lambda_k : k \in \mathbb{Z}\}$ is a closed subset of $\mathbb{R}$ consisting of a two-sided unbounded discrete sequence of eigenvalues with finite multiplicities. And the eigenspaces of D form a complete orthonormal decomposition of $L^2(M, \mathbb{S}(M))$.

Proposition 6.1 (The Weyl's Asymptotic Law) *If $m \not\equiv 3 \ (mod\ 4)$, then the growth of the Dirac eigenvalues satisfies Weyl's asymptotic law:*

$$\lim_{\Lambda \to +\infty} \frac{d_{\pm}(\Lambda)}{\Lambda^m} = C_m Vol(M, g)$$

where

$$d_+(\Lambda) = \dim_{\mathbb{C}} \Big(\bigoplus_{0 < \lambda_k \le \Lambda} \ker(D - \lambda_k) \Big),$$

$$d_-(\Lambda) = \dim_{\mathbb{C}} \Big(\bigoplus_{-\Lambda \le \lambda_k < 0} \ker(D - \lambda_k) \Big)$$

and $C_m > 0$ is a dimensional constant.

Proof For $\Lambda > 0$ let

$$N(\Lambda) = \dim \Big(\bigoplus_{|\lambda_k| \le \Lambda} \ker(D - \lambda_k) \Big).$$

be the sum of the multiplicities of eigenvalues with modulus at most Λ. Then the Schrödinger-Lichnerowicz formula for the Dirac operator yields

$$\lim_{\Lambda \to +\infty} \frac{N(\Lambda)}{\Lambda^m} = c_m \mathrm{Vol}(M, g)$$

for some positive dimensional constant c_m; see [9, Corollary 2.43], for instance.

If the dimension $m \not\equiv 3 \pmod 4$ then $\operatorname{spec}(D)$ is symmetric about the origin including the multiplicities [13, Theorem 1.3.7], hence

$$d_+(\Lambda) = d_-(\Lambda) = \frac{N(\Lambda) - \dim\ker(D)}{2}.$$

The proposition follows with $C_m = c_m/2$. □

Remark 6.2 In dimension $m \equiv 3 \pmod 4$ the spectrum $\operatorname{spec}(D)$ is not symmetric. In order to measure the lack of symmetry of $\operatorname{spec}(D)$ in this case, Atiyah, Patodi and Singer [4] introduced the so-called η-invariant of D, denoted by $\eta(D)$. In particular $\eta(D) = 0$ as soon as $\operatorname{spec}(D)$ is symmetric. However, only few η-invariants are know explicitly. It was shown in [11] that, if $m \equiv 3 \pmod 4$, for arbitrary large $\Lambda > 0$, $j \in \mathbb{N}$ and $0 < l_1 < \cdots < l_j < \Lambda$ there exists a Riemannian metric g such that $\operatorname{spec}(D_g) \cap (0, \Lambda) = \{l_1 < \cdots < l_j\}$ and $\operatorname{spec}(D_g) \cap (-\Lambda, 0) = \emptyset$. The distribution of the spectrum in these dimensions becomes complicated. For further information about spectral theory for Dirac operators and for related topics we refer to the monograph by Ginoux [13].

We now define the operator $|D - \lambda|^{\frac{1}{2}} : L^2(M, \mathbb{S}(M)) \to L^2(M, \mathbb{S}(M))$ by

$$\psi = \sum_{k \in \mathbb{Z}} a_k \eta_k \quad \mapsto \quad |D - \lambda|^{\frac{1}{2}} \psi := \sum_{k \in \mathbb{Z}} |\lambda_k - \lambda|^{\frac{1}{2}} a_k \eta_k$$

where $\eta_k \in \ker(D - \lambda_k)$ and $\int_M |\eta_k|^2 d\,\mathrm{vol}_g = 1$ for $k \in \mathbb{Z}$. The Hilbert space

$$H^{\frac{1}{2}}(M, \mathbb{S}(M)) := \left\{ \psi \in L^2(M, \mathbb{S}(M)) : |D|^{\frac{1}{2}} \psi \in L^2(M, \mathbb{S}(M)) \right\}.$$

coincides with the Sobolev space $W^{\frac{1}{2},2}(M, \mathbb{S}(M))$ (see [1, 2]). Let $P_\lambda^0 : L^2(M, \mathbb{S}(M)) \to \ker(D - \lambda)$ denote the projector. We can endow $H^{\frac{1}{2}}(M, \mathbb{S}(M))$ with the inner product

$$\langle \psi, \varphi \rangle_\lambda = \operatorname{Re}\left(|D - \lambda|^{\frac{1}{2}} \psi, |D - \lambda|^{\frac{1}{2}} \varphi \right)_2 + \operatorname{Re}(P_\lambda^0 \psi, P_\lambda^0 \varphi)_2$$

and the induced norm $\|\cdot\|_\lambda = \langle \cdot, \cdot \rangle_\lambda^{\frac{1}{2}}$, where $(\psi, \varphi)_2 = \int_M (\psi, \varphi) d\,\mathrm{vol}_g$ is the L^2-inner product on spinors. Clearly $E := H^{\frac{1}{2}}(M, \mathbb{S}(M))$ with this inner product has the orthogonal decomposition of three subspaces

$$E = E_\lambda^+ \oplus E_\lambda^0 \oplus E_\lambda^- \tag{6.2}$$

where $E_\lambda^\pm$ is the positive (resp. negative) eigenspace of $D - \lambda$, and $E_\lambda^0 = \ker(D - \lambda)$ is its kernel (which may be trivial). The dual space of E will be denoted by $E_\lambda^* = H^{-\frac{1}{2}}(M, \mathbb{S}(M))$. By $P_\lambda^\pm : E \to E_\lambda^\pm$ and $P_\lambda^0 : E \to E_\lambda^0$ we denote the orthogonal projections. If λ is clear from the context, we simply write $\psi^{\pm} = P_\lambda^{\pm}(\psi)$ and $\psi^0 = P_\lambda^0(\psi)$ for $\psi \in E$.

6.2 The Palais-Smale Condition

Equation (6.1) is the Euler-Lagrange equation of the functional

$$\mathcal{L}_\lambda(\psi) = \frac{1}{2}\int_M (D\psi, \psi) d\,\mathrm{vol}_g - \frac{\lambda}{2}\int_M |\psi|^2 d\,\mathrm{vol}_g - \frac{1}{2^*}\int_M |\psi|^{2^*} d\,\mathrm{vol}_g .$$

In the sequel, by L^q we denote the Banach space $L^q(M, \mathbb{S}(M))$ for $q \geq 1$ and by $|\cdot|_q$ we denote the usual L^q-norm.

The functional $\mathcal{L}_\lambda$ is well defined on $E = H^{\frac{1}{2}}(M, \mathbb{S}(M))$ and is of class C^1 since E embeds into $L^q(M, \mathbb{S}(M))$ for $q \in [1, 2^*]$. We write $\psi = \psi^+ + \psi^0 + \psi^- \in E = E_\lambda^+ \oplus E_\lambda^0 \oplus E_\lambda^-$ according to the decomposition (6.2). Then $\mathcal{L}_\lambda$ has the form

$$\mathcal{L}_\lambda(\psi) = \frac{1}{2}(\|\psi^+\|_\lambda^2 - \|\psi^-\|_\lambda^2) - \frac{1}{2^*}\int_M |\psi|^{2^*} d\,\mathrm{vol}_g .$$

We now investigate the Palais-Smale condition for $\mathcal{L}_\lambda$. Due to the non-compactness of the critical embedding $E = H^{\frac{1}{2}}(M, \mathbb{S}(M)) \hookrightarrow L^{2^*}(M, \mathbb{S}(M))$, one cannot expect that $\mathcal{L}_\lambda$ satisfies the Palais-Smale condition on E. We shall see that the Palais-Smale condition holds below a critical value.

Lemma 6.3 *A $(PS)_c$-sequence $(\psi_n)_n$ for $\mathcal{L}_\lambda$ is bounded, for any $c \in \mathbb{R}$. Moreover, $c = 0$ if and only if $\psi_n \to 0$.*

Proof Now let $(\psi_n)_n$ be a $(PS)_c$-sequence, then

$$\mathcal{L}_\lambda(\psi_n) = \frac{1}{2}(\|\psi_n^+\|_\lambda^2 - \|\psi_n^-\|_\lambda^2) - \frac{1}{2^*}\int_M |\psi_n|^{2^*} d\,\mathrm{vol}_g = c + o_n(1), \tag{6.3}$$

$$\nabla\mathcal{L}_\lambda(\psi_n)[\psi_n] = \|\psi_n^+\|_\lambda^2 - \|\psi_n^-\|_\lambda^2 - \int_M |\psi_n|^{2^*} d\,\mathrm{vol}_g = o(\|\psi_n\|_\lambda), \tag{6.4}$$

and

$$\begin{aligned}\nabla\mathcal{L}_\lambda(\psi_n)[\psi_n^+ - \psi_n^-] &= \|\psi_n^+ + \psi_n^-\|_\lambda^2 - \mathrm{Re}\int_M |\psi_n|^{2^*-2}(\psi_n, \psi_n^+ - \psi_n^-) d\,\mathrm{vol}_g \\ &= o(\|\psi_n^+ + \psi_n^-\|_\lambda).\end{aligned} \tag{6.5}$$

Thus, we have for every $C > 2c$ and n large:

$$C + o(\|\psi_n\|_\lambda) \geq 2\mathcal{L}_\lambda(\psi_n) - \mathcal{L}'_\lambda(\psi_n)[\psi_n] = \frac{1}{m}|\psi_n|_{2^*}^{2^*}. \tag{6.6}$$

By (6.5), we can deduce that

$$\|\psi_n^+ + \psi_n^-\|_\lambda^2 \leq \int_M |\psi_n|^{2^*-1}|\psi_n^+ - \psi_n^-| d\,\mathrm{vol}_g + o(\|\psi_n^+ + \psi_n^-\|_\lambda).$$

From this and the Hölder and Sobolev inequalities we get

$$\|\psi_n^+ + \psi_n^-\|_\lambda^2 \leq C'|\psi_n|_{2^*}^{2^*-1}\|\psi_n^+ + \psi_n^-\|_\lambda + o(\|\psi_n^+ + \psi_n^-\|_\lambda). \tag{6.7}$$

Now, by taking into account (6.6), we get

$$\|\psi_n^+ + \psi_n^-\|_\lambda^2 \leq C\big(1 + o(\|\psi_n\|_\lambda^{1-\frac{1}{2^*}})\big)\|\psi_n^+ + \psi_n^-\|_\lambda.$$

Note that $\dim E_\lambda^0 < \infty$, hence any two norms on E_λ^0 are equivalent. Therefore we have

$$\|\psi_n^0\|_\lambda^2 = |\psi_n^0|_2^2 \leq |\psi_n|_2^2 \leq C|\psi_n|_{2^*}^2 \leq C\big(1 + o(\|\psi_n\|_\lambda^{\frac{2}{2^*}})\big) \tag{6.8}$$

which implies

$$\|\psi_n\|_\lambda^2 \leq C\big(1 + o(\|\psi_n\|_\lambda^{\frac{2}{2^*}})\big) + C\big(1 + o(\|\psi_n\|_\lambda^{1-\frac{1}{2^*}})\big)\|\psi_n\|_\lambda.$$

Now the boundedness of $(\psi_n)_n$ follows from $2^* > 2$.

If $c = 0$, due to the boundedness of $(\psi_n)_n$ we can rewrite (6.6) as $o_n(1) = |\psi_n|_{2^*}^{2^*}$. The assertion then follows from (6.7) and (6.8). □

The next lemma has been proved in [14, p. 288, (5.83)]. Let

$$\mathcal{E}_0(\psi) = \frac{1}{2}\int_M (D\psi, \psi) d\,\mathrm{vol}_g - \frac{1}{2^*}\int_M |\psi|^{2^*} d\,\mathrm{vol}_g.$$

be the functional associated to

$$D\psi = |\psi|^{2^*-2}\psi \quad \text{on } M. \tag{6.9}$$

Lemma 6.4 *Let $(\psi_n)_n$ be a Palais-Smale sequence for $\mathcal{E}_0$ such that $\psi_n \rightharpoonup 0$ in E and*

$$\liminf_{n\to\infty} \int_M |\psi_n|^{2^*} d\,\mathrm{vol}_g > 0.$$

Then

$$\liminf_{n\to\infty} \mathcal{E}_0(\psi_n) = \liminf_{n\to\infty} \frac{1}{2m}|\psi_n|_{2^*}^{2^*} \geq \frac{1}{2m}\left(\frac{m}{2}\right)^m \omega_m,$$

where ω_m stands for the volume of the unit sphere (S^m, g_{S^m}).

As a consequence of Lemma 6.4, we obtain the following compactness result below the critical value

$$\gamma_{crit} := \frac{1}{2m}\left(\frac{m}{2}\right)^m \omega_m. \tag{6.10}$$

Proposition 6.5 *The functional $\mathcal{L}_\lambda$ satisfies the $(PS)_c$-condition for any $c < \gamma_{crit}$.*

Proof Let $(\psi_n)_n$ be a $(PS)_c$-sequence for $\mathcal{L}_\lambda$ on E. By Lemma 6.3, $(\psi_n)_n$ is bounded and $\psi_n \to 0$ if (and only if) $c = 0$. Now suppose $c \neq 0$ and, up to a subsequence if necessary, $\psi_n \rightharpoonup \psi_0$ in E. Then we have

$$\psi_n \to \psi_0 \quad \text{in } L^p \text{ for any } 1 \leq p < 2^* \tag{6.11}$$

and the limit spinor ψ_0 satisfies

$$D\psi_0 = \lambda\psi_0 + |\psi_0|^{2^*-2}\psi_0 \quad \text{on } M. \tag{6.12}$$

We claim that $\psi_0 \neq 0$ in E. Indeed, assume to the contrary that $\psi_0 = 0$. Then it follows from (6.11) that $(\psi_n)_n$ is also a (PS)-sequence for $\mathcal{E}_0$. Moreover,

$$\liminf_{n\to\infty} \frac{1}{2m}\int_M |\psi_n|^{2^*} d\,\mathrm{vol}_g = \liminf_{n\to\infty} \mathcal{L}_\lambda(\psi_n) - \frac{1}{2}\mathcal{L}'_\lambda(\psi_n)[\psi_n] = c > 0.$$

Therefore, by Lemma 6.4 we have

$$c = \liminf_{n\to\infty} \mathcal{L}_\lambda(\psi_n) = \liminf_{n\to\infty} \mathcal{E}_0(\psi_n) \geq \frac{1}{2m}\left(\frac{m}{2}\right)^m \omega_m$$

which contradicts the assumption $c < \frac{1}{2m}\left(\frac{m}{2}\right)^m \omega_m$.

Now let us set $\varphi_n = \psi_n - \psi_0$. Then $\varphi_n \rightharpoonup 0$ in E. By the compact embedding $E \hookrightarrow L^p$ for $1 \leq p < 2^*$, we easily get $|\varphi_n|_2 \to 0$ as $n \to \infty$. On the other hand, arguing exactly as in [14, Lemma 5.2], we obtain the Brezis-Lieb type result for the integrand of critical part, that is,

$$\int_M |\psi_n|^{2^*} d\,\mathrm{vol}_g = \int_M |\varphi_n|^{2^*} d\,\mathrm{vol}_g + \int_M |\psi_0|^{2^*} d\,\mathrm{vol}_g + o_n(1) \tag{6.13}$$

and, for arbitrary $\psi \in E$ with $\|\psi\|_\lambda \le 1$,

$$\operatorname{Re}\int_M |\psi_n|^{2^*-2}(\psi_n, \psi) = \operatorname{Re}\int_M |\varphi_n|^{2^*-2}(\varphi_n, \psi) + \operatorname{Re}\int_M |\psi_0|^{2^*-2}(\psi_0, \psi) + o_n(1). \tag{6.14}$$

Therefore, combining (6.13) and (6.14), we infer that

$$\mathcal{L}_\lambda(\psi_n) = \mathcal{E}_0(\varphi_n) + \mathcal{L}_\lambda(\psi_0) + o_n(1) \tag{6.15}$$

and

$$\nabla\mathcal{L}_\lambda(\psi_n) = \nabla\mathcal{E}_0(\varphi_n) + o_n(1), \tag{6.16}$$

where we have used (6.12), i.e. $\nabla\mathcal{L}_\lambda(\psi_0) = 0$, in the last equality.

Equations (6.15) and (6.16) imply that $(\varphi_n)_n$ is a (PS)-sequence for $\mathcal{E}_0$. If $|\varphi_n|_{2^*} \to 0$ then we easily get $\varphi_n \to 0$ in E which gives the compactness of $(\psi_n)_n$ (cf. the proof of Lemma 6.3). If $\liminf_{n\to\infty} |\varphi_n|_{2^*} > 0$, then it follows from Lemma 6.4 that

$$\liminf_{n\to\infty} \mathcal{E}_0(\varphi_n) \ge \frac{1}{2m}\Big(\frac{m}{2}\Big)^m \omega_m .$$

But this, together with (6.15), implies $\liminf_{n\to\infty} \mathcal{L}_\lambda(\psi_n) \ge \frac{1}{2m}\big(\frac{m}{2}\big)^m \omega_m$ which contradicts our assumption. Hence we must have that $(\psi_n)_n$ is compact in E. □

6.3 The Min-Max Scheme

Let us write the functional $\mathcal{L}_\lambda \in C^1(E)$ in the following form

$$\mathcal{L}_\lambda(\psi) = \frac{1}{2}\Big(\|\psi^+\|_\lambda^2 - \|\psi^-\|_\lambda^2\Big) - \mathcal{K}(\psi)$$

with

$$\mathcal{K}(\psi) = \frac{1}{2^*}\int_M |\psi|^{2^*} d\,\mathrm{vol}_g .$$

The next two lemmas collect elementary properties of $\mathcal{K}$, which are need for finding critical values. We omit the proofs since they are quite standard.

Lemma 6.6 $\mathcal{K}(0) = 0$ *and* $\frac{1}{2}\nabla\mathcal{K}(\psi)[\psi] > \mathcal{K}(\psi) > 0$ *for every* $\psi \ne 0$.

Lemma 6.7 *$\mathcal{K}$ is strictly convex, hence weakly lower semi-continuous.*

Recall the decomposition $E = E_\lambda^+ \oplus E_\lambda^0 \oplus E_\lambda^-$ and let

$$S_\lambda^+ := \{\phi \in E_\lambda^+ : \|\phi\|_\lambda = 1\}.$$

For $\phi \in S_\lambda^+$ we set

$$E_\lambda(\phi) := \mathbb{R}\phi \oplus E_\lambda^0 \oplus E_\lambda^- \quad \text{and} \quad \widehat{E}_\lambda(\phi) := \{t\phi + \chi : t \geq 0,\ \chi \in E_\lambda^0 \oplus E_\lambda^-\} \subset E_\lambda(\phi).$$

We can now explain the strategy of our arguments. Fixing $\phi \in S_\lambda^+$ we maximize $\mathcal{L}_\lambda$ on $\widehat{E}_\lambda(\phi)$. The strict convexity of $\mathcal{K}$ implies that the maximum is unique and depends continuously on ϕ. Varying ϕ the set of these maximizers defines a (topological) manifold $\mathcal{P}_\lambda$ homeomorphic to S_λ^+, and a minimizer of $\mathcal{L}_\lambda$ on this manifold will be a critical point of $\mathcal{L}_\lambda$, in fact the critical point with least energy among all critical points of $\mathcal{L}_\lambda$. Alternatively, for each $\phi \in E_\lambda^+$ the functional $E_\lambda^0 \oplus E_\lambda^- \to \mathbb{R}$, $\chi \mapsto \mathcal{L}_\lambda(\phi + \chi)$, has a unique maximum $\chi_\lambda(\phi)$. Then one can apply the standard mountain pass theorem to the reduced functional $E_\lambda^+ \to \mathbb{R}$, $\phi \mapsto \mathcal{L}_\lambda\big(\phi + \chi_\lambda(\phi)\big)$.

Lemma 6.8 *For each $\phi \in S_\lambda^+$ there exists a unique nontrivial critical point $\mu_\lambda(\phi) \in \widehat{E}_\lambda(\phi)$ of the constrained functional $\mathcal{L}_\lambda|_{\widehat{E}_\lambda(\phi)}$. Moreover the following hold:*

(a) *$\mu_\lambda(\phi)$ is the global maximum of $\mathcal{L}_\lambda|_{\widehat{E}_\lambda(\phi)}$.*
(b) *$\mu_\lambda(\phi)^+$ is bounded away from 0, i.e. $\|\mu_\lambda(\phi)^+\|_\lambda \geq \delta$ for some $\delta > 0$.*
(c) *$\mu_\lambda : S_\lambda^+ \to \widehat{E}_\lambda(\phi)$ is bounded on compact subsets of S_λ^+.*

Proof We first prove that $\sup \mathcal{L}_\lambda|_{\widehat{E}_\lambda(\phi)} > 0$ is achieved. Observe that there exists a constant $C > 0$ such that

$$C\big(|t\phi|_{2^*} + |\chi^0|_{2^*} + |\chi^-|_{2^*}\big) \leq |t\phi + \chi|_{2^*} \quad \text{for all } t\phi + \chi \in \widehat{E}_\lambda(\phi) \tag{6.17}$$

because $\mathbb{R}\phi \oplus E_\lambda^0$ is of finite dimension. This implies

$$\mathcal{L}_\lambda(t\phi + \chi) \leq \frac{t^2}{2} - \frac{1}{2}\|\chi^-\|_\lambda^2 - C'\big(|t\phi|_{2^*} + |\chi^0|_{2^*} + |\chi^-|_{2^*}\big)^{2^*}$$

on $\widehat{E}_\lambda(\phi)$, hence $\mathcal{L}_\lambda(\psi) \leq 0$ if $\|\psi\|_\lambda$ is large. On the other hand, $\sup_{\widehat{E}_\lambda(\phi)} \mathcal{L}_\lambda \geq \alpha > 0$ because $\mathcal{L}_\lambda(t\phi) = \frac{t^2}{2} + o(t^2)$ as $t \to 0$, uniformly in $\phi \in S_\lambda^+$. This together with the weak upper semi-continuity of $\mathcal{L}_\lambda$ on $\widehat{E}_\lambda(\phi)$ implies that the supremum is achieved at some $\mu_\lambda(\phi)$. It also follows that $\|\mu_\lambda(\phi)^+\|$ is bounded away from 0, and that μ_λ is bounded on compact sets.

Now we prove that any critical point $\psi \neq 0$ of $\mathcal{L}_\lambda|_{\widehat{E}_\lambda(\phi)}$ is a strict global maximum of $\mathcal{L}_\lambda|_{\widehat{E}_\lambda(\phi)}$, hence it is unique. Let $\psi \in \widehat{E}_\lambda(\phi)$ be a nontrivial critical point of $\mathcal{L}_\lambda|_{\widehat{E}_\lambda(\phi)}$ and $\varphi = (1+s)\psi + \chi \in \widehat{E}_\lambda(\phi) \setminus \{\psi\}$ be an arbitrary element of $\widehat{E}_\lambda(\phi)$; here $s \geq -1$ and $\chi \in E_\lambda^0 \oplus E_\lambda^-$. For ease of notations, let us set $G(t) = \frac{1}{2^*}t^{2^*}$ and $g(t) = t^{2^*-2}$ for $t \geq 0$, fix χ as above and set

$$\begin{aligned} h(s) :=& \frac{s^2+2s}{2} g(|\psi|)|\psi|^2 + (s+1)g(|\psi|)\operatorname{Re}(\psi, \chi) \\ &+ G(|\psi|) - G(|(1+s)\psi + \chi|) \end{aligned} \tag{6.18}$$

an elementary calculation shows:

$$\begin{aligned} \mathcal{L}_\lambda(\varphi) - \mathcal{L}_\lambda(\psi) =& \frac{1}{2}\int_M \big((D\chi, \chi) - \lambda|\chi|^2\big) d\operatorname{vol}_g \\ &+ \nabla\mathcal{K}(\psi)\Big[\frac{s^2+2s}{2}\psi + (1+s)\chi\Big] \\ &+ \mathcal{K}(\psi) - \mathcal{K}((1+s)\psi + \chi) \\ =& -\frac{1}{2}\|\chi^-\|_\lambda^2 + \int_M h(s) d\operatorname{vol}_g . \end{aligned} \tag{6.19}$$

It is sufficient to prove that the integrand $h(s)(x)$ is negative if $\psi(x) \neq 0$. Observe that $h(-1) = -\frac{1}{2}g(|\psi|)|\psi|^2 + G(|\psi|) - G(|\chi|) < 0$ and $\lim_{s\to\infty} h(s) = -\infty$. We may assume that h attains its maximum at some $s_* \in (-1, \infty)$, otherwise we are done. We distinguish s and s_* here and emphasize that s_* is a real function depending on x, and we can set $s_*(x) = -1$ provided that $h(-1)(x)$ is the extremum. We drop the argument $x \in M$ in the sequel. Then, denoted by $\varphi_* = (1+s_*)\psi + \chi$, we have:

$$0 = h'(s_*) = \big(g(|\psi|) - g(|\varphi_*|)\big)\operatorname{Re}(\psi, \varphi_*).$$

Now, if $\operatorname{Re}(\psi, \varphi_*) \neq 0$, hypothesis (f_3) implies $|\psi| = |\varphi_*| > 0$, hence for $\varphi_* \neq \psi$:

$$\begin{aligned} h(s) \leq h(s_*) &= g(|\psi|)\operatorname{Re}\Big(\psi, \frac{s_*^2+2s_*}{2}\psi + (1+s_*)\chi\Big) \\ &= -\frac{s_*^2}{2}g(|\psi|)|\psi|^2 - (1+s_*)g(|\psi|)\big(|\psi|^2 - \operatorname{Re}(\psi, \varphi_*)\big) \\ &= -\frac{1}{2}g(|\psi|)|\chi|^2 < 0 \end{aligned} \tag{6.20}$$

If $\mathrm{Re}(\psi, \varphi_*) = 0$, i.e. $\mathrm{Re}(\psi, \chi) = -(1 + s_*)|\psi|^2$ then

$$\begin{aligned} h(s) \le h(s_*) &= -\frac{(s_* + 1)^2}{2} g(|\psi|)|\psi|^2 - \frac{1}{2} g(|\psi|)|\psi|^2 + G(|\psi|) - G(|\varphi_*|) \\ &< -\frac{1}{2} g(|\psi|)|\psi|^2 + G(|\psi|) < 0. \end{aligned} \tag{6.21}$$

□

Now we consider the functional

$$\mathcal{M}_\lambda : S_\lambda^+ \to \mathbb{R}, \quad \mathcal{M}_\lambda(\phi) := \mathcal{L}_\lambda(\mu_\lambda(\phi)).$$

Proposition 6.9

(a) $\mathcal{M}_\lambda \in \mathcal{C}^1(S_\lambda^+)$ *and*

$$\nabla \mathcal{M}_\lambda(\phi)[\chi] = \|\mu_\lambda(\phi)^+\|_\lambda \nabla \mathcal{L}_\lambda(\mu_\lambda(\phi))[\chi] \quad \textit{for all } \chi \in T_\phi(S_\lambda^+).$$

(b) *If* $(\phi_n)_n$ *is a Palais-Smale sequence for* $\mathcal{M}_\lambda$ *then* $(\mu_\lambda(\phi_n))_n$ *is a Palais-Smale sequence for* $\mathcal{L}_\lambda$. *If* $(\psi_n)_n$ *is a bounded Palais-Smale sequence for* $\mathcal{L}_\lambda$ *then* $\left(\frac{1}{\|\psi_n^+\|_\lambda}\psi_n^+\right)_n$ *is a Palais-Smale sequence for* $\mathcal{M}_\lambda$.

(c) $\phi \in S_\lambda^+$ *is a critical point of* $\mathcal{M}_\lambda$ *if and only if* $\mu_\lambda(\phi)$ *is a nontrivial critical point of* $\mathcal{L}_\lambda$. *The corresponding critical values coincide.*

Proof By Lemmas 6.6–6.8 we may apply [16, Corollary 33]. □

Remark 6.10 The set $\mathcal{P}_\lambda := \{\mu_\lambda(\phi) : \phi \in S_\lambda^+\}$ is the Nehari-Pankov manifold associated to $\mathcal{L}_\lambda$. It is a topological manifold homeomorphic to S_λ^+ via the homeomorphism $\mu_\lambda : S_\lambda^+ \to \mathcal{P}_\lambda$. Neither μ_λ nor $\mathcal{P}_\lambda$ need to be of class $\mathcal{C}^1$ since $\mathcal{K}$ is not $\mathcal{C}^2$. It is therefore surprising that $\mathcal{M}_\lambda = \mathcal{L}_\lambda \circ \mu_\lambda$ is $\mathcal{C}^1$. A general discussion of the construction and the properties of the Nehari-Pankov manifold $\mathcal{P}_\lambda$ in an abstract setting can be found in [16, Chapter 4].

Remark 6.11 Observe that $E_\lambda^+ = E_{\lambda_k}^+$ for $\lambda_k \le \lambda < \lambda_{k+1}$, and $E_\lambda^+ = E_{\lambda_k}^0 \oplus E_{\lambda_k}^+$ for $\lambda_{k-1} \le \lambda < \lambda_k$, hence $E_{\lambda_k}^+ = E_{\lambda_k+\varepsilon}^+$ is a subspace of $E_{\lambda_k-\varepsilon}^+$ of codimension $d_k = \dim E_{\lambda_k}^0$ for $\varepsilon > 0$ small; see Fig. 6.1 below.

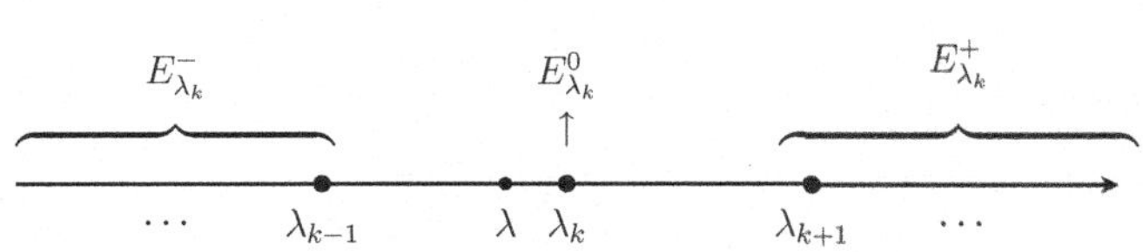

Fig. 6.1 Decomposition of $E = H^{\frac{1}{2}}(M, \mathbb{S}(M))$ for $\lambda \in (\lambda_{k-1}, \lambda_k)$

For $\lambda_{k-1} \le \lambda < \lambda_k$ the least energy solution of $\mathcal{L}_\lambda$ is obtained by minimizing $\mathcal{M}_\lambda$ on S_λ^+, equivalently by minimizing $\mathcal{L}_\lambda$ on $\mathcal{P}_\lambda \cong S_\lambda^+$. In order to obtain another solution $\widetilde{\psi}_\lambda$ for $\lambda < \lambda_k$ close to λ_k, we will construct in Lemma 6.12 and Proposition 6.14 a (topological) submanifold $\mathcal{Q}_{\lambda,k} \subset \mathcal{P}_\lambda$ that is homeomorphic to $S_{\lambda_k}^+$, hence it has codimension $d_k \ge 1$. This is a continuation of $\mathcal{P}_{\lambda_k}$, i.e. $\mathcal{Q}_{\lambda_k,k} = \mathcal{P}_{\lambda_k}$. We can then minimize $\mathcal{L}_\lambda$ on $\mathcal{Q}_{\lambda,k}$ in order to obtain a further solution $\widetilde{\psi}_\lambda$. As a consequence of our construction we expect that the S^1-orbit of $\widetilde{\psi}_\lambda$ converges to the S^1-orbit of ψ_{λ_k} as $\lambda \to \lambda_k$. This is the case if $\mathcal{L}_{\lambda_k}$ has no other minimizers on $\mathcal{P}_{\lambda_k}$ except those on the S^1-orbit of ψ_{λ_k}.

Lemma 6.12 *For $k \in \mathbb{Z}$ and $\sigma > 0$, there exists $\delta_{k,\sigma} > 0$ such that for $\lambda \in (\lambda_k - \delta_{k,\sigma}, \lambda_k]$ and $\phi \in S_{\lambda_k}^+$ with $|\phi|_2^2 \ge \sigma$ there exists a unique global maximum point $\nu_{\lambda,k}(\phi) \in \widehat{E}_{\lambda_k}(\phi)$ of the constrained functional $\mathcal{L}_\lambda|_{\widehat{E}_{\lambda_k}(\phi)}$. Moreover, $P_{\lambda_k}^+(\nu_{\lambda,k}(\phi))$ is bounded away from 0 uniformly in $\lambda \in (\lambda_k - \delta_{k,\sigma}, \lambda_k]$ and $\phi \in S_{\lambda_k}^+$.*

Proof The existence of a global maximizer of $\mathcal{L}_\lambda|_{\widehat{E}_{\lambda_k}(\phi)}$ is analogous to the one in the proof of Lemma 6.8. We only need to show the uniqueness for $\lambda < \lambda_k$ close to λ_k. Let $\psi \in \widehat{E}_{\lambda_k}(\phi)$ be a global maximum of $\mathcal{L}_\lambda|_{\widehat{E}_{\lambda_k}(\phi)}$, and observe that

$$\mathcal{L}_\lambda(\psi) \ge \inf_{S_{\lambda_k}^+} \mathcal{M}_{\lambda_k} > 0, \tag{6.22}$$

hence

$$\int_M g(|\psi|)|\psi|^2 d\,\mathrm{vol}_g = \|\psi^+\|_\lambda^2 - \|\psi^-\|_\lambda^2 = 2\mathcal{L}_\lambda(\psi) + 2\int_M G(|\psi|) d\,\mathrm{vol}_g \ge \alpha \tag{6.23}$$

and

$$\begin{aligned}\mathcal{L}_\lambda(\psi) - \frac{1}{2}\nabla\mathcal{L}_\lambda(\psi)[\psi] &= \frac{1}{2}\nabla\mathcal{K}(\psi)[\psi] - \mathcal{K}(\psi) \\ &= \int_M \frac{1}{2} g(|\psi|)|\psi|^2 - G(|\psi|) d\,\mathrm{vol}_g \ge \alpha.\end{aligned} \tag{6.24}$$

for some $\alpha > 0$. Here we still use $G(t) = \frac{1}{2^*}t^{2^*}$ and $g(t) = t^{2^*-2}$ for notation convenience. These estimates hold uniformly in $\phi \in S_{\lambda_k}^+$, $\psi \in \widehat{E}_{\lambda_k}(\phi)$ a global maximum of $\mathcal{L}_\lambda|_{\widehat{E}_{\lambda_k}(\phi)}$. Moreover, (6.17) implies

$$\begin{aligned}\sup_{\chi \in E_{\lambda_k}^0 \oplus E_{\lambda_k}^-} \mathcal{L}_\lambda(\chi) &\le \max_{\eta \in S_\lambda^+ \cap E_{\lambda_k}^0} \max_{t>0} \left[\frac{\lambda_k - \lambda}{2} t^2 |\eta|_2^2 - \frac{C}{2^*} t^{2^*} |\eta|_{2^*}^{2^*} \right] \\ &= \max_{\eta \in S_\lambda^+ \cap E_{\lambda_k}^0} \frac{(\lambda_k - \lambda)^m |\eta|_2^{2m}}{2mC^{m-1}|\eta|_{2^*}^{2^*(m-1)}} = O\big((\lambda_k - \lambda)^m\big).\end{aligned}$$

Hence we can see from (6.24) that, when for λ is close to λ_k, the projection $P^+_{\lambda_k}(\psi) = t\phi$ (some $t > 0$) plays the leading role to make $\mathcal{L}_\lambda(\psi) \geq \alpha$. Hence we have $P^+_{\lambda_k}(\psi)$ is bounded away from 0 uniformly in λ and $\phi \in S^+_{\lambda_k}$.

For $\phi \in S^+_{\lambda_k}$ with $|\phi|_2^2 \geq \sigma$, we consider an arbitrary element $\varphi = (1+s)\psi + \chi \in \widehat{E}_{\lambda_k}(\phi) \setminus \{\psi\}$, i.e. $s \geq -1$ and $\chi = \chi^0 + \chi^- \in E^0_{\lambda_k} \oplus E^-_{\lambda_k}$ with $s \neq 0$ or $\chi \neq 0$. We fix χ, define $h(s)$ as in (6.18), and deduce as in (6.19):

$$\begin{aligned}\mathcal{L}_\lambda(\varphi) - \mathcal{L}_\lambda(\psi) &= \frac{1}{2}\int_M (D\chi, \chi) d\,\mathrm{vol}_g - \frac{\lambda}{2}\int_M |\chi|^2 d\,\mathrm{vol}_g + \int_M h(s)\, d\,\mathrm{vol}_g \\ &= -\frac{1}{2}\|\chi^-\|^2_{\lambda_k} + \frac{\lambda_k - \lambda}{2}\int_M |\chi^0|^2 d\,\mathrm{vol}_g + \int_M h(s)\, d\,\mathrm{vol}_g .\end{aligned}$$

Here the quadratic part is not negative definite. In order to prove $\mathcal{L}_\lambda(\varphi) < \mathcal{L}_\lambda(\psi)$ it is sufficient to consider φ in a bounded region, i.e. $\|\varphi\|_\lambda \leq R$ for some constant $R > 0$ independent of λ, because $\mathcal{L}_\lambda(\varphi) \to -\infty$ as $\|\varphi\|_\lambda \to \infty$ uniformly for λ in a bounded interval. We also point out that $P_{\lambda_k}(\psi) = t\phi$, hence $|\psi|_2^2 \geq \tau > 0$ is bounded away from 0 as mentioned above.

Observe that (6.20) and (6.21) imply

$$h(s) \leq -\frac{1}{2}\min\left\{g(|\psi|)|\chi|^2,\ g(|\psi|)|\psi|^2 - 2G(|\psi|)\right\}.$$

Now we divide M into two parts:

$$\Omega_1 = \{x \in M : g(|\psi|)|\chi|^2 \geq g(|\psi|)|\psi|^2 - 2G(|\psi|)\},$$

and

$$\Omega_2 = \{x \in M : g(|\psi|)|\chi|^2 < g(|\psi|)|\psi|^2 - 2G(|\psi|)\}.$$

Case 1 $\int_{\Omega_1} |\psi|^2 d\,\mathrm{vol}_g \geq \frac{\tau}{2}$.
In this case we have

$$\int_{\{x\in\Omega_1 : |\psi(x)| \geq s_0\}} |\psi|^2 d\,\mathrm{vol}_g \geq \frac{\tau}{4}$$

for $s_0 > 0$ small, hence there is a constant $c_0 > 0$ such that

$$\int_{\Omega_1} g(|\psi|)|\psi|^2 - 2G(|\psi|)\, d\,\mathrm{vol}_g \geq \int_{\{x\in\Omega_1 :\, |\psi| \geq s_0\}} c_0|\psi|^2 \geq \frac{c_0\tau}{4}.$$

Then $\mathcal{L}_\lambda(\varphi) \leq \mathcal{L}_\lambda(\psi) - \frac{c_0\tau}{8}$ for $\lambda_k - \lambda$ is small. Recall that we only need to consider φ, hence χ^0 in a bounded region.

Case 2 $\int_{\Omega_2} |\psi|^2 d\,\mathrm{vol}_g \geq \frac{\tau}{2}$.
Here we observe that

$$\frac{\tau}{2} \leq \int_{\Omega_2} |\psi|^2 d\,\mathrm{vol}_g \leq \left(\int_{\Omega_2} |\psi|^{2^*-2} d\,\mathrm{vol}_g\right)^{1-\theta} \left(\int_{\Omega_2} |\psi|^{2^*} d\,\mathrm{vol}_g\right)^{\theta}$$

with $\theta = \frac{m-2}{m-1}$. Now the $H^{1/2}$-boundedness of ψ and $g(t) = t^{2^*-2}$ imply that

$$\int_{\Omega_2} g(|\psi|) d\,\mathrm{vol}_g \geq \tau > 0 \tag{6.25}$$

is bounded away from 0. We now claim that the quadratic form

$$\chi \mapsto -\|\chi^-\|^2_{\lambda_k} - \int_{\Omega_2} g(|\psi|)|\chi|^2 d\,\mathrm{vol}_g$$

is negative definite on $E^0_{\lambda_k} \oplus E^-_{\lambda_k}$, uniformly in ϕ. Indeed, since $E^0_{\lambda_k}$ is finite dimensional and the nodal set of any eigenspinor in $E^0_{\lambda_k}$ is of measure zero [5], it follows from (6.25) that there exists a positive constant $c_0 > 0$ such that

$$c_0|\chi^0|^2_2 \leq \int_M g(|\psi|)|\chi^0|^2 d\,\mathrm{vol}_g \leq |\chi^0|^2_\infty \int_M g(|\psi|) d\,\mathrm{vol}_g$$

The constant c_0 can be chosen independently of ψ because global maximizers of $\mathcal{L}_\lambda|_{\widehat{E}_{\lambda_k}(\phi)}$ are bounded in $H^{1/2}$. Now we conclude:

$$\mathcal{L}(\varphi) - \mathcal{L}(\psi) \leq -\frac{1}{2}\|\chi^-\|^2_{\lambda_k} + \frac{\lambda_k - \lambda}{2}\int_M |\chi^0|^2 d\,\mathrm{vol}_g - \frac{1}{2}\int_{\Omega_2} g(|\psi|)|\chi|^2 d\,\mathrm{vol}_g < 0$$

provided $\lambda_k - \lambda$ is small enough.

□

For $\sigma > 0$ and $\lambda_k - \delta_{k,\sigma} < \lambda \leq \lambda_k$, we have a map

$$\nu_{\lambda,k} = \nu_{\lambda,k,\sigma} : S^+_{\lambda_k,\sigma} = \left\{\phi \in S^+_{\lambda_k} : |\phi|^2_2 > \sigma\right\} \to E.$$

Observe that $\nu_{\lambda,k,\sigma}(\phi) = \nu_{\lambda,k,\sigma'}(\phi)$ for $|\phi|^2_2 > \sigma > \sigma'$ which justifies the notation $\nu_{\lambda,k}$.

Remark 6.13

(a) The map $\nu_{\lambda,k} : S^+_{\lambda_k,\sigma} \to E$ is an embedding, and its image $\mathcal{Q}_{\lambda,k,\sigma} := \nu_{\lambda,k}(S^+_{\lambda_k,\sigma})$ is a topological manifold homeomorphic to $S^+_{\lambda_k,\sigma}$. We claim that $\mathcal{Q}_{\lambda,k,\sigma} \subset \mathcal{P}_\lambda$. Consider $\phi \in S^+_{\lambda_k,\sigma}$, and let $\nu_{\lambda,k}(\phi) = t\phi + \chi^0 + \chi^- \in \mathbb{R}\phi \oplus E^0_{\lambda_k} \oplus E^-_{\lambda_k}$. This is a maximizer of $\mathcal{L}_\lambda$ on $\widehat{E}_{\lambda_k}(\phi)$, hence on $\mathbb{R}^+(t\phi +$

$\chi^0) \oplus E^-_{\lambda_k} = \widehat{E}_\lambda\left(\frac{1}{\|t\phi+\chi^0\|_\lambda}(t\phi+\chi^0)\right) \subset \widehat{E}_{\lambda_k}(\phi)$. It follows that $\nu_{\lambda,k}(\phi) = \mu_\lambda\left(\frac{1}{\|t\phi+\chi^0\|_\lambda}(t\phi+\chi^0)\right)$.

(b) The non-ground-state solution $\widetilde{\psi}_\lambda$ will be a minimizer of $\mathcal{L}_\lambda$ on $\mathcal{Q}_{\lambda,k,\sigma}$. It is an interesting problem to determine the range of $\lambda < \lambda_k$ so that one can define $\mathcal{Q}_{\lambda,k,\sigma}$, for instance, whether it can be defined up to or below λ_{k-1}. This will depend on the geometry of M.

For $\sigma > 0$ and $\lambda \in (\lambda_k - \delta_{k,\sigma}, \lambda_k]$ we consider the functional

$$\mathcal{N}_{\lambda,k} : S^+_{\lambda_k,\sigma} \to \mathbb{R}, \quad \mathcal{N}_{\lambda,k}(\phi) = \mathcal{L}_\lambda(\nu_{\lambda,k}(\phi)).$$

We point out that $\nu_{\lambda_k,k} = \mu_{\lambda_k}$ and $\mathcal{N}_{\lambda_k,k} = \mathcal{M}_{\lambda_k}$. The arguments from [16, Corollary 33] imply the following proposition.

Proposition 6.14 *Fix $\sigma > 0$ and $\lambda \in (\lambda_k - \delta_{k,\sigma}, \lambda_k]$.*

(a) $\mathcal{N}_{\lambda,k} \in C^1(S^+_{\lambda_k,\sigma})$ *and*

$$\nabla\mathcal{N}_{\lambda,k}(\phi)[\chi] = \left\|P^+_{\lambda_k}\nu_{\lambda,k}(\phi)\right\|_{\lambda_k} \nabla\mathcal{L}_\lambda(\nu_{\lambda,k}(\phi))[\chi] \quad \text{for all } \chi \in T_\phi(S^+_{\lambda_k,\sigma}).$$

(b) *If $(\phi_n)_n$ is a Palais-Smale sequence for $\mathcal{N}_{\lambda,k}$ in $S^+_{\lambda_k,\sigma}$ then $(\nu_{\lambda,k}(\phi_n))_n$ is a Palais-Smale sequence for $\mathcal{L}_\lambda$.*

(c) *$\phi \in S^+_{\lambda_k,\sigma}$ is a critical point of $\mathcal{N}_{\lambda,k}$ if and only if $\nu_{\lambda,k}(\phi)$ is a nontrivial critical point of $\mathcal{L}_\lambda$. The corresponding critical values coincide.*

Now we show that $\inf\mathcal{N}_{\lambda,k} > \inf\mathcal{M}_{\lambda_k}$ is achieved for $\lambda \le \lambda_k$ close to λ_k, and is non-increasing and continuous in λ, provided $\inf\mathcal{M}_{\lambda_k} < \gamma_{crit}$. In Sect. 6.5 we shall prove that $\inf\mathcal{M}_{\lambda_k} < \gamma_{crit}$ holds for all $k \ge 1$.

Lemma 6.15

(a) *For all $\phi \in S^+_{\lambda_k}$ there holds $\nu_{\lambda,k}(\phi) \to \mu_{\lambda_k}(\phi)$ as $\lambda \nearrow \lambda_k$. The convergence is uniform on compact subsets.*

(b) *If $\inf\mathcal{M}_{\lambda_k} < \gamma_{crit}$, then there exists $\sigma > 0$ such that*

$$\inf_{S_{\lambda_k,\sigma}} \mathcal{N}_{\lambda,k} < \gamma_{crit} \quad \text{for all } \lambda \in (\lambda_k - \delta_{k,\sigma}, \lambda_k],$$

and the infimum is achieved.

(c) *The function $(\lambda_k - \delta_{k,\sigma}, \lambda_k] \to \mathbb{R}$, $\lambda \mapsto \inf\mathcal{N}_{\lambda,k}$, is non-increasing and continuous.*

Proof

(a) We fix $\phi \in S^+_{\lambda_k}$ and consider $\psi_\lambda = \nu_{\lambda,k}(\phi)$. Observe that $\psi_\lambda = t_\lambda\phi + \chi_\lambda$ with $t_\lambda > 0$ and $\chi_\lambda \in E^0_{\lambda_k} \oplus E^-_{\lambda_k}$. By the boundedness of $\{\psi_\lambda\}$, we have $\psi_\lambda \rightharpoonup \psi_0$

with $t_\lambda \to t_0$ and $\chi_\lambda \rightharpoonup \chi_0$ as $\lambda \nearrow \lambda_k$. Then the weak lower semi-continuity of $|\cdot|_2$ and $\mathcal{K}$ on $\widehat{E}_k(\phi)$ implies

$$\limsup_{\lambda\nearrow\lambda_k} \mathcal{N}_{\lambda,k}(\phi) \le \frac{1}{2}\int_M (D\psi_0, \psi_0) d\,\mathrm{vol}_g - \frac{\lambda_k}{2}\int_M |\psi_0|^2 d\,\mathrm{vol}_g - \mathcal{K}(\psi_0) \le \mathcal{M}_{\lambda_k}(\phi).$$

On the other hand, the monotonicity of $\mathcal{L}_\lambda$ with respect to λ yields

$$0 \le \mathcal{N}_{\lambda,k}(\phi) - \mathcal{M}_{\lambda_k}(\phi) \le \frac{\lambda_k - \lambda}{2}|\psi_\lambda|_2^2 \to 0 \quad \text{as } \lambda \nearrow \lambda_k.$$

It follows that $\psi_0 = \mu_{\lambda_k}(\phi)$ and

$$\lim_{\lambda\nearrow\lambda_k} \|P^-_{\lambda_k}\chi_\lambda\|^2_\lambda = \|P^-_{\lambda_k}\chi_0\|^2_{\lambda_k}.$$

Since $E^0_{\lambda_k}$ is of finite dimension, we have $\psi_\lambda \to \mu_{\lambda_k}(\phi)$ as $\lambda \nearrow \lambda_k$.

(b) Since $c_{\lambda_k} := \inf \mathcal{M}_{\lambda_k} < \gamma_{crit}$, the infimum is achieved and the set of minimizers

$$A := A_{\lambda_k} := \{\phi \in S^+_{\lambda_k} : \mathcal{M}_{\lambda_k}(\phi) = c_{\lambda_k}\}$$

is compact because the (PS)-condition holds at this level. Let $U_\rho(A) \subset S^+_{\lambda_k}$ be the ρ-neighborhood of A with respect to $\|\cdot\|_{\lambda_k}$. We choose $\rho > 0$ so that the L^2-norm is bounded away from 0 on $U := U_\rho(A)$, that is $U \subset S_{\lambda_k,\sigma}$ for some $\sigma > 0$. Using again the Palais-Smale condition below γ_{crit} there exists $\alpha > 0$ with $c_{\lambda_k} + 3\alpha < \gamma_{crit}$ and such that

$$\inf \mathcal{M}_{\lambda_k}(S^+_{\lambda_k} \setminus U) \ge c_{\lambda_k} + 2\alpha.$$

The sets $B := \mu_{\lambda_k}(A)$ and $V := \mu_{\lambda_k}(U)$ are contained in the Nehari-Pankov manifold $\mathcal{P}_{\lambda_k}$ from Remark 6.13. By definition there holds

$$\inf \mathcal{L}_{\lambda_k}\big(\mathcal{P}_{\lambda_k} \setminus V\big) \ge c_{\lambda_k} + 2\alpha \quad \text{and} \quad \inf \mathcal{L}_{\lambda_k}(A) = c_{\lambda_k}.$$

Using the monotonicity of $\mathcal{L}_\lambda$ with respect to λ we obtain for $\lambda < \lambda_k$ close to λ_k

$$\inf \mathcal{L}_\lambda\big(\mathcal{Q}_{\lambda,k,\sigma} \setminus V_\lambda\big) \ge c_{\lambda_k} + 2\alpha \quad \text{and} \quad \inf \mathcal{L}_{\lambda_k}(B_\lambda) \le c_{\lambda_k} + \alpha$$

where $V_\lambda = \nu_{\lambda,k}(U)$ and $B_\lambda = \nu_{\lambda,k}(A)$. This implies

$$\inf \mathcal{N}_{\lambda,k}\big(S^+_{\lambda_k,\sigma} \setminus U\big) \geq c_{\lambda_k} + 2\alpha \quad \text{and} \quad \inf \mathcal{N}_{\lambda,k}(A) \leq c_{\lambda_k} + \alpha.$$

Then it follows that $\mathcal{N}_{\lambda,k}$ achieves its infimum on $S_{\lambda_k,\sigma}$ in the set U.

(c) The function $\inf \mathcal{N}_{\lambda,k}$ is non-increasing in λ because $\mathcal{L}_\lambda(\psi)$ is decreasing with respect to λ. For a given $\lambda \in (\lambda_k - \delta_{k,\sigma}, \lambda_k]$ there exists a spinor field $\phi \in S^+_{\lambda_k}$ such that $\mathcal{N}_{\lambda,k}(\phi) = \inf \mathcal{N}_{\lambda,k}$. Then we have for $\lambda' < \lambda$:

$$\inf \mathcal{N}_{\lambda,k} \leq \inf \mathcal{N}_{\lambda',k} \leq \mathcal{N}_{\lambda',k}(\phi) \to \mathcal{N}_{\lambda,k}(\phi) = \inf \mathcal{N}_{\lambda,k} \quad \text{as } \lambda' \nearrow \lambda,$$

hence $\lambda \mapsto \inf \mathcal{N}_{\lambda,k}$ is continuous from the left. In order to prove continuity from the right at $\lambda \in (\lambda_k - \delta_{k,\sigma}, \lambda_k)$, consider a sequence $\lambda'_n \searrow \lambda$ and let $\phi_n \in S^+_{\lambda_k}$ be a minimizer of $\mathcal{N}_{\lambda'_n,k}$. Then $\psi_n := \nu_{\lambda'_n,k}(\phi_n)$ is a critical point of $\mathcal{L}_{\lambda'_n}$ and $\mathcal{L}_{\lambda'_n}(\psi_n) = \mathcal{N}_{\lambda'_n,k}(\phi_n) = \inf \mathcal{N}_{\lambda'_n,k} \leq \inf \mathcal{N}_{\lambda,k}$, hence $(\psi_n)_n$ is a Palais-Smale sequence for $\mathcal{L}_\lambda$ at a level $c = \lim_{n\to\infty} \inf \mathcal{N}_{\lambda'_n,k} \leq \inf \mathcal{N}_{\lambda,k} < \gamma_{crit}$. It follows that $\psi_n \to \psi$ along a subsequence, and ψ is a critical point of $\mathcal{L}_\lambda$ at the level c. Equivalently, $\phi_n \to \phi \in S^+_{\lambda_k}$ with $\psi = \nu_{\lambda,k}(\phi)$, and ϕ is a critical point of $\mathcal{N}_{\lambda,k}$. This implies $c \geq \inf \mathcal{N}_{\lambda,k}$, and the continuity from the right follows.

□

6.4 Test Spinors, Auxiliary Estimates and Annihilation of the Kernels

Our proof relies on the construction of a test spinor on M in order to show $\inf \mathcal{M}_\lambda < \gamma_{crit}$ under the conditions of Theorem 3.34 mentioned in Chap. 3. The test spinor comes from a spinor on $\mathbb{R}^m$ being cut-off and transplanted to M so that it has support in a small neighborhood of an arbitrary point $p_0 \in M$. We first need to recall a construction from the paper [3] of Ammann et al.

To begin with we fix $\psi_0 \in \mathbb{S}_m$ with $|\psi_0| = m^{\frac{m-1}{2}}$ arbitrarily, set $\mu(x) = \frac{1}{1+|x|^2}$ for $x \in \mathbb{R}^m$ and define

$$\psi(x) = \mu(x)^{\frac{m}{2}}(1 - x) \cdot \psi_0$$

see [3, Section 5.1]. Then it is standard to verify that

$$D_{\mathbb{R}^m}\psi = m\mu\psi \tag{6.26}$$

and

$$|\psi| = m^{\frac{m-1}{2}}\mu^{\frac{m-1}{2}} = \left(\frac{m}{1+|x|^2}\right)^{\frac{m-1}{2}}. \tag{6.27}$$

We choose $\delta < i(M)/2$ where $i(M) > 0$ is the injectivity radius of M. Let $\eta : \mathbb{R}^m \to \mathbb{R}$ be a smooth cut-off function satisfying $\eta(x) = 1$ if $|x| \le \delta$ and $\eta(x) = 0$ if $|x| \ge 2\delta$. Now we define $\psi_\varepsilon : \mathbb{R}^m \to \mathbb{S}_m$ by

$$\varphi_\varepsilon(x) = \eta(x)\psi_\varepsilon(x) \quad \text{where} \quad \psi_\varepsilon(x) = \varepsilon^{-\frac{m-1}{2}}\psi(x/\varepsilon). \tag{6.28}$$

In order to transplant the test spinor on M, we recall the Bourguignon-Gauduchon-trivialization [10]. Here we fix $p_0 \in M$ arbitrarily, and let $(x_1, \dots, x_m)$ be the normal coordinates given by the exponential map

$$\exp_{p_0} : \mathbb{R}^m \cong T_{p_0}M \supset U \to V \subset M, \quad x \mapsto p = \exp_{p_0}(x).$$

For $p \in V$ let $G(p) = (g_{ij}(p))_{ij}$ denote the corresponding metric at p. Since $G(p)$ is symmetric and positive definite, the square root $B(p) = (b_{ij}(p))_{ij}$ of $G(p)^{-1}$ is well defined, symmetric and positive definite. It can be thought of as linear isometry

$$B(p) : (\mathbb{R}^m \cong T_{\exp_{p_0}^{-1}(p)}U, g_{\mathbb{R}^m}) \to (T_pV, g).$$

We obtain an isomorphism of $SO(m)$-principal bundles:

$$\begin{array}{ccc} P_{SO}(U, g_{\mathbb{R}^m}) & \xrightarrow{\ \phi\ } & P_{SO}(V, g) \\ \downarrow & & \downarrow \\ T_{p_0}M \supset U & \xrightarrow{\ \exp_{p_0}\ } & V \subset M \end{array}$$

where $\phi(y_1, \dots, y_m) = (By_1, \dots, By_m)$ for an oriented frame $(y_1, \dots, y_m)$ on U. Notice that ϕ commutes with the right action of $SO(m)$, hence it induces an isomorphism of spin structures:

$$\begin{array}{ccc} Spin(m) \times U = P_{Spin}(U, g_{\mathbb{R}^m}) & \longrightarrow & P_{Spin}(V, g) \subset P_{Spin}(M) \\ \downarrow & & \downarrow \\ T_{p_0}M \supset U & \xrightarrow{\ \exp_{p_0}\ } & V \subset M \end{array}$$

Thus we obtain an isomorphism between the spinor bundles $\mathbb{S}(U)$ and $\mathbb{S}(V)$:

$$\mathbb{S}(U) := P_{Spin}(U, g_{\mathbb{R}^m}) \times_\rho \mathbb{S}_m \longrightarrow \mathbb{S}(V) := P_{Spin}(V, g) \times_\rho \mathbb{S}_m \subset \mathbb{S}(M) \tag{6.29}$$

where $(\rho, \mathbb{S}_m)$ is the complex spin representation.

Setting $e_i = B(\partial_i) = \sum_j b_{ij}\partial_j$ we obtain an orthonormal frame $(e_1, \dots, e_m)$ of (TV, g). In order to simplify the notation, we use ∇ and $\bar{\nabla}$, respectively, for the

Levi-Civita connections on $(TU, g_{\mathbb{R}^m})$ and (TV, g) and for the natural lifts of these connections to the spinor bundles $\mathbb{S}(U)$ and $\mathbb{S}(V)$, respectively. For the Clifford multiplications on these bundles, we shall write "·" in both cases, that is,

$$e_i \cdot \bar{\psi} = B(\partial_i) \cdot \bar{\psi} = \overline{\partial_i \cdot \psi}.$$

Now a spinor $\varphi \in \Gamma(\mathbb{S}(U))$ corresponds via the isomorphisms (6.29) to a spinor $\bar{\phi} \in \Gamma(\mathbb{S}(V))$. In particular, since the spinors $\varphi_\varepsilon \in \Gamma(\mathbb{S}(U))$ from (6.28) have compact support in U they correspond to spinors $\bar{\varphi}_\varepsilon \in \Gamma(\mathbb{S}(M))$ with compact support in V. These are not quite our test spinors, because they do not lie in $\mathcal{P}_\lambda$. In the rest of this section we prove some estimates that will be needed in order to control the critical values.

We write D and $\bar{D}$ for the Dirac operators acting on $\Gamma(\mathbb{S}(U))$ and $\Gamma(\mathbb{S}(V))$, respectively. By [3, Proposition 3.2] there holds

$$\bar{D}\bar{\varphi}_\varepsilon = \overline{D\varphi_\varepsilon} + W \cdot \bar{\varphi}_\varepsilon + X \cdot \bar{\varphi}_\varepsilon + \sum_{i,j} (b_{ij} - \delta_{ij}) \overline{\partial_i \cdot \nabla_{\partial_j} \varphi_\varepsilon} \tag{6.30}$$

with $W \in \Gamma(Cl(TV))$ and $X \in \Gamma(TV)$ given

$$W = \frac{1}{4} \sum_{\substack{i,j,k \\ i \neq j \neq k \neq i}} \sum_{\alpha,\beta} b_{i\alpha}(\partial_\alpha b_{j\beta}) b^{-1}_{\beta k} e_i \cdot e_j \cdot e_k,$$

and

$$X = \frac{1}{4} \sum_{i,k} (\bar{\Gamma}^i_{ik} - \bar{\Gamma}^k_{ii}) e_k = \frac{1}{2} \sum_{i,k} \bar{\Gamma}^i_{ik} e_k;$$

here $(b^{-1}_{ij})_{ij}$ denotes the inverse matrix of B, and $\bar{\Gamma}^k_{ij} := g(\bar{\nabla}_{e_j} e_j, e_k)$. In the sequel we identify $x \in \mathbb{R}^m$ with $p = \exp_{p_0} x \in M$ for notational convenience. As remarked in [3, 14], observing that $B = (G^{-1})^{\frac{1}{2}}$ and $G = I + O(|x|^2)$ as $|x| \to 0$, we deduce

$$b_{ij} = \delta_{ij} + O(|x|^2), \quad W = O(|x|^3) \quad \text{and} \quad X = O(|x|) \quad \text{as } |x| \to 0. \tag{6.31}$$

In the sequel we use the notation $f_\varepsilon \lesssim g_\varepsilon$ for two functions f_ε and g_ε, when there exists a constant $C > 0$ independent of ε such that $f_\varepsilon \le C g_\varepsilon$.

Lemma 6.16 *Let $\bar{\varphi}_\varepsilon \in \mathbb{S}(V)$ be as above and set $\bar{R}_\varepsilon := \bar{D}\bar{\varphi}_\varepsilon - |\bar{\varphi}_\varepsilon|^{2^*-2}\bar{\varphi}_\varepsilon$. Then*

$$\|\bar{\varphi}_\varepsilon\|_{E^*_\lambda} \lesssim \begin{cases} \varepsilon^{\frac{1}{2}} & \text{if } m = 2, \\ \varepsilon |\ln \varepsilon|^{\frac{2}{3}} & \text{if } m = 3, \\ \varepsilon & \text{if } m \ge 4, \end{cases}$$

and

$$\|\bar{R}_\varepsilon\|_{E_\lambda^*} \lesssim \begin{cases} \varepsilon^{\frac{m-1}{2}} & \text{if } 2 \le m \le 4, \\ \varepsilon^2 |\ln \varepsilon|^{\frac{3}{5}} & \text{if } m = 5, \\ \varepsilon^2 & \text{if } m \ge 6. \end{cases}$$

Proof For $\psi \in E$ with $\|\psi\| \le 1$ and ε small we have

$$\begin{aligned} \left| \int_M (\bar{\varphi}_\varepsilon, \psi) d\,\mathrm{vol}_g \right| &= \left| \int_{B_{2\delta}(p_0)} (\bar{\varphi}_\varepsilon, \psi) d\,\mathrm{vol}_g \right| \\ &\le \left(\int_{B_{2\delta}(p_0)} |\bar{\varphi}_\varepsilon|^{\frac{2m}{m+1}} d\,\mathrm{vol}_g \right)^{\frac{m+1}{2m}} |\psi|_{2^*} \\ &\lesssim \left(\int_{|x| \le 2\delta} |\psi_\varepsilon|^{\frac{2m}{m+1}} dx \right)^{\frac{m+1}{2m}} \\ &\lesssim \left(\varepsilon^{\frac{2m}{m+1}} \int_0^{\frac{2\delta}{\varepsilon}} \frac{r^{m-1}}{(1+r^2)^{\frac{m(m-1)}{m+1}}} dr \right)^{\frac{m+1}{2m}} \\ &\lesssim \begin{cases} \varepsilon^{\frac{1}{2}} & \text{if } m = 2, \\ \varepsilon |\ln \varepsilon|^{\frac{2}{3}} & \text{if } m = 3, \\ \varepsilon & \text{if } m \ge 4, \end{cases} \end{aligned} \tag{6.32}$$

This implies the estimate on $\|\bar{\varphi}_\varepsilon\|_{E_\lambda^*}$.

In order to estimate $\bar{R}_\varepsilon$, (6.26) and (6.27) yield

$$\begin{aligned} D\varphi_\varepsilon &= \nabla\eta \cdot \psi_\varepsilon + \eta D\psi_\varepsilon \\ &= \nabla\eta \cdot \psi_\varepsilon + |\varphi_\varepsilon|^{2^*-2}\varphi_\varepsilon + (\eta - \eta^{2^*-1})|\psi_\varepsilon|^{2^*-2}\psi_\varepsilon. \end{aligned} \tag{6.33}$$

Using (6.30), we obtain

$$\bar{R}_\varepsilon = A_1 + A_2 + A_3 + A_4 + A_5 + A_6$$

where

$$\begin{aligned} A_1 &= \overline{\nabla\eta \cdot \psi_\varepsilon}, \\ A_2 &= (\eta - \eta^{2^*-1})|\bar{\psi}_\varepsilon|^{2^*-2}\bar{\psi}_\varepsilon, \\ A_3 &= \eta W \cdot \bar{\psi}_\varepsilon, \\ A_4 &= \eta X \cdot \bar{\psi}_\varepsilon, \end{aligned}$$

$$A_5 = \eta \sum_{i,j} (b_{ij} - \delta_{ij}) \overline{\partial_i \cdot \nabla_{\partial_j} \psi_\varepsilon},$$

$$A_6 = \sum_{i,j} (b_{ij} - \delta_{ij}) \partial_j \eta \, \overline{\partial_i \cdot \psi_\varepsilon}.$$

In the following estimates we use that the support of η is contained in $B_{2\delta}(0) \subset \mathbb{R}^m$. Analogous to (6.32), using (6.27) and (6.31) we estimate:

$$\begin{aligned}
\|A_1\|_{E_\lambda^*} &\lesssim \left(\int_{B_{2\delta}(p_0)} \left| \overline{\nabla \eta \cdot \psi_\varepsilon} \right|^{\frac{2m}{m+1}} d\,\mathrm{vol}_g \right)^{\frac{m+1}{2m}} \\
&\lesssim \left(\int_{\delta \le |x| \le 2\delta} |\psi_\varepsilon|^{\frac{2m}{m+1}} dx \right)^{\frac{m+1}{2m}} \\
&\lesssim \left(\varepsilon^{\frac{2m}{m+1}} \int_{\frac{\delta}{\varepsilon}}^{\frac{2\delta}{\varepsilon}} \frac{r^{m-1}}{(1+r^2)^{\frac{m(m-1)}{m+1}}} dr \right)^{\frac{m+1}{2m}} \lesssim \varepsilon^{\frac{m-1}{2}}
\end{aligned}$$

$$\begin{aligned}
\|A_2\|_{E_\lambda^*} &\lesssim \left(\int_{B_{2\delta}(p_0)} \left(\eta - \eta^{\frac{m+1}{m-1}} \right)^{\frac{2m}{m+1}} |\bar{\psi}_\varepsilon|^{\frac{2m}{m-1}} d\,\mathrm{vol}_g \right)^{\frac{m+1}{2m}} \\
&\lesssim \left(\int_{\delta \le |x| \le 2\delta} |\psi_\varepsilon|^{\frac{2m}{m-1}} dx \right)^{\frac{m+1}{2m}} \lesssim \left(\int_{\frac{\delta}{\varepsilon}}^{\frac{2\delta}{\varepsilon}} \frac{r^{m-1}}{(1+r^2)^m} dr \right)^{\frac{m+1}{2m}} \\
&\lesssim \varepsilon^{\frac{m+1}{2}}
\end{aligned}$$

$$\begin{aligned}
\|A_3\|_{E_\lambda^*} &\lesssim \left(\int_{B_{2\delta}(p_0)} |W|^{\frac{2m}{m+1}} |\bar{\psi}_\varepsilon|^{\frac{2m}{m+1}} d\,\mathrm{vol}_g \right)^{\frac{m+1}{2m}} \lesssim \left(\int_{|x| \le 2\delta} |x|^{\frac{6m}{m+1}} |\psi_\varepsilon|^{\frac{2m}{m+1}} dx \right)^{\frac{m+1}{2m}} \\
&\lesssim \left(\varepsilon^{\frac{8m}{m+1}} \int_0^{\frac{2\delta}{\varepsilon}} \frac{r^{\frac{6m}{m+1}+m-1}}{(1+r^2)^{\frac{m(m-1)}{m+1}}} dr \right)^{\frac{m+1}{2m}} \lesssim \begin{cases} \varepsilon^{\frac{m-1}{2}} & \text{if } 2 \le m \le 8 \\ \varepsilon^4 |\ln \varepsilon|^{\frac{5}{9}} & \text{if } m = 9 \\ \varepsilon^4 & \text{if } m \ge 10 \end{cases}
\end{aligned}$$

$$\begin{aligned}
\|A_4\|_{E_\lambda^*} &\lesssim \left(\int_{B_{2\delta}(p_0)} |X|^{\frac{2m}{m+1}} |\bar{\psi}_\varepsilon|^{\frac{2m}{m+1}} d\,\mathrm{vol}_g \right)^{\frac{m+1}{2m}} \lesssim \left(\int_{|x| \le 2\delta} |x|^{\frac{2m}{m+1}} |\psi_\varepsilon|^{\frac{2m}{m+1}} dx \right)^{\frac{m+1}{2m}} \\
&\lesssim \left(\varepsilon^{\frac{4m}{m+1}} \int_0^{\frac{2\delta}{\varepsilon}} \frac{r^{\frac{2m}{m+1}+m-1}}{(1+r^2)^{\frac{m(m-1)}{m+1}}} dr \right)^{\frac{m+1}{2m}} \lesssim \begin{cases} \varepsilon^{\frac{m-1}{2}} & \text{if } 2 \le m \le 4 \\ \varepsilon^2 |\ln \varepsilon|^{\frac{3}{5}} & \text{if } m = 5 \\ \varepsilon^2 & \text{if } m \ge 6 \end{cases}
\end{aligned}$$

$$\|A_5\|_{E_\lambda^*} \lesssim \left(\int_{|x|\le 2\delta} |x|^{\frac{4m}{m+1}} |\nabla\psi_\varepsilon|^{\frac{2m}{m+1}} dx\right)^{\frac{m+1}{2m}} \lesssim \left(\varepsilon^{\frac{4m}{m+1}} \int_0^{\frac{2\delta}{\varepsilon}} \frac{r^{\frac{4m}{m+1}+m-1}}{(1+r^2)^{\frac{m^2}{m+1}}} dr\right)^{\frac{m+1}{2m}}$$

$$\le \left(\varepsilon^{\frac{4m}{m+1}} \int_0^{\frac{2\delta}{\varepsilon}} \frac{r^{\frac{2m}{m+1}+m-1}}{(1+r^2)^{\frac{m(m-1)}{m+1}}} dr\right)^{\frac{m+1}{2m}} \lesssim \begin{cases} \varepsilon^{\frac{m-1}{2}} & \text{if } 2 \le m \le 4 \\ \varepsilon^2 |\ln\varepsilon|^{\frac{3}{5}} & \text{if } m = 5 \\ \varepsilon^2 & \text{if } m \ge 6 \end{cases}$$

Here we used the inequality $|\nabla\psi(x)| \lesssim \mu(x)^{\frac{m}{2}}$ and the same estimate as for $\|A_4\|_{E_\lambda^*}$. Finally there holds:

$$\|A_6\|_{E_\lambda^*} \lesssim \left(\int_{|x|\le 2\delta} |x|^{\frac{6m}{m+1}} |\psi_\varepsilon|^{\frac{2m}{m+1}} dx\right)^{\frac{m+1}{2m}} \lesssim \begin{cases} \varepsilon^{\frac{m-1}{2}} & \text{if } 2 \le m \le 8 \\ \varepsilon^4 |\ln\varepsilon|^{\frac{5}{9}} & \text{if } m = 9 \\ \varepsilon^4 & \text{if } m \ge 10 \end{cases}$$

Here we used $|\nabla\eta(x)| \lesssim |x|$ and the same estimate as for $\|A_3\|_{E_\lambda^*}$.

From these estimates we finally obtain:

$$\|\bar{R}_\varepsilon\|_{E_\lambda^*} \lesssim \begin{cases} \varepsilon^{\frac{m-1}{2}} & \text{if } 2 \le m \le 4 \\ \varepsilon^2 |\ln\varepsilon|^{\frac{3}{5}} & \text{if } m = 5 \\ \varepsilon^2 & \text{if } m \ge 6 \end{cases}$$

□

Lemma 6.17 *Let $\bar{\varphi}_\varepsilon \in \mathbb{S}(V)$ be as above and let ω_m stand for the volume of the standard sphere S^m. Then*

$$\int_M |\bar{\varphi}_\varepsilon|^2 d\,\mathrm{vol}_g \gtrsim \begin{cases} \varepsilon|\ln\varepsilon| + O(\varepsilon) & \text{if } m = 2 \\ \varepsilon + O(\varepsilon^2) & \text{if } m \ge 3 \end{cases} \tag{6.34}$$

and

$$\frac{1}{2}\int_M (\bar{D}\bar{\varphi}_\varepsilon, \bar{\varphi}_\varepsilon) d\,\mathrm{vol}_g - \frac{1}{2^*}\int_M |\bar{\varphi}_\varepsilon|^{2^*} d\,\mathrm{vol}_g$$

$$\le \frac{1}{2m}\left(\frac{m}{2}\right)^m \omega_m + \begin{cases} O(\varepsilon) & \text{if } m = 2, \\ O(\varepsilon^2|\ln\varepsilon|) & \text{if } m = 3, \\ O(\varepsilon^2) & \text{if } m \ge 4, \end{cases} \tag{6.35}$$

Proof For the first estimate, by taking into account $d\,\mathrm{vol}_g = d\,\mathrm{vol}_{\mathbb{R}^m} + O(|x|^2)$ in normal coordinates at p_0 we have

$$
\begin{aligned}
\int_M |\bar{\varphi}_\varepsilon|^2 d\,\mathrm{vol}_g &= \int_{B_{2\delta}(p_0)} |\bar{\varphi}_\varepsilon|^2 d\,\mathrm{vol}_g \\
&= \int_{|x|\le\delta} |\psi_\varepsilon|^2 dx + O\Big(\int_{\delta<|x|\le 2\delta} |\psi_\varepsilon|^2 dx\Big) \\
&\quad + O\Big(\int_{|x|\le 2\delta} |x|^2 |\psi_\varepsilon|^2 dx\Big) \\
&= \varepsilon m^{m-1}\omega_{m-1}\int_0^{\frac{\delta}{\varepsilon}} \frac{r^{m-1}}{(1+r^2)^{m-1}}dr + O\bigg(\varepsilon\int_{\frac{\delta}{\varepsilon}}^{\frac{2\delta}{\varepsilon}} \frac{r^{m-1}}{(1+r^2)^{m-1}}dr\bigg) \\
&\quad + O\bigg(\varepsilon^3\int_0^{\frac{2\delta}{\varepsilon}} \frac{r^{m+1}}{(1+r^2)^{m-1}}dr\bigg) \\
&= A(\varepsilon) + O(\varepsilon^{m-1}) + \begin{cases} O(\varepsilon^{m-1}) & \text{if } m = 2, 3, \\ O(\varepsilon^3|\ln\varepsilon|) & \text{if } m = 4, \\ O(\varepsilon^3) & \text{if } m \ge 5 \end{cases} \\
&= A(\varepsilon) + \begin{cases} O(\varepsilon^{m-1}) & \text{if } m = 2, 3, \\ O(\varepsilon^3|\ln\varepsilon|) & \text{if } m = 4, \\ O(\varepsilon^3) & \text{if } m \ge 5, \end{cases}
\end{aligned}
$$

where

$$
A(\varepsilon) = \begin{cases} \varepsilon\big(\ln(\varepsilon^2+\delta^2) - 2\ln\varepsilon\big)m^{m-1}\omega_{m-1} & \text{if } m = 2 \\ \varepsilon m^{m-1}\omega_{m-1}\int_0^\infty \frac{r^{m-1}}{(1+r^2)^{m-1}}dr & \text{if } m \ge 3 \end{cases}
$$

This implies (6.34).

Now we come to (6.35). Analogously to the arguments in Lemma 6.16, we shall use (6.30) and (6.33) in order to get

$$
\int_M (\bar{D}\bar{\varphi}_\varepsilon, \bar{\varphi}_\varepsilon) d\,\mathrm{vol}_g = J_1 + J_2 + \cdots + J_7
$$

where

$$
J_1 = \mathrm{Re}\int_M \eta\cdot(\overline{\nabla\eta\cdot\psi_\varepsilon}, \bar{\psi}_\varepsilon) d\,\mathrm{vol}_g
$$

$$
\begin{aligned}
J_2 &= \int_M |\bar{\varphi}_\varepsilon|^{2^*} d\,\mathrm{vol}_g \\
J_3 &= \int_M (\eta - \eta^{2^*-1}) \cdot \eta \cdot |\bar{\psi}_\varepsilon|^{2^*} d\,\mathrm{vol}_g \\
J_4 &= \mathrm{Re} \int_M \eta^2 \cdot (W \cdot \bar{\psi}_\varepsilon, \bar{\psi}_\varepsilon) d\,\mathrm{vol}_g \\
J_5 &= \mathrm{Re} \int_M \eta^2 \cdot (X \cdot \bar{\psi}_\varepsilon, \bar{\psi}_\varepsilon) d\,\mathrm{vol}_g \\
J_6 &= \mathrm{Re} \sum_{i,j} \int_M \eta^2 (b_{ij} - \delta_{ij}) (\overline{\partial_i \cdot \nabla_{\partial_j} \psi_\varepsilon}, \bar{\psi}_\varepsilon) d\,\mathrm{vol}_g \\
J_7 &= \mathrm{Re} \sum_{i,j} \int_M \eta \cdot (b_{ij} - \delta_{ij}) \partial_j \eta \cdot (\overline{\partial_i \cdot \psi_\varepsilon}, \bar{\psi}_\varepsilon) d\,\mathrm{vol}_g\,.
\end{aligned}
$$

We have:

$$J_1 = 0$$

$$
\begin{aligned}
J_2 &= \int_{|x| \le \delta} |\psi_\varepsilon|^{2^*} dx + O\left(\int_{\delta < |x| \le 2\delta} |\psi_\varepsilon|^{2^*} dx \right) + O\left(\int_{|x| \le 2\delta} |x|^2 |\psi_\varepsilon|^{2^*} dx \right) \\
&= m^m \omega_{m-1} \int_0^{\frac{\delta}{\varepsilon}} \frac{r^{m-1}}{(1+r^2)^m} dr + O\left(\int_{\frac{\delta}{\varepsilon}}^{\frac{2\delta}{\varepsilon}} \frac{r^{m-1}}{(1+r^2)^m} dr \right) \\
&\quad + O\left(\varepsilon^2 \int_0^{\frac{2\delta}{\varepsilon}} \frac{r^{m+1}}{(1+r^2)^m} dr \right) \\
&= m^m \omega_{m-1} \int_0^\infty \frac{r^{m-1}}{(1+r^2)^m} dr + O(\varepsilon^m) + \begin{cases} O(\varepsilon^2 |\ln \varepsilon|), & \text{if } m = 2 \\ O(\varepsilon^2), & \text{if } m \ge 3 \end{cases} \\
&= m^m \omega_{m-1} \int_0^\infty \frac{r^{m-1}}{(1+r^2)^m} dr + \begin{cases} O(\varepsilon^2 |\ln \varepsilon|) & \text{if } m = 2 \\ O(\varepsilon^2) & \text{if } m \ge 3 \end{cases}
\end{aligned}
$$

$$J_3 \lesssim \int_{\delta \le |x| \le 2\delta} |\psi_\varepsilon|^{2^*} dx \lesssim \varepsilon^m$$

$$J_4 \lesssim \int_{|x| \le 2\delta} |x|^3 |\psi_\varepsilon|^2 dx \lesssim \varepsilon^4 \int_0^{\frac{2\delta}{\varepsilon}} \frac{r^{m+2}}{(1+r^2)^{m-1}} dr \lesssim \begin{cases} \varepsilon^{m-1} & \text{if } 2 \le m \le 4 \\ \varepsilon^4 |\ln \varepsilon| & \text{if } m = 5 \\ \varepsilon^4 & \text{if } m \ge 6 \end{cases}$$

$$J_5 = 0$$

$$J_6 \lesssim \int_{|x|\le 2\delta} |x|^2 |\nabla\psi_\varepsilon| \cdot |\psi_\varepsilon| dx \lesssim \varepsilon^2 \int_0^{\frac{2\delta}{\varepsilon}} \frac{r^{m+1}}{(1+r^2)^{m-\frac{1}{2}}} dr$$
$$\lesssim \begin{cases} \varepsilon & \text{if } m = 2 \\ \varepsilon^2 |\ln\varepsilon| & \text{if } m = 3 \\ \varepsilon^2 & \text{if } m \ge 4 \end{cases}$$

Here we used the inequality $|\nabla\psi(x)| \lesssim \mu(x)^{\frac{m}{2}}$.

$$J_7 \lesssim \int_{|x|\le 2\delta} |x|^3 |\psi_\varepsilon|^2 dx \lesssim \begin{cases} \varepsilon^{m-1} & \text{if } 2 \le m \le 4 \\ \varepsilon^4 |\ln\varepsilon| & \text{if } m = 5 \\ \varepsilon^4 & \text{if } m \ge 6 \end{cases}$$

Here we used the inequality $|\nabla\eta(x)| \lesssim x$ and the same estimate as for J_4.

Combining these estimates we deduce that

$$\frac{1}{2}\int_M (\bar{D}\bar{\varphi}_\varepsilon, \bar{\varphi}_\varepsilon) d\,\mathrm{vol}_g - \frac{1}{2^*}\int_M |\bar{\varphi}_\varepsilon|^{2^*} d\,\mathrm{vol}_g$$
$$\le \frac{1}{2} m^{m-1} \omega_{m-1} \int_0^\infty \frac{r^{m-1}}{(1+r^2)^m} dr + \begin{cases} O(\varepsilon) & \text{if } m = 2 \\ O(\varepsilon^2 |\ln\varepsilon|) & \text{if } m = 3 \\ O(\varepsilon^2) & \text{if } m \ge 4 \end{cases}$$

Finally (6.35) follows upon taking into account that

$$\omega_m = \int_{\mathbb{R}^m} \left(\frac{2}{1+|x|^2}\right)^m dx = 2^m \omega_{m-1} \int_0^\infty \frac{r^{m-1}}{(1+r^2)^m} dr.$$

□

By virtue of Lemma 6.16 and 6.17, our next step is to apply Theorem 3.34 (with index $\alpha \equiv \lambda_k$) and to obtain the inequality

$$\inf \mathcal{M}_{\lambda_k} < \gamma_{crit} \tag{6.36}$$

which plays the fundamental role in the compactness condition of Proposition 6.9 and Lemma 6.15. And in order to do this, let us consider the following procedure which annihilates the subspace E_λ^0. Since we have

$$\mathcal{L}_\lambda(\psi) = \frac{1}{2}\left(\|\psi^+\|_\lambda^2 - \|\psi^-\|_\lambda^2\right) - \frac{1}{2^*}\int_M |\psi|^{2^*} d\,\mathrm{vol}_g$$

for all $\psi \in E$. Let

$$T_\lambda : L^{2^*} = L^{2^*}(M, \mathbb{S}(M)) \to E_\lambda^0 = \ker(D - \lambda), \quad T_\lambda(\psi) = \arg\min_{\phi \in E_\lambda^0} |\psi - \phi|_{2^*}^{2^*},$$

i.e. $T_\lambda(\psi) \in E_\lambda^0$ is the best approximation of $\psi \in L^{2^*}$ in E_λ^0. This exists and is unique because the L^{2^*} norm is uniformly convex and E_λ^0 is finite-dimensional. Of course, T_λ is not linear in general, and $T_\lambda \equiv 0$ when $E_\lambda^0 = \{0\}$. Clearly we have $|\psi|_{2^*}^{2^*} \geq |\psi - T_\lambda(\psi)|_{2^*}^{2^*}$ and therefore

$$\mathcal{L}_\lambda(\psi) \leq \widetilde{\mathcal{E}}_\lambda(\psi) := \mathcal{L}_\lambda(\psi - T_\lambda(\psi)) = \frac{1}{2}\big(\|\psi^+\|_\lambda^2 - \|\psi^-\|_\lambda^2\big) - \frac{1}{2^*}\int_M |\psi - T_\lambda(\psi)|^{2^*} d\,\mathrm{vol}_g\,. \tag{6.37}$$

This implies

$$\inf \mathcal{M}_\lambda = \inf_{\phi \in S_\lambda^+} \max_{\psi \in \widehat{E}_\lambda(\phi)} \mathcal{L}_\lambda(\psi) \leq \inf_{\phi \in S_\lambda^+} \max_{\psi \in \widehat{E}_\lambda(\phi)} \widetilde{\mathcal{E}}_\lambda(\psi). \tag{6.38}$$

We need to collect some properties of T_λ and of the functionals

$$\mathcal{F}_\lambda : L^{2^*} \to \mathbb{R}, \quad \mathcal{F}_\lambda(\psi) = \frac{1}{2^*}|\psi - T_\lambda(\psi)|_{2^*}^{2^*}$$

and

$$\widetilde{\mathcal{E}}_\lambda : E \to \mathbb{R}, \quad \widetilde{\mathcal{E}}_\lambda(\psi) = \mathcal{L}_\lambda(\psi - T_\lambda(\psi)) = \frac{1}{2}(\|\psi^+\|_\lambda^2 - \|\psi^-\|_\lambda^2) - \mathcal{F}_\lambda(\psi).$$

Lemma 6.18

(a) *For $\psi \in L^{2^*}$, $\phi \in E_\lambda^0$ and $t \in \mathbb{R}$ there holds: $T_\lambda(t\psi) = tT_\lambda(\psi)$, $T_\lambda(\psi + \phi) = T_\lambda(\psi) + \phi$ and $\mathcal{F}_\lambda(\psi + \phi) = \mathcal{F}_\lambda(\psi)$.*
(b) *T_λ is of class C^1 on $L^{2^*} \setminus E_\lambda^0$. Moreover, $\nabla T_\lambda(\psi)[\psi] = T_\lambda(\psi)$ for all $\psi \in L^{2^*} \setminus E_\lambda^0$.*
(c) *$\mathcal{F}_\lambda$ is of class C^2 on $L^{2^*} \setminus E_\lambda^0$ with derivative*

$$\nabla \mathcal{F}_\lambda(\psi)[\phi] = \mathrm{Re} \int_M |\psi - T_\lambda(\psi)|^{2^*-2}(\psi - T_\lambda(\psi), \phi) d\,\mathrm{vol}_g\,.$$

(d) *$\mathcal{F}_\lambda$ is convex.*
(e) *If $\psi \in \widetilde{E} = E_\lambda^+ \oplus E_\lambda^-$ is a critical point of $\widetilde{\mathcal{E}}_\lambda$ then $\psi - T_\lambda(\psi) \in E$ is a critical point of $\mathcal{L}_\lambda$.*

Proof (a) is trivial. For the proof of (b) we fix $\psi \in L^{2^*} \setminus E_\lambda^0$ and consider the map $\mathcal{F}_{\lambda,\psi} : E_\lambda^0 \to \mathbb{R}$ defined by $\mathcal{F}_{\lambda,\psi}(\phi) := \frac{1}{2^*}|\psi - \phi|_{2^*}^{2^*}$. Observe that $\phi = T_\lambda(\psi)$

is the unique solution of $\nabla \mathcal{F}_{\lambda,\psi}(\phi) = 0$ because $\mathcal{F}_{\lambda,\psi}$ is of class $\mathcal{C}^2$ and strictly convex. A simple computation yields for $\phi, w \in E_\lambda^0$:

$$\nabla^2 \mathcal{F}_{\lambda,\psi}(\phi)[w,w] \geq \int_M |\psi - \phi|^{2^*-2} \cdot |w|^2 d\,\mathrm{vol}_g\,. \tag{6.39}$$

We claim that there exists $\delta > 0$ such that

$$\nabla^2 \mathcal{F}_{\lambda,\psi}(T_\lambda(\psi))[w,w] \geq \delta \|w\|_\lambda^2 \quad \text{for all } w \in E_\lambda^0. \tag{6.40}$$

Assuming (6.40), it follows that $\nabla^2 \mathcal{F}_{\lambda,\psi}(T_\lambda(\psi)) : E_\lambda^0 \to E_\lambda^0$ is invertible. Therefore we can apply the implicit function theorem to $\nabla \mathcal{F}_{\lambda,\psi}(T_\lambda(\psi)) = 0$ and see that $\nabla T_\lambda(\psi) : L^{2^*} \to E_\lambda^0$ exists on a neighborhood of ψ in $L^{2^*} \setminus E_\lambda^0$. Moreover, $\nabla T_\lambda(\psi)[\psi] = T_\lambda(\psi)$ follows from (a).

It remains to show (6.40). Arguing by contradiction, we find a sequence $\{w_n\} \subset E_\lambda^0$ such that $\|w_n\|_\lambda = 1$ and $\nabla^2 \mathcal{F}_{\lambda,\psi}(T_\lambda(\psi))[w_n, w_n] \to 0$. By using (6.39) and the fact E_λ^0 is finite-dimensional, we deduce that there exists $w_0 \in E_\lambda^0$ with $\|w_0\|_\lambda = 1$ such that $w_n \to w_0$ along a subsequence, and $\int_M |\psi - T_\lambda(\psi)|^{2^*-2} |w_0|^2 d\,\mathrm{vol}_g = 0$. Since $w_0 \in E_\lambda^0$ is an eigenspinor, Bär's theorem [5] yields that the nodal set of w_0 has measure zero. This implies $\psi - T_\lambda(\psi) = 0$ in L^{2^*} which is impossible.

In order to prove (c) observe that $\nabla \mathcal{F}_\lambda(\psi)$ is trivial on E_λ^0 by (a). This implies for $\phi \in L^{2^*}$:

$$\begin{aligned}\nabla \mathcal{F}_\lambda(\psi)[\phi] &= \mathrm{Re} \int_M |\psi - T_\lambda(\psi)|^{2^*-2} \big(\psi - T_\lambda(\psi), \phi - \nabla T_\lambda(\psi)[\phi]\big) d\,\mathrm{vol}_g \\ &= \mathrm{Re} \int_M |\psi - T_\lambda(\psi)|^{2^*-2} (\psi - T_\lambda(\psi), \phi) d\,\mathrm{vol}_g\,.\end{aligned}$$

Now $\mathcal{F}_\lambda$ is $\mathcal{C}^2$ because T_λ is $\mathcal{C}^1$.

(d) For $\psi_0, \psi_1 \in L^{2^*}$ and $\alpha \in [0,1]$ we obtain, using the definition of T_λ:

$$\begin{aligned}&\big|(1-\alpha)\psi_0 + \alpha\psi_1 - T_\lambda\big((1-\alpha)\psi_0 + \alpha\psi_1\big)\big|_{2^*}^{2^*} \\ &\quad \leq \big|(1-\alpha)\psi_0 + \alpha\psi_1 - \big((1-\alpha)T_\lambda(\psi_0) + \alpha T_\lambda(\psi_1)\big)\big|_{2^*}^{2^*} \\ &\quad \leq (1-\alpha)\big|\psi_0 - T_\lambda(\psi_0)\big|_{2^*}^{2^*} + \alpha\big|\psi_1 - T_\lambda(\psi_1)\big|_{2^*}^{2^*}\end{aligned}$$

(e) Let $\psi \in E_\lambda^+ \oplus E_\lambda^-$ be a critical point of $\widetilde{\mathcal{E}}_\lambda$. Then we have for arbitrary $\phi^+ \in E_\lambda^+$ and $\phi^- \in E_\lambda^-$:

$$0 = \nabla \widetilde{\mathcal{E}}_\lambda(\psi)[\phi^+ + \phi^-] = \langle \psi^+, \phi^+ \rangle_\lambda - \langle \psi^-, \phi^- \rangle_\lambda - \nabla \mathcal{F}_\lambda(\psi)[\phi^+ + \phi^-]. \tag{6.41}$$

Since $\nabla\mathcal{F}_\lambda(\psi)[\phi^0] \equiv 0$ for any $\phi^0 \in E^0_\lambda$, it follows from (6.41) that, setting $\phi = \phi^+ + \phi^0 + \phi^-$,

$$\nabla\mathcal{L}_\lambda(\psi - T_\lambda(\psi))[\phi] = \langle \psi^+, \phi^+\rangle_\lambda - \langle \psi^-, \phi^-\rangle_\lambda - \nabla\mathcal{F}_\lambda(\psi)[\phi] = 0.$$

Since ϕ^+, ϕ^0 and ϕ^- are arbitrary, we see that $\psi - T_\lambda(\psi)$ is a critical point of $\mathcal{L}_\lambda$. □

The second derivative of $\mathcal{F}_\lambda$ is given by

$$\begin{aligned}\nabla^2\mathcal{F}_\lambda(\psi)[\phi,\chi] = \mathrm{Re}\int_M &|\psi - T_\lambda(\psi)|^{2^*-2}(\chi - \nabla T_\lambda(\psi)[\chi], \phi)d\,\mathrm{vol}_g\\ &+\frac{2}{m-1}\int_M |\psi - T_\lambda(\psi)|^{2^*-4}\,\mathrm{Re}(\psi - T_\lambda(\psi), \chi - \nabla T_\lambda(\psi)[\chi])\\ &\cdot \mathrm{Re}(\psi - T_\lambda(\psi), \phi)d\,\mathrm{vol}_g\,.\end{aligned}$$

Using $\nabla T_\lambda(\psi)[\psi] = T_\lambda(\psi)$ and that $\nabla^2\mathcal{F}_\lambda(\psi)[\phi,\chi] = 0$ for $\psi \in L^{2^*}\setminus E^0_\lambda$, $\phi \in L^{2^*}$ and $\chi \in E^0_\lambda$, an elementary calculation yields for $\psi \in L^{2^*}\setminus E^0_\lambda$ and $\phi \in L^{2^*}$:

$$\begin{aligned}&\big(\nabla^2\mathcal{F}_\lambda(\psi)[\psi,\psi] - \nabla\mathcal{F}_\lambda(\psi)[\psi]\big)\\ &\quad+2\big(\nabla^2\mathcal{F}_\lambda(\psi)[\psi,\phi] - \nabla^2\mathcal{F}_\lambda(\psi)[\phi]\big) + \nabla^2\mathcal{F}_\lambda(\psi)[\phi,\phi]\\ &\geq \frac{2}{m+1}\int_M |\psi - T_\lambda(\psi)|^{2^*}d\,\mathrm{vol}_g > 0\end{aligned}$$

Lemma 6.18 implies that $\widetilde{\mathcal{E}}_\lambda(\psi+\phi) = \widetilde{\mathcal{E}}_\lambda(\psi)$ for $\psi \in E$ and $\phi \in E^0_\lambda$. Therefore we only need to consider $\widetilde{\mathcal{E}}_\lambda$ on $\widetilde{E}_\lambda = E^+_\lambda \oplus E^-_\lambda$, so that the functional $\widetilde{\mathcal{E}}$ satisfies all the hypotheses of Theorem 3.34.

6.5 The Bifurcation Diagram

In this section, the main purpose is to apply the localized energy estimate, in particular those discussed in Sect. 3.3, to study the problem (6.1) concerning the bifurcation from the spectrum. Our first result can be state as follows.

Theorem 6.19

(1) *The spinorial Brezis-Nirenberg problem* (6.1) *has a least energy solution* ψ_λ *for every* $\lambda > 0$. *The function*

$$\mathbb{R}^+ \to \mathbb{R}^+, \quad \lambda \mapsto \mathcal{L}_\lambda(\psi_\lambda),$$

is continuous and nonincreasing on each interval $[\lambda_k, \lambda_{k+1})$, $k \geq 1$, *and on* $(0, \lambda_1)$. *Moreover,* $\psi_\lambda \to 0$ *hence* $\mathcal{L}_\lambda(\psi_\lambda) \searrow 0$ *as* $\lambda \nearrow \lambda_k$.

(2) *For* $k \geq 1$ *there exists* $a_k \in (\lambda_{k-1}, \lambda_k)$ *such that* (6.1) *has for* $\lambda \in (a_k, \lambda_k)$ *a solution* $\widetilde{\psi}_\lambda$ *so that the map*

$$(a_k, \lambda_{k+1}) \to \mathbb{R}^+, \qquad \lambda \mapsto \begin{cases} \mathcal{L}_\lambda(\widetilde{\psi}_\lambda) & \text{if } \lambda < \lambda_k \\ \mathcal{L}_\lambda(\psi_\lambda) & \text{if } \lambda \geq \lambda_k \end{cases}$$

is continuous and non-increasing.

Remark 6.20 By Theorem 6.19 (and its proof later, see also Lemma 6.15), the least energy solution ψ_λ for $\lambda \in [\lambda_k, \lambda_{k+1})$ can be continued to $\lambda \in (a_k, \lambda_{k+1})$ in the sense that the energy, and the corresponding min-max description, changes continuously. We do not know whether the solutions ψ_λ and $\widetilde{\psi}_\lambda$ depend continuously on λ. Since $\mathcal{L}_\lambda(\widetilde{\psi}_\lambda) \to \mathcal{L}_{\lambda_k}(\psi_{\lambda_k}) > 0$ and $\mathcal{L}_\lambda(\psi_\lambda) \to 0$ as $\lambda \nearrow \lambda_k$, we see that $\widetilde{\psi}_\lambda$ is different from ψ_λ. From a variational point of view ψ_λ is a minimizer of $\mathcal{L}_\lambda$ on the Nehari-Pankov manifold $\mathcal{P}_\lambda$, $\lambda \in [a_k, \lambda_k)$, whereas $\widetilde{\psi}_\lambda$ minimizes $\mathcal{L}_\lambda$ on a submanifold $\mathcal{Q}_{\lambda,k}$ of $\mathcal{P}_\lambda$ of codimension $d_k = \dim_{\mathbb{R}} \ker(D - \lambda_k)$, $\lambda \in (a_k, \lambda_k)$. $\mathcal{Q}_{\lambda,k}$ is a continuation of $\mathcal{P}_{\lambda_k}$ for $\lambda < \lambda_k$ close to λ_k.

Next we state several multiplicity results. Observe that given a solution ψ of (6.1) and $\zeta \in S^1 \subset \mathbb{C}$ then $\zeta\psi$ is also a solution. Therefore we only count S^1-orbits of solutions in our multiplicity results. Clearly the solutions ψ_λ and $\widetilde{\psi}_\lambda$ from Theorems 6.19 lie on different S^1-orbits because they lie on different energy levels. As above we set $d_k = \dim_{\mathbb{R}} \ker(D - \lambda_k)$.

Theorem 6.21 *For* $k \in \mathbb{Z}$ *there exists* $b_k < \lambda_k$ *such that the following holds. For each* $\lambda \in (b_k, \lambda_k)$ *problem* (6.1) *has at least* $\frac{d_k}{2}$ S^1*-orbits of solutions* $\psi_{\lambda,j}$, $j = 1, \dots, \frac{d_k}{2}$. *These satisfy* $\psi_{\lambda,j} \to 0$ *as* $\lambda \nearrow \lambda_k$.

This theorem follows from [8, Theorem (3.1)]. That the bifurcation is subcritical is a consequence of the sign of our nonlinearity. If the manifold M has an additional symmetry group Γ inducing an action of Γ on $H^{\frac{1}{2}}(M, \mathbb{S}(M))$ so that $\mathcal{L}_\lambda$ is also invariant under this action then one should count $S^1 \times \Gamma$-orbits of critical points in Theorem 6.21.

Combining Theorems 6.19 and 6.21 we obtain a multiplicity result for solutions of (6.1) provided $\lambda < \lambda_k$ is close to λ_k. If we order the $\psi_{\lambda,j}$ by $\mathcal{L}_\lambda(\psi_{\lambda,j}) \leq \mathcal{L}_\lambda(\psi_{\lambda,j+1})$ then the least energy solution ψ_λ corresponds to $\psi_{\lambda,1}$. As a consequence of our results, for $\max\{a_k, b_k\} < \lambda < \lambda_k$ we have $\frac{d_k}{2} + 1$ S^1-orbits of solutions, namely the S^1-orbits of the high energy solutions $\widetilde{\psi}_\lambda$ from Theorem 6.19, and the $\frac{d_k}{2}$ S^1-orbits of low energy solutions from Theorem 6.21 which bifurcate from the trivial branch $\{(\lambda, 0) : \lambda \in \mathbb{R}\}$.

Theorem 6.19 essentially yields that, for $k \geq 1$ the least energy solution $\psi_\lambda = \psi_{\lambda,1}$ bifurcating from λ_{k+1} can be continued a bit below λ_k, turning into the "bound state" $\widetilde{\psi}_\lambda$ for $\lambda < \lambda_{k-1}$. If $d_k > 2$ then it is of course an interesting problem whether the bound states $\psi_{\lambda,j}$, $j = 2, \ldots, d_k$, bifurcating from λ_{k+1} can also be extended to below λ_k, in the sense of Remark 6.20. Our last result and its proof suggest that all solutions bifurcating from λ_k can be extended to $\lambda \in (\lambda_k - \nu, \lambda_k)$ with

$$\nu := \frac{m}{2}\left(\frac{\omega_m}{\text{Vol}(M, g)}\right)^{\frac{1}{m}} > 0. \tag{6.42}$$

This holds for all $k \in \mathbb{Z}$.

Theorem 6.22 *For $\lambda \in \mathbb{R}$, the problem* (6.1) *has at least*

$$\ell(\lambda) = \sum_{\lambda < \lambda_k < \lambda + \nu} \dim_{\mathbb{C}} \ker(D - \lambda_k)$$

distinct S^1-orbits of solutions $\psi_{\lambda,j}$, $j = 1, \ldots, \ell(\lambda)$.

The Weyl formula and the symmetry of the eigenvalues of D in dimension $m \not\equiv 3 \pmod 4$ imply for $\Lambda > 0$ that

$$d_+(\Lambda) := \dim_{\mathbb{C}}\left(\bigoplus_{0 < \lambda_k \le \Lambda} \ker(D - \lambda_k)\right) \sim c_M \Lambda^m$$

and

$$d_-(\Lambda) := \dim_{\mathbb{C}}\left(\bigoplus_{-\Lambda \le \lambda_k < 0} \ker(D - \lambda_k)\right) \sim c_M \Lambda^m$$

where $c_M > 0$ is a constant depending on $m = \dim(M)$ and $\text{vol}(M, g)$; see Proposition 6.1. From this we immediately obtain the following result.

Proposition 6.23 *If $m \not\equiv 3 \pmod 4$ then $\ell(\lambda) \to \infty$ as $|\lambda| \to \infty$, hence the number of S^1-orbits of solutions of* (6.1) *becomes unbounded as $|\lambda| \to \infty$.*

Remark 6.24

(a) If $m \equiv 3 \pmod 4$ the Schrödinger-Lichnerowicz formula only yields the following lower bound for the number $n(\lambda)$ of S^1-orbits of solutions of (6.1):

$$n(\lambda) + n(-\lambda) \ge \sum_{\lambda_k \in (\lambda, \lambda+\nu) \cup (-\lambda-\nu, -\lambda)} \frac{d_k}{2} \to \infty \quad \text{as } \lambda \to \infty.$$

(b) We recall that Theorem 6.19 do not yield any solution of (6.1) with $\lambda = \lambda_k$, $k \le 0$, whereas Theorems 6.21, 6.22, and Proposition 6.23 do not distinguish between the sign of λ or whether λ lies in $\text{spec}(D)$.

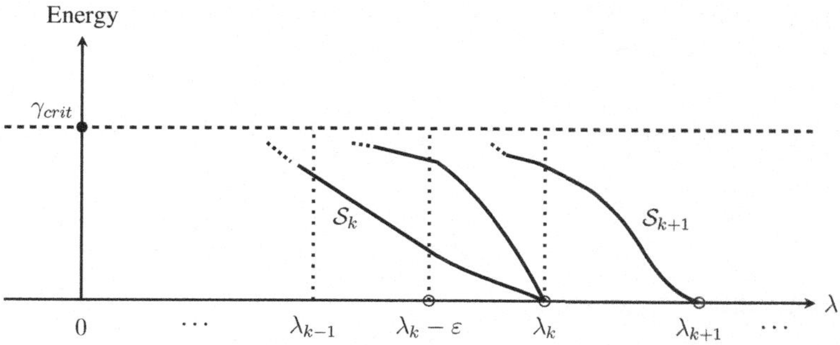

Fig. 6.2 Least energy branches $\mathcal{S}_k$, $\mathcal{S}_{k+1}$ plus one high energy branch bifurcating from λ_k

Setting

$$\mathcal{S}_k := \{(\lambda, \mathcal{L}_\lambda(\psi_\lambda)) : \lambda_{k-1} \le \lambda < \lambda_k\} \cup \{(\lambda, \mathcal{L}_\lambda(\widetilde{\psi}_\lambda)) : \lambda < \lambda_{k-1} \text{ close to } \lambda_{k-1}\}$$

we can visualize the aforementioned results about energy branches bifurcating from λ_k and λ_{k+1} in the following figure.

In Fig. 6.2 the dotted line $\lambda = \lambda_k - \varepsilon$ intersects at least two energy branches of solutions for $\varepsilon > 0$ small; the intersection with $\mathcal{S}_k$ gives us a least energy solution. The branch $\mathcal{S}_k$ of least energy solutions bifurcating from λ_k continues to λ_{k-1} and below; similarly for the branch $\mathcal{S}_{k+1}$ of least energy solutions bifurcating from λ_{k+1} sketched in Fig. 6.2. We do not know whether one of the high energy branches continues up to λ_{k-1}.

Now we address the main inequality

$$\inf \mathcal{M}_{\lambda_k} < \gamma_{crit}$$

using our test spinor $\bar{\varphi}_\varepsilon$ in (6.28). Actually, we will prove a stronger result for all $\lambda \in [\lambda_k, \lambda_{k+1})$. To start with, we have

$$|\bar{\varphi}_\varepsilon|_1 = \int_M |\bar{\varphi}_\varepsilon| d\,\mathrm{vol}_g \lesssim \varepsilon^{\frac{m-1}{2}} \quad \text{and} \quad |\bar{\varphi}_\varepsilon|_{2^*-1}^{2^*-1} = \int_M |\bar{\varphi}_\varepsilon|^{2^*-1} d\,\mathrm{vol}_g \lesssim \varepsilon^{\frac{m-1}{2}} \tag{6.43}$$

As an immediate consequence of these two estimates, jointly with the L^{2^*}-boundedness of $\bar{\varphi}_\varepsilon$ (see in the proof of Lemma 6.17 the estimate of J_2), we have that $\bar{\varphi}_\varepsilon \rightharpoonup 0$ in $L^{2^*}(M, \mathbb{S}(M))$. By Lemma 6.18 (c), we can get that

$$\mathrm{Re} \int_M |\bar{\varphi}_\varepsilon - T_\lambda(\bar{\varphi}_\varepsilon)|^{2^*-2} (\bar{\varphi}_\varepsilon - T_\lambda(\bar{\varphi}_\varepsilon), T_\lambda(\bar{\varphi}_\varepsilon)) d\,\mathrm{vol}_g \equiv 0 \tag{6.44}$$

for all $\varepsilon > 0$. And hence, we can deduce from the boundedness of $T_\lambda(\bar{\varphi}_\varepsilon)$ in E_λ^0 and $\dim E_\lambda^0 < \infty$ that $T_\lambda(\bar{\varphi}_\varepsilon) \to 0$ as $\varepsilon \to 0$ (since E_λ^0 is of finite dimension, this convergence will be valid in any norm). Furthermore, we can also deduce from the definition of $\bar{\varphi}_\varepsilon$ and (6.44) that

$$\int_{M\setminus B_{2\delta}(p_0)} |T_\lambda(\bar{\varphi}_\varepsilon)|^{2^*} d\,\mathrm{vol}_g \le \int_M |\bar{\varphi}_\varepsilon - T_\lambda(\bar{\varphi}_\varepsilon)|^{2^*-2}|T_\lambda(\bar{\varphi}_\varepsilon)|^2 d\,\mathrm{vol}_g$$
$$= \mathrm{Re}\int_M |\bar{\varphi}_\varepsilon - T_\lambda(\bar{\varphi}_\varepsilon)|^{2^*-2}(\bar{\varphi}_\varepsilon, T_\lambda(\bar{\varphi}_\varepsilon))d\,\mathrm{vol}_g\,.$$

This implies that

$$|T_\lambda(\bar{\varphi}_\varepsilon)|_\infty^{2^*} \lesssim \int_{M\setminus B_{2\delta}(p_0)} |T_\lambda(\bar{\varphi}_\varepsilon)|^{2^*} d\,\mathrm{vol}_g$$
$$\lesssim \int_M \Big(|\bar{\varphi}_\varepsilon|^{2^*-1}|T_\lambda(\bar{\varphi}_\varepsilon)| + |T_\lambda(\bar{\varphi}_\varepsilon)|^{2^*-1}|\bar{\varphi}_\varepsilon|\Big)d\,\mathrm{vol}_g$$
$$\le |T_\lambda(\bar{\varphi}_\varepsilon)|_\infty \int_M |\bar{\varphi}_\varepsilon|^{2^*-1}d\,\mathrm{vol}_g + |T_\lambda(\bar{\varphi}_\varepsilon)|^{2^*-1}\int_M |\bar{\varphi}_\varepsilon| d\,\mathrm{vol}_g\,.$$

Then, using the facts $2^* > 2$ and $|T_\lambda(\bar{\varphi}_\varepsilon)|_\infty \to 0$ as $\varepsilon \to 0$, we conclude from (6.43) that

$$|T_\lambda(\bar{\varphi}_\varepsilon)|_\infty \lesssim \varepsilon^{\frac{(m-1)^2}{2(m+1)}} \tag{6.45}$$

And thus we get

$$\Big|\int_M |\bar{\varphi}_\varepsilon - T_\lambda(\bar{\varphi}_\varepsilon)|^{2^*} - |\bar{\varphi}_\varepsilon|^{2^*} d\,\mathrm{vol}_g\Big|$$
$$\lesssim \Big|\int_M |\bar{\varphi}_\varepsilon - \theta T_\lambda(\bar{\varphi}_\varepsilon)|^{2^*-2}(\bar{\varphi}_\varepsilon - \theta T_\lambda(\bar{\varphi}_\varepsilon), T(\bar{\varphi}_\varepsilon))\, d\,\mathrm{vol}_g\Big|$$
$$\le \int_M |\bar{\varphi}_\varepsilon - \theta T_\lambda(\bar{\varphi}_\varepsilon)|^{2^*-1}\cdot |T_\lambda(\bar{\varphi}_\varepsilon)|\, d\,\mathrm{vol}_g$$
$$\lesssim \int_M |\bar{\varphi}_\varepsilon|^{2^*-1}|T_\lambda(\bar{\varphi}_\varepsilon)| + |T_\lambda(\bar{\varphi}_\varepsilon)|^{2^*} d\,\mathrm{vol}_g$$
$$\lesssim |T_\lambda(\bar{\varphi}_\varepsilon)|_\infty \int_M |\bar{\varphi}_\varepsilon|^{2^*-1}d\,\mathrm{vol}_g + |T_\lambda(\bar{\varphi}_\varepsilon)|_\infty^{2^*} \lesssim \varepsilon^{\frac{m(m-1)}{m+1}}$$

where $0 < \theta < 1$. Moreover, for any $\psi \in E$ with $\|\psi\|_\lambda \le 1$, we can deduce from (6.43) and (6.45) that

$$\begin{aligned}
&\left| \int_M |\bar\varphi_\varepsilon - T_\lambda(\bar\varphi_\varepsilon)|^{2^*-2}(\bar\varphi_\varepsilon - T_\lambda(\bar\varphi_\varepsilon), \psi) - |\bar\varphi_\varepsilon|^{2^*-2}(\bar\varphi_\varepsilon, \psi) d\,\mathrm{vol}_g \right| \\
&\qquad \le (2^*-1)\int_M |\bar\varphi_\varepsilon - \theta T_\lambda(\bar\varphi_\varepsilon)|^{2^*-2} \cdot |T_\lambda(\bar\varphi_\varepsilon)| \cdot |\psi| d\,\mathrm{vol}_g \\
&\qquad \lesssim \int_M |\bar\varphi_\varepsilon|^{2^*-2} \cdot |T_\lambda(\bar\varphi_\varepsilon)| \cdot |\psi| + |T_\lambda(\bar\varphi_\varepsilon)|^{2^*-1} \cdot |\psi| d\,\mathrm{vol}_g \\
&\qquad \lesssim |T_\lambda(\bar\varphi_\varepsilon)|_\infty \cdot \left(\int_M |\bar\varphi_\varepsilon|^{\frac{4m}{m^2-1}} d\,\mathrm{vol}_g \right)^{\frac{m+1}{2m}} + |T_\lambda(\bar\varphi_\varepsilon)|_\infty^{2^*-1} \lesssim \varepsilon^{\frac{m-1}{2}}.
\end{aligned}$$

Therefore $(\widetilde\varphi_\varepsilon := \bar\varphi_\varepsilon - T_\lambda(\bar\varphi_\varepsilon))_\varepsilon$ is a (PS)-sequence for $\widetilde{\mathcal{E}}_\lambda$ on E. Combining all these facts and denoting $(\mathcal{J}_\lambda, \eta_\lambda)$ the reduction couple associated to $\widetilde{\mathcal{E}}_\lambda$ on E_λ^+ (in the sense of Proposition 3.29), and applying Theorem 3.34, we finally obtain

$$\max_{t>0} \mathcal{J}_\lambda(t\bar\varphi_\varepsilon^+) \le \frac{1}{2m}\left(\frac{m}{2}\right)^m \omega_m + \begin{cases} -\lambda C\varepsilon + O(\varepsilon^{\frac{3}{2}}) & \text{if } m = 3, \\ -\lambda C\varepsilon + O(\varepsilon^2) & \text{if } m \ge 4 \end{cases}$$

for some constant $C > 0$ depending only on the dimension and $\lambda \in [\lambda_k, \lambda_{k+1})$. And if $\lambda \in (\lambda_k, \lambda_{k+1})$, that is $E_\lambda^0 = \{0\}$, we have

$$\max_{t>0} \mathcal{J}_\lambda(t\bar\varphi_\varepsilon^+) \le \frac{1}{2m}\left(\frac{m}{2}\right)^m \omega_m + \begin{cases} -\lambda C\varepsilon|\ln\varepsilon| + O(\varepsilon) & \text{if } m = 2, \\ -\lambda C\varepsilon + O(\varepsilon^2|\ln\varepsilon|^{\frac{4}{3}}) & \text{if } m = 3, \\ -\lambda C\varepsilon + O(\varepsilon^2) & \text{if } m \ge 4. \end{cases}$$

Now (6.38) implies, setting $\phi_\varepsilon = \frac{1}{\|\bar\varphi_\varepsilon^+\|_\lambda}\bar\varphi_\varepsilon^+$,

$$\inf \mathcal{M}_\lambda \le \max_{\psi \in \widehat{E}_\lambda(\phi_\varepsilon)} \widetilde{\mathcal{E}}_\lambda(\psi) = \max_{t>0} \mathcal{J}_\lambda(t\bar\varphi_\varepsilon^+) < \frac{1}{2m}\left(\frac{m}{2}\right)^m \omega_m = \gamma_{crit}$$

for $\varepsilon > 0$ small. This, jointly with Lemma 6.15, concludes the proof of Theorem 6.19.

To see Theorem 6.22, let us begin with the functional $\mathcal{R}_\lambda : \widetilde{E}_\lambda \setminus \{0\} \to \mathbb{R}$ defined by

$$\mathcal{R}_\lambda(\psi) = \frac{\int_M (D\psi, \psi) - \lambda|\psi|^2 d\,\mathrm{vol}_g}{\left(\int_M |\psi - T_\lambda(\psi)|^{2^*} d\,\mathrm{vol}_g\right)^{\frac{2}{2^*}}} = \frac{\|\psi^+\|_\lambda^2 - \|\psi^-\|_\lambda^2}{|\psi - T_\lambda(\psi)|_{2^*}^2}$$

Then we have

$$\nabla\mathcal{R}_\lambda(\psi)[\varphi] = \frac{2}{A_\lambda(\psi)^{\frac{2}{2^*}}}\Big(\mathrm{Re}\int_M ((D-\lambda)\psi,\varphi)d\,\mathrm{vol}_g - \frac{\mathcal{R}_\lambda(\psi)}{2^*}A_\lambda(\psi)^{\frac{2-2^*}{2^*}}\cdot\nabla A_\lambda(\psi)[\varphi]\Big)$$

where

$$A_\lambda(\psi) = \int_M |\psi - T_\lambda(\psi)|^{2^*}d\,\mathrm{vol}_g$$

For any $\phi \in E_\lambda^+ \setminus \{0\}$ and any $c > 0$ the set $\{\chi \in E_\lambda^- : \mathcal{R}_\lambda(\phi+\chi) \geq c\}$ is strictly convex and bounded because

$$\chi \mapsto \|\phi\|_\lambda^2 - \|\chi\|_\lambda^2 - \lambda|\phi+\chi|_2^2 - c\big|\phi+\chi - T_\lambda(\phi+\chi)\big|_{2^*}^{2^*}$$

is strictly concave and anti-coercive on E_λ^-. Therefore the map $\chi \mapsto \mathcal{R}_\lambda(\phi+\chi)$ has a unique maximum point $\chi_\phi \in E_\lambda^-$. Now, let us define

$$\mathcal{S}_\lambda(\phi) = \mathcal{R}_\lambda(\phi+\chi_\phi)$$

Lemma 6.25 $\mathcal{S}_\lambda(\phi) = \big(2m\mathcal{J}_\lambda(\phi)\big)^{\frac{1}{m}}$ *for* $\phi \in \widetilde{\mathcal{P}}_\lambda := \{\phi \in E_\lambda^+\setminus\{0\} : \nabla\mathcal{J}_\lambda(\phi)[\phi] = 0\}$.

Proof Recall that $\mathcal{J}_\lambda(\phi) = \widetilde{\mathcal{E}}_\lambda(\phi+\eta_\lambda(\phi))$ for $\phi \in E_\lambda^+$. If we chose especially an element $\phi \in \widetilde{\mathcal{P}}_\lambda$, then

$$0 = \int_M \big(D(\phi+\eta_\lambda(\phi)), \phi+\eta_\lambda(\phi)\big)d\,\mathrm{vol}_g - \lambda|\phi+\eta_\lambda(\phi)|_2^2 - \big|\phi+\eta_\lambda(\phi) - T_\lambda(\phi+\eta_\lambda(\phi))\big|_{2^*}^{2^*}$$

Hence, with this ϕ, we have $\mathcal{J}_\lambda(\phi) = \mathcal{J}_\lambda(\phi) - \frac{1}{2}\nabla\mathcal{J}_\lambda(\phi)[\phi] = \frac{1}{2m}\big|\phi+\eta_\lambda(\phi) - T_\lambda(\phi+\eta_\lambda(\phi))\big|_{2^*}^{2^*}$.

On the other hand, we find that

$$\mathcal{R}_\lambda'(\phi+\eta_\lambda(\phi))[\chi] \equiv 0 \quad \text{for all}\,\chi \in E_\lambda^-.$$

This together with the fact $\mathcal{R}_\lambda(\phi + \eta_\lambda(\phi)) > 0$ yields that $\chi_\phi = \eta_\lambda(\phi)$ for $\phi \in \widetilde{\mathcal{P}}_\lambda$. And this in turn implies

$$\mathcal{S}_\lambda(\phi) = \left(\int_M \left| \phi + \eta_\lambda(\phi) - T_\lambda(\phi + \eta_\lambda(\phi)) \right|^{2^*} \right)^{1-\frac{2}{2^*}} = \left(2m \mathcal{J}_\lambda(\phi) \right)^{\frac{1}{m}}$$

□

Theorem 6.22 follows from the next Proposition. Recall the definition of

$$\nu = \frac{m}{2} \left(\frac{\omega_m}{\mathrm{Vol}(M, g)} \right)^{\frac{1}{m}}$$

from (6.42).

Proposition 6.26 *For $\lambda \in \mathbb{R}$ the functional $\mathcal{M}_\lambda$ admits at least*

$$\ell = \ell(\lambda) = \dim_{\mathbb{C}} \left(\bigoplus_{\lambda < \lambda_k < \lambda + \nu} \ker(D - \lambda_k) \right).$$

distinct S^1-orbits of critical points.

Proof Let $\gamma = \gamma_{S^1}$ denote the S^1-genus, i.e. for a topological space $X \neq \emptyset$ on which the group S^1 acts continuously without fixed points, $\gamma(X)$ is the infimum over all $k \in \mathbb{N}$ such that there exist finite subgroups $H_1, \ldots, H_k \subset S^1$ and a continuous equivariant map $X \to S^1/H_1 * \cdots * S^1/H_k$ where $*$ denotes the join. If no such k exists this means $\gamma(X) = \infty$. The S^1-genus has properties analogous to the Krasnoselski genus for spaces with an action of $\mathbb{Z}_2$; see [6, 7]. Now we define

$$\beta_j := \inf_{\substack{A \subset S_\lambda^+ \\ \gamma(A) \geq j}} \max_{\phi \in A} \mathcal{M}_\lambda(\phi) \quad \text{for } j \geq 1.$$

where we only consider S^1-invariant subsets $A \subset S_\lambda^+$. Clearly we have $\beta_j \leq \beta_{j+1}$ for all $j \geq 1$. If

$$X_j \subset \bigoplus_{\lambda < \lambda_k < \lambda + \nu} \ker(D - \lambda_k)$$

is any complex j-dimensional subspace, $1 \leq j \leq \ell$, then $X_j \cap S_\lambda^+$ is equivariantly homeomorphic to the unit sphere in X_j, hence $\gamma(X_j \cap S_\lambda^+) = \dim_{\mathbb{C}}(X_j)$. Now (6.37) and Lemma 6.25 imply

$$\beta_j \leq \max_{\phi \in X_j \cap S_\lambda^+} \mathcal{M}_\lambda(\phi) \leq \max_{\phi \in X_j \cap \widetilde{\mathcal{P}}_\lambda} \mathcal{J}_\lambda(\phi) = \frac{1}{2m} \max_{\phi \in X_j \cap \widetilde{\mathcal{P}}_\lambda} \mathcal{S}_\lambda(\phi)^m.$$

Observe that, for any $\phi \in X_j \cap \widetilde{\mathcal{P}}_\lambda$,

$$\|\phi\|_\lambda^2 - \|\chi_\phi\|_\lambda^2 < \|\phi\|_\lambda^2 \le \nu|\phi|_2^2 < \nu|\phi + \chi_\phi - T_\lambda(\phi + \chi_\phi)|_2^2$$

where the last inequality follows from the fact that $\phi \in E_\lambda^+$, $\chi_\phi \in E_\lambda^-$ and $T_\lambda(\phi + \chi_\phi) \in E_\lambda^0$ are orthogonal in L^2. Finally, using Hölder's inequality, we find for $j = 1, \ldots, \ell$:

$$\beta_j < \frac{\nu^m}{2m} \cdot \max_{\phi \in X_j \cap \widetilde{\mathcal{P}}_\lambda} \left(\frac{|\phi + \chi_\phi - T_\lambda(\phi + \chi_\phi)|_2^2}{|\phi + \chi_\phi - T_\lambda(\phi + \chi_\phi)|_{2^*}^2} \right)^m \le \nu^m \cdot \frac{\mathrm{Vol}(M, g)}{2m} = \gamma_{crit}.$$

Since the Palais-Smale condition holds below γ_{crit} each β_j, $j = 1, \ldots, \ell$, is a critical value, and if $\beta_j = \beta_{j+1}$ for some j then $\mathcal{M}_\lambda$ has infinitely many S^1-orbits of critical points at the levelβ_j; see [7, Theorem 2.19]. □

References

1. Adams, R.: Sobolev Spaces. Academic Press, New York (1975)
2. Ammann, B.: A variational problem in conformal spin geometry. Habilitationsschrift, Universität Hamburg (2003)
3. Ammann, B., Grossjean, J.-F., Humbert, E., Morel, B.: A spinorial analogue of Aubin's inequality. Math. Z. **260**, 127–151 (2008)
4. Atiyah, M.F., Patodi, V.K., Singer, I.M.: Spectral asymmetry and Riemannian geometry I, II and III. Math. Proc. Camb. Philos. Soc. **77**, 43–69 (1975); **78**, 405–432 (1975); **79**, 71–99 (1975)
5. Bär, C.: On nodal sets for Dirac and Laplace operators. Commun. Math. Phys. **188**(3), 709–721 (1997)
6. Bartsch, T.: On the genus of representation spheres. Comment. Math. Helv. **65**(1), 85–95 (1990)
7. Bartsch, T.: Topological Methods for Variational Problems with Symmetries. Lect. Notes Math., vol. 1560. Springer, Berlin-Heidelberg (1993)
8. Bartsch, T., Clapp, M.: Bifurcation theory for symmetric potential operators and the equivariant cup-length. Math. Z. **204**(3), 341–356 (1990)
9. Berline, N., Getzler, E., Vergne, M.: Heat Kernels and Dirac Operators. Grundlehren Text Editions. Springer, Berlin (2003)
10. Bourguignon, J.-P., Gauduchon, P.: Spineurs, opérateurs de Dirac et variations de métriques. Commun. Math. Phys. **144**(3), 581–599 (1992)
11. Dahl, M.: Prescribing eigenvalues of the Dirac operator. Manuscr. Math. **118**(2), 191–199 (2005)
12. Friedrich, T.: Dirac Operators in Riemannian Geometry. Grad. Stud. Math., vol. 25. American Mathematical Society, Providence (2000)
13. Ginoux, N.: The Dirac Spectrum. Lecture Notes in Math., vol. 1976. Springer, Berlin (2009)
14. Isobe, T.: Nonlinear Dirac equations with critical nonlinearities on compact spin manifolds. J. Funct. Anal. **260**, 253–307 (2011)
15. Lawson, H.B., Michelson, M.L.: Spin Geometry. Princeton University Press, Princeton (1989)
16. Szulkin, A., Weth, T.: The method of Nehari manifold. In: Handbook of Nonconvex Analysis and Applications, pp. 597–632. Int. Press, Somerville (2010)

Chapter 7
Isometrically Embedded Sphere with Prescribed Mean Curvature

This chapter is devoted to the study of the prescribing mean curvature problem. After recalling some basic notions and facts, we apply the localization method to find existence results.

7.1 Spinorial Weierstrass Representation

In this section we recall some well-known concepts in Riemannian geometry. In the presentation we will be as concise as possible, in order to arrive soon to the spinorial Yamabe type equation. We refer for example to [21, 22, 28, 29, 34–36], for detailed derivations of the geometric quantities, their motivation and applications.

7.1.1 Overview

The fact that a constant mean curvature (cmc) surface in $\mathbb{R}^3$ yields a spinor satisfying a linear Dirac equation is a modification of results by Weierstrass and has already been known for a long time. Some of the ideas can be trace back to the work of [19]. Abresch presented it in a talk in Luminy, which was cited by many geometrists, but—unfortunately—was never published. For other references, we refer to [28, 34–36] and works by Richter, Pinkall, Kamberov, Bobenko and their collaborators.

For general cases, let us consider (N, h) a 3-dimensional oriented Riemannian manifold with a fixed spin structure and (M, g) an oriented surface isometrically immersed into N. It follows directly that the spin structure of N induces a spin structure on M. Moreover, the spinor bundle $\mathbb{S}(N)$ of the ambient space N yields by

© The Author(s), under exclusive license to Springer Nature Singapore Pte Ltd. 2025
Y. Ding, T. Xu, *Localization Approaches in Strongly Indefinite Problems*, Series in Contemporary Mathematics 6, https://doi.org/10.1007/978-981-97-9539-0_7

restriction the spinor bundle $\mathbb{S}(M)$ of the surface M. Over M this bundle decompose into

$$\mathbb{S}(M) = \mathbb{S}(M)^+ \oplus \mathbb{S}(M)^-$$

where the subbundles $\mathbb{S}(M)^\pm$ are given by

$$\mathbb{S}(M)^\pm = \{\psi \in \mathbb{S}(M) : i \cdot e_1 \cdot e_2 \cdot \psi = \pm\psi\}.$$

Here $\{e_1, e_2\}$ denotes an oriented orthonormal frame in TM and $X \cdot \psi$ stands for the Clifford multiplication of a spinor $\psi \in \mathbb{S}(M)$ by a vector $X \in TM$. Since in the 3-dimensional spin representation the relation

$$e_1 \cdot e_2 = e_3$$

holds, we ca replace the Clifford product $e_1 \cdot e_2$ by the normal vector e_3 of $M \hookrightarrow N$, i.e. $\mathbb{S}(M)^\pm = \{\psi \in \mathbb{S}(M) : i \cdot e_3 \cdot \psi = \pm\psi\}$. For a spinor field Ψ defined on the 3-manifold N, its restriction $\psi = \Psi|_M$ is a spinor field on M and decomposes into $\psi = \psi_+ + \psi_-$ with

$$\psi_+ = \frac{1}{2}(\psi + i \cdot e_3 \cdot \psi), \quad \psi_- = \frac{1}{2}(\psi - i \cdot e_3 \cdot \psi).$$

If denoted by ∇^N and ∇^M hte covariant derivatives in the spinor bundles $\mathbb{S}(N)$ and $\mathbb{S}(M)$, respectively. For any vector $X \in TM$, we have the well-known formula (see [14])

$$\nabla^N_X \Psi = \nabla^M_X \psi - \frac{1}{2}\nabla_X e_3 \cdot e_3 \cdot \psi$$

The endomorphism $X \mapsto \nabla_X e_3$ coincides with the second fundamental form $\mathrm{II} : TM \to TM$ of the submanifold $M \hookrightarrow N$. Since II is symmetric, the Clifford product $e_1 \cdot \mathrm{II}(e_1) + e_2 \cdot \mathrm{II}(e_2)$ is a scalar and equals $-2H$, where H stands for the mean curvature of the surface M in N. Notice that the Dirac operator D_g of M is given by the formula

$$D_g\psi = e_1 \cdot \nabla^M_{e_1}\psi + e_2 \cdot \nabla^M_{e_2}\psi$$

can now be expressed by the covariant derivative ∇^N and the mean curvature vector:

$$e_1 \cdot \nabla^N_{e_1}\Psi + e_2 \cdot \nabla^N_{e_2}\Psi = D_g\psi + H \cdot e_3 \cdot \psi$$

Suppose that the spinor field Ψ on N is a real Killing spinor, i.e., there exists $\lambda \in \mathbb{R}$ such that for any tangent vector $\tau \in TN$ the derivative of Ψ in the direction of τ is given by the Clifford multiplication

$$\nabla_\tau^N \Psi = \lambda \cdot \tau \cdot \Psi.$$

For the restriction $\psi = \Psi|_M$, we have immediately

$$D_g \psi = -2\lambda\psi - H \cdot e_3 \cdot \psi$$

Using the decomposition $\psi = \psi_+ + \psi_-$, the above equation is equivalent to the pair of equations

$$D_g\psi_+ = (-2\lambda - iH)\psi_- \quad \text{and} \quad D_g\psi_- = (-2\lambda + iH)\psi_+.$$

As a first special case, we have

Proposition 7.1 *Let M be a minimal surface in N. Then the restriction $\psi = \Psi|_M$ of any real Killing spinor Ψ on N is an eigenspinor of constant length on M, i.e., there holds $D_g\psi = -2\lambda\psi$ and $|\psi|_g \equiv 1$.*

On the other hand, if Ψ is a parallel spinor (i.e., $\lambda = 0$) on N, then we obtain

$$D_g\psi_+ = -iH\psi_- \quad \text{and} \quad D_g\psi_- = iH\psi_+$$

And if we introduce the spinor field $\bar{\psi} = \psi_+ - i\psi_-$, a direct computation shows

$$D_g\bar{\psi} = H\bar{\psi}. \tag{7.1}$$

Moreover, the spinor field $\bar{\psi}$ is still of constant length.

The previous discussion corresponds to the spinorial Weierstrass representation, and Friedrich has written a very nice article [21] in which it is beautifully explained: in fact, it will be also true if one uses (7.1) to induce a immersion of M into an 3-dimensional manifold carrying a Killing spinor.

Theorem 7.2 (The Spinorial Weierstrass Representation) *Let (M, g) be an oriented 2-dimensional Riemannian manifold and $H : M \to \mathbb{R}$ a smooth function. Then there is a correspondence between the following data:*

(1) *An isometric immersion $(\widetilde{M}, g) \to \mathbb{R}^3$ of the universal covering $\widetilde{M}$ into the Euclidean space $\mathbb{R}^3$ with mean curvature H.*
(2) *A solution φ with constant length $|\varphi| = 1$ of the Dirac equation $D_g\varphi = H\varphi$.*
(3) *A pair $(\varphi, \mathcal{E})$ consisting of a symmetric endomorphism $\mathcal{E}$ such that $\mathrm{tr}(\mathcal{E}) = -H$ and a spinor field φ satisfying the equation $\nabla_X\varphi = \mathcal{E}(X) \cdot_g \varphi$.*

7.1.2 A Spinorial Yamabe Type Equation

For a given Riemannian surface (M, g) with a fixed spin structure σ, in order to obtain a constant length spinor field satisfying

$$D_g\varphi = H\varphi, \quad |\varphi|_g \equiv 1 \tag{7.2}$$

it seems very interesting to consider a conformal change of the Eq. (7.2). That is, we may begin with the following nonlinear partial differential equation:

$$D_g\psi = H|\psi|_g^2\psi \quad \text{on } M. \tag{7.3}$$

Suppose we have a spinor field ψ that satisfies the above equation and that never vanishes. Then one may introduce a conformal metric $\tilde{g} = |\psi|_g^4 g$ on M. Through the well-known formula for the Dirac operator under a conformal change of the metrics (see for instance [23, Proposition 1.3.10]), we obtain a new spinor field $\varphi := \iota(|\psi|_g^{-1}\psi)$ which satisfies

$$D_{\tilde{g}}\varphi = H\varphi, \quad |\varphi|_{\tilde{g}} \equiv 1,$$

where $\iota : (\mathbb{S}(M), g) \to (\mathbb{S}(M), \tilde{g})$ is a fiberwise isometry. And thus we find an isometric immersion $(M, \tilde{g}) \hookrightarrow (\mathbb{R}^3, g_{\mathbb{R}^3})$ that realizes H as prescribed mean curvature.

Equation (7.3) is usually referred as the *spinorial Yamabe type equation*, since it can be seen as a spinorial counterpart of the prescribed scalar curvature equation posed by Yamabe in which the Laplace-Beltrami operator is involved. It follows from the variational framework in Chap. 6 that the structure of (7.3) is variational, and the presence of the exponent in the nonlinearity makes the study of (7.3) a non-compact variational problem. Hence, the analytic study of (7.3) is rather difficult.

In recent years, the solvability of Eq. (7.3) on spin manifolds has been investigated by several mathematicians. For the case $H \equiv constant > 0$, B. Ammann together with his collaborators published a serious of works (see [3–6]) concerning the existence of solutions to the following constraint problem on general dimensions

$$D\psi = \lambda_{min}^+(M, g, \sigma)|\psi|^{\frac{2}{m-1}}\psi, \quad \|\psi\|_{L^{2^*}(M)} = 1, \tag{7.4}$$

where $\lambda_{min}^+(M, g, \sigma) > 0$ is the Bär-Hijazi-Lott invariant of the spin manifold (M, g) (see [23]) and $2^* = \frac{2m}{m-1}$, $m \geq 2$. By assuming

$$\lambda_{min}^+(M, g, \sigma) < \lambda_{min}^+(S^m, g_{S^m}, \sigma_{S^m}) = \frac{m}{2}\omega_m^{1/m}, \tag{7.5}$$

where (S^m, g_{S^m}) denotes the standard sphere equipped with the canonical metric and ω_m stands for the volume of S^m, the results in [3, 5] assert that (7.4) has a solution and, particularly for dimension two, this leads to constant mean curvature immersion of surfaces into $\mathbb{R}^3$. Although the inequality (7.5) is known for may spin manifolds (see for example [6]), whether this assumption holds generally for a compact Riemannian spin surface of genus ≥ 1 is still a open problem. For the case $H \neq constant$, one may write $H = \lambda^+_{min}(M, g, \sigma) \cdot Q$ for some smooth function $Q : M \to \mathbb{R}$. Then, in [32], Raulot obtained a existence result under a similar assumption as (7.5), that is,

$$\lambda^+_{min}(M, g, \sigma) \cdot (\max Q)^{\frac{m-1}{m}} < \lambda^+_{min}(S^m, g_{S^m}, \sigma_{S^m}). \tag{7.6}$$

And Theorem 12 of [32] affirms the existence of a solution to the equation

$$D\psi = \lambda^+_{min}(M, g, \sigma)Q(\xi)|\psi|^{\frac{2}{m-1}}\psi, \qquad \int_M Q(\xi)|\psi|^{2^*} d\,\mathrm{vol}_g = 1.$$

However, Theorem 12 of [32] does not apply for any Riemannian surface and for S^m, $m \geq 2$, due to the break down of the assumption (7.6) in dimension two and for the spheres.

We are going to obtain here the existence result starting from the sphere S^2, by posing sufficient conditions on the mean curvature function H. The results we want to discuss here are the following twofold: counting zeros of a solution of (7.3) and existence of a solution of (7.3).

Theorem 7.3 (Xu [39]) *On a closed Riemannian surface (M, g) of genus γ, with a fixed spin structure, suppose that $H : M^2 \to (0, \infty)$ is a smooth function. Let ψ be a solution of the equation*

$$D_g\psi = H|\psi|^2_g\psi \quad on\ M^2. \tag{7.7}$$

Then the number of zeros of ψ is at most

$$\gamma - 1 + \frac{\int_{M^2} H^2|\psi|^4_g d\,\mathrm{vol}_g}{4\pi}.$$

Remark 7.4 We note that, by the regularity arguments in [25] (see also [37]), a weak solution to Eq. (7.7) is smooth provided that H is smooth. Then it follows from Bär's theorem (cf. [11]) that the zero set of ψ should be discrete. Theorem 7.3 may be used to derive upper estimates for the number of nodal points. In order to explain this we write $H_{max} = \max_{M^2} H$ and we suppose that ψ satisfies the energy estimate

$$\int_{M^2} H|\psi|^4_g d\,\mathrm{vol}_g < \frac{8\pi}{H_{max}} \tag{7.8}$$

and M^2 has genus 0 (i.e. M^2 is topologically a sphere), then we have

$$\#\psi^{-1}(0) \le -1 + \frac{\int_{M^2} H^2|\psi|_g^4 d\,\mathrm{vol}_g}{4\pi} \le -1 + \frac{H_{max}\int_{M^2} H|\psi|_g^4 d\,\mathrm{vol}_g}{4\pi} < 1.$$

And thus ψ will have no zero at all.

After having studied the zero set, we discuss the existence of a (weak) solution to (7.7) on the 2-sphere in the next theorem. In order to state the result, we introduce the following notation: For a function $H \in C^2(S^2)$, we set $H_{max} := \max_{\xi\in S^2} H(\xi)$ and $H_{min} := \min_{\xi\in S^2} H(\xi)$. We then denote the set of maximum points as $\mathcal{H} := \{\xi \in S^2 : H(\xi) = H_{max}\}$, and $\mathcal{H}_\delta := \{\xi \in S^2 : \mathrm{dist}_{g_{S^2}}(\xi, \mathcal{H}) < \delta\}$ its δ-neighborhood for $\delta > 0$. Our criteria of the function H will be formulated as

(H) *$\mathcal{H}$ is not contractible in its δ-neighborhood $\mathcal{H}_\delta$, for some small $\delta > 0$, but there exists $d \in \left(\max\left\{\frac{1}{2}H_{max}, H_{min}\right\}, H_{max}\right)$ such that $\mathcal{H}$ is contractible in $\{\xi \in S^2 : H(\xi) \ge d\}$. There is no critical value of H in the interval (d, H_{max}), and if $\xi \in S^2$ is a critical point of H with $H(\xi) = d$ then the Hessian of H at ξ is positive definite.*

And then, combining Theorem 7.3, the result is as follows

Theorem 7.5 (Xu [39]) *If $H : S^2 \to (0, \infty)$ is a C^2 function satisfying condition (H), then there is a solution ψ to Eq.* (2.12) *with the energy estimate*

$$\frac{4\pi}{H_{max}} < \int_{S^2} H|\psi|^4_{g_{S^2}} d\,\mathrm{vol}_{g_{S^2}} < \frac{8\pi}{H_{max}}.$$

Particularly, if H is smooth, then the solution ψ has no zero at all, i.e. the nodal set $\psi^{-1}(0)$ is empty.

Remark 7.6 By virtue of the spinorial Weierstraß representation, Theorem 7.5 implies the existence of an isometric immersion $\Pi : (S^2, |\psi|^4_{g_{S^2}} g_{S^2}) \to (\mathbb{R}^3, g_{\mathbb{R}^3})$. In particular, the pull-back of the Euclidean volume form under this immersion will be $\Pi^*(d\mu) = |\psi|^4_{g_{S^2}} d\,\mathrm{vol}_{g_{S^2}}$. Notice that the Willmore functional for this immersion Π, denoted by $\mathcal{W}(\Pi)$, appears as the integral squared norm of the mean curvature H. Hence, we have the estimate

$$\mathcal{W}(\Pi) = \int_{S^2} H^2|\psi|^4_{g_{S^2}} d\,\mathrm{vol}_{g_{S^2}} \le H_{max}\int_{S^2} H|\psi|^4_{g_{S^2}} d\,\mathrm{vol}_{g_{S^2}} < 8\pi.$$

Due to Li-Yau's inequality [30, Theorem 6], the immersion Π covers points in $\mathbb{R}^3$ at most once. And thus Π is in fact an isometric embedding.

We note that the class of functions H that satisfy the hypothesis (H) is dense, in C^1-topology, in the space of positive smooth functions (see Fig. 7.1 for a local visual representation of this type of functions). Then Theorem 7.5 and Remark 7.6 together

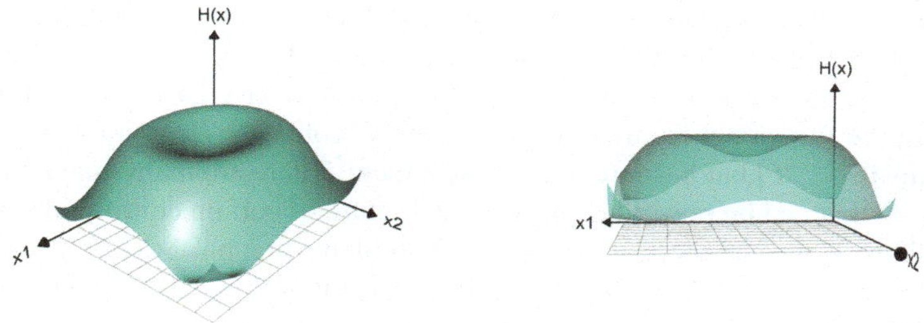

Fig. 7.1 A possible shape of function H's, in local coordinates, which satisfies the criteria (H)

immediately imply the next corollary, which provides an answer to Question 1-3 mentioned in Chap. 2: Sect. 2.4. For ease of notation, we will denote $[g_{S^2}] = \{f^2 g_{S^2} : f \in C^\infty(S^2), f > 0\}$ the standard conformal class of g_{S^2}.

Corollary 7.7 *Let C_+^∞ denote the class of positive smooth functions on S^2 and let*

$$\mathscr{H} = \{H \in C_+^\infty : H \equiv constant \text{ or } H \text{ satisfies condition } (H)\}.$$

Then $\mathscr{H}$ has the following properties:

(1) *$\mathscr{H}$ is dense in C_+^∞ with respect to the C^1-topology;*
(2) *for any $H \in \mathscr{H}$, there exists a conformal metric $g_H \in [g_{S^2}]$ and an isometric immersion $\Pi_H : (S^2, g_H) \to (\mathbb{R}^3, g_{\mathbb{R}^3})$ realizing H as its mean curvature;*
(3) *the Willmore energy of Π_H satisfies*

$$\mathcal{W}(\Pi_H) = \int_{S^2} H^2 d\,\mathrm{vol}_{g_H} < 8\pi,$$

and thus Π_H is an isometric embedding.

We point out that, in [26], T. Isobe also considered a similar problem. He proved that when the function H is very close to a constant, that is $H(\xi) = 1 + \varepsilon Q(\xi)$ on S^2 with Q being a specific Morse function and $\varepsilon > 0$ is small, then there exists a branched immersion $\Pi_{\varepsilon,Q} : S^2 \to \mathbb{R}^3$ that realizes its mean curvature as $1 + \varepsilon Q$. Here, by branched immersion we mean that $\Pi_{\varepsilon,Q}$ is an immersion except at a discrete set of points. The method in [26] is also based on the spinorial Weierstrass representation and the nonlinear Dirac equation (2.12).

Another result concerning this problem is due to M. Anderson [9]. He proved that for any positive function $H : S^2 \to (0, \infty)$ in the class $C^{k,\alpha}$, $k \geq 0$ and $\alpha \in (0, 1)$, there exists a branched immersion $\Pi : S^2 \to \mathbb{R}^3$ and an positive affine function ℓ on S^2 such that the mean curvature of Π is prescribed by $H + \ell$. In this setting the function $H > 0$ lies arbitrarily in the class $C^{k,\alpha}$, and thus it can be far away from a constant. However, the affine function ℓ cannot be prescribed.

Note that, in Corollary 7.7, the mean curvature is accurately prescribed by the functions in the class $\mathscr{H}$ and, particularly, the presence of branch points is ruled out. Hence, comparing with [9, 26], our approach is in a different perspective. The class $\mathscr{H}$ has its generality in the sense that only a "volcano"-shaped assumption is required and, furthermore, such a class of functions can give an explicit estimate of the Willmore functional from above. However, Corollary 7.7 can not cover any of the results in [9, 26] but has intersection with them. There are many functions lying outside $\mathscr{H}$ that can be prescribed as mean curvature of certain conformal isometric embeddings. However, there are still several problems that have to be solved.

Last but not least, we want to remind the readers that Eq. (2.12) for an arbitrarily given H does not necessarily admit a non-trivial solution. As was shown in [7], there is an obstruction to the existence of a non-trivial solution to Eq. (2.12). In fact, if there exists a non-trivial solution ψ to Eq. (2.12), then H must satisfy

$$\int_{S^2} (\nabla_X H)|\psi|^4_{g_{S^2}} d\,\mathrm{vol}_{g_{S^2}} = 0 \tag{7.9}$$

for any conformal vector field X on S^2. In particular, $\nabla_X H$ must change sign on S^2. The obstruction (7.9) is quite similar to the Kazdan-Warner obstruction for Gaussian and scalar curvature equations (see [27]).

7.2 Variational Settings

We will consider the following semilinear Dirac equation with a critical exponent in generic dimension $m \geq 2$:

$$D_{g_{S^m}} \psi = H|\psi|^{\frac{2}{m-1}}_{g_{S^m}} \psi \quad \text{on } (S^m, g_{S^m}) \tag{7.10}$$

under the generalized assumption that $H \in C^2(S^m)$ satisfies

($H*$) *$\mathcal{H}$ is not contractible in its δ-neighborhood $\mathcal{H}_\delta$, for some small $\delta > 0$, but there exists $d \in \left(\max\left\{2^{-\frac{1}{m-1}} H_{max}, H_{min}\right\}, H_{max}\right)$ such that $\mathcal{H}$ is contractible in $\left\{\xi \in S^m : H(\xi) \geq d\right\}$. There is no critical value of H in the interval (d, H_{max}), and if $\xi \in S^m$ is a critical point of H with $H(\xi) = d$ then the Hessian of H at ξ is positive definite.*

Here, analogous to the 2-sphere case, $H_{max} := \max_{\xi \in S^m} H(\xi)$ and $H_{min} := \min_{\xi \in S^m} H(\xi)$, $\mathcal{H} := \left\{\xi \in S^m : H(\xi) = H_{max}\right\}$ is the collection of maximum points and $\mathcal{H}_\delta := \left\{\xi \in S^m : \mathrm{dist}_{g_{S^m}}(\xi, \mathcal{H}) < \delta\right\}$ is its δ-neighborhood for $\delta > 0$. Inspired by the idea of Yamabe [40] and Aubin [10] when they deal with the Yamabe problem, here we mainly focus on a specific regularization of Eq. (7.10), namely, a family of related equations with subcritical exponent

$$D_{g_{S^m}} \psi = H|\psi|^{p-2}_{g_{S^m}} \psi \quad \text{on } (S^m, g_{S^m}) \tag{7.11}$$

where $p \in (2, \frac{2m}{m-1})$. If one considers the variational formulation where solutions of Eq. (7.11) are obtained as critical points of the energy functional

$$\mathcal{L}_p(\psi) := \frac{1}{2}\int_{S^m} (D\psi, \psi) d\,\mathrm{vol}_{g_{S^m}} - \frac{1}{p}\int_{S^m} H|\psi|^p d\,\mathrm{vol}_{g_{S^m}},$$

then the main ingredient of this paper lies in the compactness of a family of critical points ψ_p as $p \to \frac{2m}{m-1}$ of $\mathcal{L}_p$. One way to show the compactness is to apply the regularity arguments to get uniform $C^{1,\alpha}$-boundedness for ψ_p, and hence by Arzelà-Ascoli theorem, up to a subsequence, the spinor field ψ_p converges to a solution of Eq. (7.10). This has been shown when the energy $\mathcal{L}_p(\psi_p)$ is below a so-called blow-up threshold (see [3, 32]). However, in our case, one must allow for blow-up phenomenon, and the situation becomes much more complicated. Whether or not one can obtain a uniform $C^{1,\alpha}$-boundedness for these ψ_p is not clear. Here our strategy is to use build a concentration-compactness alternative for the sequence $\{\psi_p\}$. To rule out the blow-up phenomenon, one only needs to clarify that concentration will not occur, and this will be done in Sect. 7.5.

From now on, let us restrict ourselves to the case $M = S^m$ with the standard metric g_{S^m}. For simplicity, we denote $D = D_{g_{S^m}}$ the associated Dirac operator on the spinor bundle $\mathbb{S}(S^m)$. Let $Spec(D)$ be its spectrum. It is well-known that D is essentially self-adjoint in $L^2(S^m, \mathbb{S}(S^m))$ and has compact resolvents (see [22, 23, 29]). Furthermore, we have

$$Spec(D) = \left\{ \pm\left(\frac{m}{2} + j\right) : j = 0, 1, 2, \dots \right\}$$

and, for each j, the eigenvalues $\pm\left(\frac{m}{2} + j\right)$ have the multiplicity

$$2^{[\frac{m}{2}]}\binom{m+j-1}{j}$$

(see [33]). This concludes that the spectrum of the Dirac operator is discrete. For notation convenience, we can write $Spec(D) = \{\lambda_k\}_{k\in\mathbb{Z}\setminus\{0\}}$ with

$$\lambda_k = \frac{k}{|k|}\left(\frac{m}{2} + |k| - 1\right).$$

An application of the classical spectral theory implies that the eigenspaces of D form a complete orthonormal decomposition of $L^2(S^m, \mathbb{S}(S^m))$, that is,

$$L^2(S^m, \mathbb{S}(S^m)) = \overline{\bigoplus_{\lambda\in Spec(D)} \ker(D - \lambda I)}.$$

Let us denote $m(\lambda_k)$ the multiplicity of $\lambda_k \in Spec(D)$. It is clear that $L^2(S^m, \mathbb{S}(S^m))$ possesses a normalized Hilbert basis consisting of eigenspinors of D, that is, $\{\eta_{k,j}\}$ for $k \in \mathbb{Z} \setminus \{0\}$ and $1 \le j \le m(\lambda_k)$ with corresponding eigenvalues λ_k. Hence $D\eta_{k,j} = \lambda_k \eta_{k,j}$. We can then define the operator $|D|^{1/2} : L^2(S^m, \mathbb{S}(S^m)) \to L^2(S^m, \mathbb{S}(S^m))$ by

$$|D|^{1/2}\psi = \sum_{k\in\mathbb{Z}\setminus\{0\}} \sum_{j=1}^{m(\lambda_k)} |\lambda_k|^{1/2} a_{k,j}\eta_{k,j},$$

for $\psi = \sum a_{k,j}\eta_{k,j} \in L^2(S^m, \mathbb{S}(S^m))$. Let us set

$$H^{1/2}(S^m, \mathbb{S}(S^m))$$
$$:= \left\{ \psi = \sum a_{k,j}\eta_{k,j} \in L^2(S^m, \mathbb{S}(S^m)) : \sum_{k\in\mathbb{Z}\setminus\{0\}} \sum_{j=1}^{m(\lambda_k)} |\lambda_k||a_{k,j}|^2 < \infty \right\}.$$

The space $H^{1/2}(S^m, \mathbb{S}(S^m))$ coincides with the Sobolev space $W^{\frac{1}{2},2}(S^m, \mathbb{S}(S^m))$ (see [1, 3]), and hence we have the Sobolev embeddings: $H^{1/2}(S^m, \mathbb{S}(S^m)) \hookrightarrow L^q(S^m, \mathbb{S}(S^m))$ for $2 \le q \le \frac{2m}{m-1}$. In the sequel, we endow $E := H^{1/2}(S^m, \mathbb{S}(S^m))$ with the inner product

$$\langle \psi, \varphi \rangle = \mathrm{Re}\left(|D|^{1/2}\psi, |D|^{1/2}\varphi \right)_2$$

and the induced norm $\|\cdot\| = \langle \cdot, \cdot \rangle^{1/2}$, where $(\psi, \varphi)_2 = \int_{S^m} (\psi, \varphi) d\,\mathrm{vol}_{g_{S^m}}$ is the L^2-inner product on spinors. Let us mention here that $\|\cdot\|$ is equivalent to the graph norm for the operator $|D|^{1/2}$, and hence $(E, \|\cdot\|)$ is complete and $C^\infty(S^m, \mathbb{S}(S^m))$ is dense in E. Furthermore, E induces a splitting $E = E^+ \oplus E^-$ with

$$E^+ := \overline{\mathrm{span}\{\eta_{k,j}\}_{k>0}} \quad \text{and} \quad E^- := \overline{\mathrm{span}\{\eta_{k,j}\}_{k<0}}, \tag{7.12}$$

where the closure is taken in the $\|\cdot\|$-topology. It is then clear that these are orthogonal subspaces of E on which the action $\int_{S^m} (D\psi, \psi) d\,\mathrm{vol}_{g_{S^m}}$ is positive or negative. In the sequel, with respect to this decomposition, we will write $\psi = \psi^+ + \psi^-$ for any $\psi \in E$. The dual space of E will be denoted by $E^* := H^{-\frac{1}{2}}(S^m, \mathbb{S}(S^m))$.

As was mentioned before, we consider nonlinear Dirac equations of the following form:

$$D\psi = H|\psi|^{p-2}\psi \quad \text{on } S^m, \tag{7.13}$$

where $H : S^m \to (0, +\infty)$ is a C^2 function, $p \in (2, 2^*]$ and $2^* = \frac{2m}{m-1}$.

Equation (7.13) has a variational structure: ψ is a solution to (7.13) if and only if ψ is a critical point of $\mathcal{L}_p$ defined by

$$\begin{aligned}\mathcal{L}_p(\psi) &= \frac{1}{2}\int_{S^m}(D\psi,\psi)d\,\mathrm{vol}_{g_{S^m}} - \frac{1}{p}\int_{S^m} H|\psi|^p d\,\mathrm{vol}_{g_{S^m}}\\ &= \frac{1}{2}\left(\|\psi^+\|^2 - \|\psi^-\|^2\right) - \frac{1}{p}\int_{S^m} H|\psi|^p d\,\mathrm{vol}_{g_{S^m}}\end{aligned}$$

for $\psi = \psi^+ + \psi^- \in E = E^+ \oplus E^-$.

In order to simplify the notation, we will occasionally use the notation $\mathcal{L}$ instead of $\mathcal{L}_{2^*}$ in the sequel. And for first and second derivatives, we write $\mathcal{L}'_p(\psi)[\varphi]$ for the derivative of $\mathcal{L}_p$ at ψ applied to φ and, similarly, we write $\mathcal{L}''_p(\psi)[\varphi,\phi]$ for the second derivative of $\mathcal{L}_p$ at ψ applied to φ and ϕ. By L^p we denote the Banach space $L^p(S^m, \mathbb{S}(S^m))$ for $p \geq 1$ and by $|\cdot|_p$ we denote the usual L^p-norm.

Since both the negative and positive parts of $Spec(D)$ are unbounded, the quadratic form in $\mathcal{L}_p$ is of strongly indefinite type. We present here again the reduction procedure of strongly indefinite functionals:

Proposition 7.8

(1) *There exists a C^1 map $h_p : E^+ \to E^-$ such that for $\psi \in E$:*

$$\mathcal{L}'_p(\psi)[v] = 0 \quad \forall v \in E^- \quad \Longrightarrow \quad \psi^- = h_p(\psi^+). \tag{7.14}$$

Moreover, $h_p(u)$ maximizes $\mathcal{L}_p(u + v)$ over all $v \in E^-$ for $u \in E^+$.

(2) *The functional $I_p : E^+ \to \mathbb{R}$, $I_p(u) = \mathcal{L}_p(u + h_p(u))$, satisfies*

$$I'_p(u) = 0 \quad \Longrightarrow \quad \mathcal{L}'_p(u + h_p(u)) = 0.$$

(3) *For every $u \in E^+ \setminus \{0\}$, the map $I_{p,u} : [0,\infty) \to \mathbb{R}$, $I_{p,u}(t) = I_p(tu)$, is of class C^2 and satisfies*

$$I'_{p,u}(t) = 0,\ t > 0 \quad \Longrightarrow \quad I''_{p,u}(t) < 0. \tag{7.15}$$

Moreover $I_{p,u}(0) = I'_{p,u}(0) = 0$ and $I''_{p,u}(0) > 0$.

Remark 7.9 Here some remarks are in order. In Chap. 6 we have set up a general variational framework to study the equation $D\psi = \lambda\psi + |\psi|^{2^*-2}\psi$ and its variations on an arbitrary spin manifold M, where $\lambda \in \mathbb{R}$ is a real parameter. Proposition 7.8 can be viewed as a direct consequence of Proposition 3.13 or 3.29, applied to the functional $\mathcal{L}_p$. The appearance of the positive function H and the subcritical exponent p in the functional $\mathcal{L}_p$ are rather harmless since they do not affect the strict convexity of the super-quadratic part. Hence there is no obstacle to obtain Proposition 7.8.

In the sequel, we shall call (I_p, h_p) the reduction couple for $\mathcal{L}_p$ on E^+. From Proposition 7.8, it follows that critical points of I_p and $\mathcal{L}_p$ are in one-to-one correspondence via the injective map $u \mapsto u + h_p(u)$ from E^+ to E. The Nehari-Pankov manifold for $\mathcal{L}_p$ is then defined as

$$\mathcal{N}_p = \{u \in E^+ \setminus \{0\} : I'_p(u)[u] = 0\} \tag{7.16}$$

so that the reduced functional I_p is bounded from below on $\mathcal{N}_p$. By Proposition 7.8 (3), we have that $\mathcal{N}_p$ is a C^1 manifold of codimension 1 in E^+ and that $\mathcal{N}_p$ is a natural constraint for the problem of finding non-trivial critical points of I_p on E^+. Indeed, one may consider the auxiliary functional $\mathcal{K}_p(u) = I'_p(u)[u]$ and can compute that $\mathcal{K}'_p(u)[u] = I'_p(u)[u] + I''_p(u)[u, u]$. Then, for $u \in \mathcal{N}_p$, one derives $\mathcal{K}_p(u) = I'_{p,u}(1) = 0$ and $\mathcal{K}'_p(u)[u] = I''_{p,u}(1) < 0$, which suggests $\mathcal{N}_p$ to be a C^1 manifold. Moreover, according to the Lagrange multiplier rule, if $u \in \mathcal{N}_p$ is a constrained critical point of I_p, then there exists $\lambda \in \mathbb{R}$ such that

$$I'_p(u)[v] = \lambda \mathcal{K}'_p(u)[v] = \lambda\big(I'_p(u)[v] + I''_p(u)[u, v]\big) \qquad \forall v \in E^+.$$

Letting $v = u$, so that $I'_{p,u}(1) = I'_p(u)[u] = 0$, we can conclude from the above relation and the fact $I''_{p,u}(1) = I''_p(u)[u, u] < 0$ that we must have $\lambda = 0$, and hence u is indeed a critical point of I_p in E^+. We also mention that, for $u \in E^+ \setminus \{0\}$, the function $I_{p,u}(t)$ attains its unique positive critical point at $t = t(p, u) > 0$ which realizes its maximum (correspondingly, $t(p, u)u \in \mathcal{N}_p$ is called the projection of u on $\mathcal{N}_p$), and hence we have $\mathcal{N}_p$ is homeomorphic to the unit sphere of E^+ and

$$I_p(u) = \max_{t>0} I_p(tu) = \max_{\psi \in \operatorname{span}\{u\} \oplus E^-} \mathcal{L}_p(\psi) \quad \forall u \in \mathcal{N}_p.$$

In the sequel, it is convenient to consider the reduced functional I_p restricted to the constraint $\mathcal{N}_p$. However, an additional difficulty may arise in the variational arguments: it is nearly impossible to calculate the value of $I_p(u)$ at any point $u \in \mathcal{N}_p$. Even if we choose an explicit spinor $\psi \in E \setminus \{0\}$, we cannot directly calculate its projection ψ^+ in E^+ and the corresponding projection $t(p, \psi^+)\psi^+$ in $\mathcal{N}_p$, moreover, we know almost nothing about the images of $h_p(t\psi^+)$ for $t > 0$. We state, in the next lemma, a result about an upper bound estimate for the reduced functional I_p on $\mathcal{N}_p$ in the directions of "almost critical points" of I_p that will play a significant role our argument.

Lemma 7.10 *Let $p_n \nearrow 2^*$ be a converging sequence and suppose that there exists a sequence of spinor fields $\{\psi_n\} \subset E$ such that*

$$c_1 \le \mathcal{L}_{p_n}(\psi_n) \le c_2 \quad \textit{and} \quad \mathcal{L}'_{p_n}(\psi_n) \to 0 \tag{7.17}$$

for some constants $c_1, c_2 > 0$, *as* $n \to \infty$. *Then* $\{\psi_n\}$ *is bounded in* E *such that*

$$\|\psi_n^- - h_{p_n}(\psi_n^+)\| = O\big(\|\mathcal{L}'_{p_n}(\psi_n)\|_{E^*}\big)$$

and

$$\max_{t>0} I_{p_n}(t\psi_n^+) \le \mathcal{L}_{p_n}(\psi_n) + O\big(\|\mathcal{L}'_{p_n}(\psi_n)\|^2_{E^*}\big).$$

Lemma 7.10 is obtained by applying Theorem 3.21. The idea in the proof is built upon the strict concavity of $I_{p,u}(t)$ around its maximum point (i.e. (7.15)) and an application of the Inverse Function Theorem to the auxiliary functional $\mathcal{K}_{p_n}(u) = I'_{p_n}(u)[u]$ near the point ψ_n^+. As an immediate consequence of Lemma 7.10, we have

Corollary 7.11 *Let* $p_n \nearrow 2^*$ *as* $n \to \infty$ *and* $c_1, c_2 > 0$. *For any* $\theta > 0$, *there exists* $\alpha > 0$ *such that for all large* n *and* $\psi \in E$ *satisfying*

$$c_1 \le \mathcal{L}_{p_n}(\psi) \le c_2 \quad \textit{and} \quad \|\mathcal{L}'_{p_n}(\psi)\|_{E^*} \le \alpha$$

there holds

$$\max_{t>0} I_{p_n}(t\psi^+) \le \mathcal{L}_{p_n}(\psi) + \theta.$$

For $p \in (2, 2^*]$, let us set

$$\tau_p := \inf\big\{I_p(u) : u \in \mathcal{N}_p\big\}. \tag{7.18}$$

As a consequence of the following estimate

$$I_p(u) \ge \mathcal{L}_p(u) = \frac{1}{2}\|u\|^2 - \frac{1}{p}\int_{S^m} H|u|^p d\,\mathrm{vol}_{g_{S^m}} \ge \frac{1}{2}\|u\|^2 - C\|u\|^p, \quad \forall u \in E^+$$

where $C > 0$ is a positive constant depending on H_{max} and the Sobolev embedding $E \hookrightarrow L^p$, it turns out immediately that τ_p is lower bounded by a positive constant.

Lemma 7.12 *Let* $p_n \nearrow 2^*$ *as* $n \to \infty$, *then* $\tau_{p_n} \to \tau_{2^*}$ *as* $n \to \infty$.

Proof For arbitrary $\varepsilon > 0$ and $k \in \mathbb{N}$, by the Ekeland variational principle [20, 31, 38], we can first choose $u \in \mathcal{N}_{2^*}$ such that $\tau_{2^*} \le I_{2^*}(u) < \tau_{2^*} + \frac{1}{k}$ and $\|I'_{2^*}(u)\| \le \frac{1}{k}$.

Let $s_n > 0$ be such that $s_n u \in \mathcal{N}_{p_n}$. It follows from the arguments in [17, see page 182] that $\{s_n\}$ must be bounded (otherwise, $I_{p_n}(s_n u) \to -\infty$ as $s_n \to +\infty$). And hence $\tau_{p_n} \le I_{p_n}(s_n u)$ is bounded from above.

Note that $\|I'_{2^*}(u)\| = \|\mathcal{L}'_{2^*}(u + h_{2^*}(u))\|$ (see for instance [17, Lemma 2.2]) and that

$$\begin{aligned}
&\mathcal{L}'_{2^*}(u + h_{2^*}(u))[\varphi] - \mathcal{L}'_{p_n}(u + h_{2^*}(u))[\varphi] \\
&= \mathrm{Re}\int_{S^m} H\big(1 - |u + h_{2^*}(u)|^{2^*-p_n}\big)|u + h_{2^*}(u)|^{p_n-2}(u + h_{2^*}(u), \varphi)d\,\mathrm{vol}_{g_{S^m}} \\
&= o_n(1)\|\varphi\|
\end{aligned}$$

for all $\varphi \in E$, we derive that $\|\mathcal{L}'_{p_n}(u + h_{2^*}(u))\| \le \frac{2}{k}$ for all large n. And thus, by applying Corollary 7.11, we have

$$\tau_{p_n} \le I_{p_n}(s_n u) = \max_{s>0} I_{p_n}(su) \le \mathcal{L}_{p_n}(u + h_{2^*}(u)) + \varepsilon \tag{7.19}$$

provided that k is fixed large enough. Since the function $p \mapsto \int_{S^m} H|u + h_{2^*}(u)|^p d\,\mathrm{vol}_{g_{S^m}}$ is continuous, we have that $\mathcal{L}_{p_n}(u+h_{2^*}(u)) = \mathcal{L}_{2^*}(u+h_{2^*}(u))+ o_n(1) = I_{2^*}(u) + o_n(1)$. Therefore, by (7.19) and the arbitrariness of ε, we have $\limsup_{n\to\infty} \tau_{p_n} \le \tau_{2^*}$.

On the other hand, we can choose $u_n \in \mathscr{N}_{p_n}$ such that $I_{p_n}(u_n) \le \tau_{p_n} + \frac{1}{n}$. Let $t_n > 0$ be such that $t_n u_n \in \mathscr{N}_{2^*}$. Then it follows that $\{t_n\}$ is also bounded. And we can derive that

$$\tau_{2^*} \le I_{2^*}(t_n u_n) = \mathcal{L}_{2^*}(t_n u_n + h_{2^*}(t_n u_n)) \le \mathcal{L}_{p_n}(t_n u_n + h_{2^*}(t_n u_n)) + o_n(1), \tag{7.20}$$

where in the last inequality we have used the fact

$$\begin{aligned}
\int_{S^m} H|\psi|^{p_n} d\,\mathrm{vol}_{g_{S^m}} &\le \Big(\int_{S^m} H|\psi|^{2^*} d\,\mathrm{vol}_{g_{S^m}}\Big)^{\frac{p_n}{2^*}} \Big(\int_{S^m} H d\,\mathrm{vol}_{g_{S^m}}\Big)^{\frac{2^*-p_n}{2^*}} \\
&= \int_{S^m} H|\psi|^{2^*} d\,\mathrm{vol}_{g_{S^m}} + o_n(1)
\end{aligned}$$

when $\|\psi\|$ is bounded independent of n. Hence, we see from (7.20) that $\tau_{2^*} \le I_{p_n}(u_n) + o_n(1) \le \tau_{p_n} + \frac{1}{n} + o_n(1)$ as $n \to \infty$, which completes the proof. □

We conclude this subsection by formulating the exact value of τ_{2^*}, that we will need in the sequel, i.e.

$$\tau_{2^*} = \frac{1}{2m(H_{max})^{m-1}}\Big(\frac{m}{2}\Big)^m \omega_m \tag{7.21}$$

where ω_m stands for the volume of (S^m, g_{S^m}). To see this, we first need to show that $\tau_{2^*} \geq \frac{1}{2m(H_{max})^{m-1}}(\frac{m}{2})^m \omega_m$. It can be seen from the fact $H(\xi) \leq H_{max}$ for $\xi \in S^m$ that

$$\mathcal{L}_{2^*}(\psi) \geq \mathcal{L}_{max}(\psi) := \frac{1}{2}(\|\psi^+\|^2 - \|\psi^-\|^2) - \frac{H_{max}}{2^*}\int_{S^m} |\psi|^{2^*} d\,\mathrm{vol}_{g_{S^m}} \tag{7.22}$$

for all $\psi \in E$. And critical points of $\mathcal{L}_{max}$ correspond to solutions of

$$D\psi = H_{max}|\psi|^{2^*-2}\psi \quad \text{on } S^m. \tag{7.23}$$

Since $H_{max} > 0$, by normalizing Eq. (7.23), we turn to consider

$$D\psi = |\psi|^{2^*-2}\psi \quad \text{on } S^m. \tag{7.24}$$

Then solutions of Eq. (7.23) and (7.24) are in one-to-one correspondence via the map

$$\psi \mapsto (H_{max})^{\frac{m-1}{2}}\psi. \tag{7.25}$$

Recall that $\mathbb{R}^m$ and $S^m \setminus \{N\}$ (where $N \in S^m$ is the North pole) are conformally equivalent, it follows from [25, Proposition 4.1] and (7.25) that if ψ is a nontrivial solution to Eq. (7.23) then

$$\mathcal{L}_{max}(\psi) \geq \frac{1}{2m(H_{max})^{m-1}}(\frac{m}{2})^m \omega_m. \tag{7.26}$$

Since it is well-known that Eq. (7.24) has a specific solution, which is a Killing spinor to the constant $-1/2$ (see e.g. [8, Section 5.3]), we can see from (7.25) that the equality in (7.26) is achieved. Furthermore, notice that Proposition 7.8 can be applied to the functional $\mathcal{L}_{max}$. Let us denote I_{max} the reduced functional for $\mathcal{L}_{max}$ on E^+ and $\mathscr{N}_{max}$ the associated Nehari-Pankov manifold. Then, by (7.22), we have $I_{2^*}(u) \geq I_{max}(u)$ for all $u \in E^+$ and, particularly,

$$\max_{t>0} I_{2^*}(tu) \geq I_{max}(u) \quad \text{for all } u \in \mathscr{N}_{max}.$$

Jointly with the definition of τ_{2^*} and (7.26), we obtain

$$\tau_{2^*} \geq \frac{1}{2m(H_{max})^{m-1}}(\frac{m}{2})^m \omega_m.$$

The other inequality requires a delicate upper bound estimate for τ_{2^*}. In fact, by a careful choice of the parameters, Lemma 7.19 in Sect. 7.5 provides a proof, and one may also refer to a similar result in [8].

7.3 A Concentration-Compactness Alternative

Again, we choose $\{p_n\}$ to be a strictly increasing sequence such that $\lim_{n\to\infty} p_n = 2^*$. In what follows, we shall collect some asymptotic properties of solutions to the equation

$$D\psi = H|\psi|^{p_n-2}\psi \quad \text{on } S^m \tag{7.27}$$

with a specific energy constraints, the proofs will be postponed in the Appendix. First of all, an alternative result comes as follows:

Proposition 7.13 *Suppose* $\{\psi_n\} \subset E$ *is a sequence such that*

$$\tau_{2^*} \le \mathcal{L}_{p_n}(\psi_n) \le 2\tau_{2^*} - \theta \quad \textit{and} \quad \mathcal{L}'_{p_n}(\psi_n) \to 0, \quad \textit{as } n \to \infty \tag{7.28}$$

for some constant $\theta > 0$*. Then, up to a subsequence, either* $\psi_n \rightharpoonup 0$ *in* E *or* ψ_n *converges in* E *to a non-trivial solution* ψ_0 *of Eq.* (7.10)

In order to characterize the non-compact case (which corresponds to the weakly convergence $\psi_n \rightharpoonup 0$ in E) in the above alternatives, let us introduce the following notations. Let $Crit[H] = \{\xi \in S^m : \nabla H(\xi) = 0\}$ denote the critical set of the function H. Let $\xi \in V \subset S^m$ be an arbitrary point, we denote $\exp_\xi : U \subset T_\xi S^m \cong \mathbb{R}^m \to V, x = (x_1, \dots, x_n) \mapsto y = \exp_\xi x$ the exponential map and $(x_1, \dots, x_m)$ means the normal coordinates. For a sequence of points $\{a_n\} \subset S^m$ and an arbitrary decreasing sequence $R_n \searrow 0$, as $n \to \infty$, we define the rescaled geodesic normal coordinates near each a_n via $\mu_n(x) = \exp_{a_n}(R_n x)$. Denoting $B_R^0 = \{x \in \mathbb{R}^m : |x| < R\}$, where $|\cdot|$ is the Euclidean norm in $\mathbb{R}^m$, we have a conformal equivalence $(B_R^0, R_n^{-2}\mu_n^* g_{S^m}) \cong (B_{R_n R}(a_n), g_{S^m}) \subset S^m$ for all large n. For ease of notation, we set $g_n = R_n^{-2}\mu_n^* g_{S^m}$. By using the trivialization of the spinor bundle constructed by Bourguignon and Gauduchon [16], we see that the coordinate map μ_n induces a spinor bundle identification $\overline{(\mu_n)_*} : \mathbb{S}_x(B_R^0, g_n) \to \mathbb{S}_{\mu_n(x)}(B_{R_n R}(a_n), g_{S^m})$ for each n. Moreover, due to the compactness of S^m, we may assume without loss of generality that $a_n \to a \in S^m$ as $n \to \infty$. Then, for all n large, the aforementioned identifications depend smoothly in a_n. Particularly, we have the following blow-up result (the proof is postponed in the Appendix).

Proposition 7.14 *Let* $\{\psi_n\} \subset E$ *satisfy* (7.28)*. If* $\{\psi_n\}$ *does not contain any compact subsequence. Then, up to a subsequence if necessary, there exist a sequence* $\{a_n\} \subset S^m$, $a_n \to a \in S^m$, *as* $n \to \infty$*, a sequence of radius* $R_n \searrow 0$*, a real number* $\lambda \in (2^{-\frac{1}{m-1}}, 1]$ *and a non-trivial solution* ϕ_0 *of the equation*

$$D_{g_{\mathbb{R}^m}}\phi = \lambda H(a)|\phi|^{2^*-2}\phi \quad \textit{on } \mathbb{R}^m$$

such that

$$\lim_{n\to\infty} R_n^{\frac{m-1}{2}(2^*-p_n)} = \lambda$$

and

$$\psi_n = R_n^{-\frac{m-1}{2}} \eta(\cdot)\overline{(\mu_n)_*} \circ \phi_0 \circ \mu_n^{-1} + o_n(1) \quad \text{in } E$$

as $n \to \infty$*, where* $\eta \in C^\infty(S^m)$ *is a cut-off function such that* $\eta \equiv 1$ *on* $B_r(a)$ *and* $\operatorname{supp} \eta \subset B_{2r}(a)$*, some* $r > 0$*. Moreover, we have*

$$\mathcal{L}_{p_n}(\psi_n) \geq \frac{1}{2m(\lambda H(a))^{m-1}} \Big(\frac{m}{2}\Big)^m \omega_m + o_n(1), \quad \text{as } n \to \infty$$

where ω_m *stands for the volume of* (S^m, g_{S^m})*. If* $\mathcal{L}'_{p_n}(\psi_n) \equiv 0$ *for all* n *in (7.28), then* $a \in Crit[H]$ *and* $\lambda = 1$.

Remark 7.15

(1) Propositions 7.13 and 7.14 are closely linked. More precisely, together with (7.21), we can conclude that, if $\{\psi_n\}$ is a sequence of critical points satisfying the energy estimates in (7.28) and does not contain any compact subsequence, then (up to a subsequence) the associated blow-up point $a \in Crit[H]$ satisfies $H(a) \geq 2^{-\frac{1}{m-1}} H_{max} + \theta'$, for some $\theta' > 0$.

(2) It is worth pointing out that, for $H \not\equiv constant$, the ground state energy solutions of (7.27) (that is, the solutions with energy τ_{p_n}) should blow up when $n \to \infty$. To see this, let us consider generically $\psi_n \in E$ such that $\mathcal{L}_{p_n}(\psi_n) \to \tau_{2^*}$ and $\mathcal{L}'_{p_n}(\psi_n) \to 0$ as $n \to \infty$. By Proposition 7.13, it follows that either (a) $\{\psi_n\}$ converges strongly to $\psi_0 \neq 0$ which satisfies Eq. (7.10) or (b) the sequence $\{\psi_n\}$ blows-up.

If (a) holds, then $\mathcal{L}_{2^*}(\psi_0) = \tau_{2^*}$. To rule out this case, let us first recall the estimate (7.22), where the functional $\mathcal{L}_{max}$ is introduced. As before, we denote (I_{max}, h_{max}) the reduction couple for $\mathcal{L}_{max}$ on E^+ and $\mathscr{N}_{max}$ the associated Nehari-Pankov manifold. Then (7.22) implies that $I_{2^*}(u) \geq I_{max}(u)$ for all $u \in E^+$. Since $\mathcal{L}_{2^*}(\psi_0) = \tau_{2^*}$ and $\mathcal{L}'_{2^*}(\psi_0) = 0$, we have $\psi_0^+ \in \mathscr{N}_{2^*}$, $I_{2^*}(\psi_0^+) = \tau_{2^*}$ and $I'_{2^*}(\psi_0^+) = 0$. Let $t_0 > 0$ be such that $t_0\psi_0^+ \in \mathscr{N}_{max}$. We claim that $t_0 \neq 1$. If, by contradiction, we have $t_0 = 1$. Then $\psi_0^+ \in \mathscr{N}_{2^*} \cap \mathscr{N}_{max}$. By (7.26), we find $I_{max}(\psi_0^+) = \tau_{2^*}$ and $I'_{max}(\psi_0^+) = 0$ (this is because τ_{2^*} equals to the lowest energy level for I_{max} on the constraint $\mathscr{N}_{max}$). Hence, we have

$$\tau_{2^*} = \frac{1}{2}\big(\|\psi_0^+\|^2 - \|h_{max}(\psi_0^+)\|^2\big) - \frac{H_{max}}{2^*}\int_{S^m} |\psi_0^+ + h_{max}(\psi_0^+)|^{2^*} d\,\mathrm{vol}_{g_{S^m}} \tag{7.29}$$

and $\widetilde{\psi}_0 := \psi_0^+ + h_{max}(\psi_0^+)$ is a nontrivial solution of Eq. (7.23). Then, by the regularity result for weak solutions to Eq. (7.23) (see e.g. [18, 25, 37]) and the *Weak Unique Continuation Property* for Dirac type operators (see [15, Theorem 2.1]), we find that the zero set of $\widetilde{\psi}_0$ does not contain any nonempty open set. And therefore it follows directly from (7.29) that

$$\begin{aligned}\tau_{2^*} &= \frac{1}{2}\big(\|\psi_0^+\|^2 - \|h_{max}(\psi_0^+)\|^2\big) - \frac{H_{max}}{2^*}\int_{S^m} |\psi_0^+ + h_{max}(\psi_0^+)|^{2^*} d\,\mathrm{vol}_{g_{S^m}} \\ &< \frac{1}{2}\big(\|\psi_0^+\|^2 - \|h_{max}(\psi_0^+)\|^2\big) - \frac{1}{2^*}\int_{S^m} H|\psi_0^+ + h_{max}(\psi_0^+)|^{2^*} d\,\mathrm{vol}_{g_{S^m}} \\ &= \mathcal{L}_{2^*}(\widetilde{\psi}_0) \le I_{2^*}(\psi_0^+) = \tau_{2^*}\end{aligned}$$

which is a contradiction, proving the claim. Let's say $t_0 \neq 1$. Then we can infer

$$\begin{aligned}\tau_{2^*} &= I_{2^*}(\psi_0^+) > I_{2^*}(t_0\psi_0^+) \ge I_{max}(t_0\psi_0^+) \\ &\ge \inf_{u\in\mathcal{N}_{max}} I_{\max}(u) = \frac{1}{2m(H_{max})^{m-1}}\Big(\frac{m}{2}\Big)^m\omega_m = \tau_{2^*},\end{aligned}$$

which is again a contradiction. Thus we see that only (b) can occur.

7.4 Analysis on the Nodal Set

In what follows, we shall study the nodal set of solutions to the following nonlinear Dirac equation with critical exponent

$$D\psi = H|\psi|^{2^*-2}\psi \quad \text{on } S^m \tag{7.30}$$

And let us begin with the following lemma which can be viewed as a generalization of [5, Lemma 4.1] for the case $H \not\equiv constant$.

Lemma 7.16 *Let (M, g) be an arbitrary Riemannian spin m-manifold (not necessary complete or compact). Assume that there is a spinor ψ of constant length 1 and with $D_g\psi = H\psi$ for a real-valued smooth function $H : M \to \mathbb{R}$, and let $\{e_1, \dots, e_m\}$ be a local orthonormal frame field. Then*

$$\mathrm{Scal}_g = \frac{4(m-1)}{m}H^2 - 4\sum_{k=1}^{m}|\nabla_{e_k}^H\psi|^2$$

where Scal_g is the scalar curvature of M with respect to g, $\nabla_X^H\psi := \nabla_X\psi + \frac{H}{m}X\cdot_g\psi$ for $X \in \Gamma(TM)$ is an induced covariant derivative and $\cdot_g$ denotes the Clifford multiplication.

Proof The algebraic properties of Clifford multiplication imply that ∇^H is metric in the sense that

$$X(\varphi_1, \varphi_2) = (\nabla^H_X \varphi_1, \varphi_2) + (\varphi_1, \nabla^H_X \varphi_2)$$

for all $\varphi_1, \varphi_2 \in \Gamma(\mathbb{S}(M))$ and $X \in \Gamma(TM)$.

Locally, let $-\Delta^H = -\sum_{k=1}^m \nabla^H_{e_k}\nabla^H_{e_k}$ denote the Laplace operator corresponds to ∇^H. Then we have

$$\begin{aligned} -\operatorname{Re}(\Delta^H\psi, \psi) &= -\operatorname{Re}\sum_{k=1}^m (\nabla^H_{e_k}\nabla^H_{e_k}\psi, \psi) \\ &= -\operatorname{Re}\sum_{k=1}^m \left[e_k(\nabla^H_{e_k}\psi, \psi) - (\nabla^H_{e_k}\psi, \nabla^H_{e_k}\psi)\right] \\ &= -\operatorname{Re}\sum_{k=1}^m e_k(\nabla^H_{e_k}\psi, \psi) + \sum_{k=1}^m |\nabla^H_{e_k}\psi|^2. \end{aligned} \tag{7.31}$$

Since $(\nabla^H_{e_k}\psi, \psi) = (\nabla_{e_k}\psi, \psi) + \frac{H}{m}\cdot(e_k \cdot_g \psi, \psi)$, we have

$$\operatorname{Re}(\nabla^H_{e_k}\psi, \psi) = \operatorname{Re}(\nabla_{e_k}\psi, \psi) = \frac{1}{2}\operatorname{Re}\left\langle \operatorname{grad}|\psi|^2, e_k\right\rangle = 0.$$

Thus (7.31) gives us

$$-\operatorname{Re}(\Delta^H\psi, \psi) = \sum_{k=1}^m |\nabla^H_{e_k}\psi|^2. \tag{7.32}$$

On the other hand, by the twisted version of the Schrödinger-Lichnerowicz formula (see for instance [22, Chapter 5]), we have

$$\left(D - \frac{H}{m}\right)^2 = -\Delta^H + \frac{\operatorname{Scal}_g}{4} + \frac{1-m}{m^2}H^2.$$

And therefore, by $D\psi = H\psi$, $|\psi| \equiv 1$ and (7.32), we obtain

$$\begin{aligned} \frac{(m-1)^2H^2}{m^2} &= -\operatorname{Re}(\Delta^H\psi, \psi) + \frac{\operatorname{Scal}_g}{4} + \frac{1-m}{m^2}H^2 \\ &= \sum_{k=1}^m |\nabla^H_{e_k}\psi|^2 + \frac{\operatorname{Scal}_g}{4} + \frac{1-m}{m^2}H^2, \end{aligned}$$

which completes the proof. □

The following theorem is due to Bär [11]:

Theorem 7.17 *Let (M, g) be compact and connected spin m-manifold and let φ be a solution of*

$$D\varphi = P\varphi$$

where P is a smooth endomorphism. Then the zero set of φ has at most Hausdorff dimension $m-2$. And if $m = 2$, then the zero set is discrete.

Unfortunately, in general, Eq. (7.30) does not satisfy Bär's theorem because $2^* - 2 = \frac{2}{m-1} \notin \mathbb{Z}$ for $m \geq 4$ and $H|\psi|^{2^*-2}$ could not be smooth even for $m = 3$. However, in dimension 2, we have $2^* = 4$ and in this case solutions of Eq. (7.30) and the corresponding $P = H|\psi|^2$ are smooth enough provided that $H \in C^\infty$. Thus, applying Theorem 7.17, we have the nodal set of a solution ψ is a discrete subset.

Proposition 7.18 *On a compact boundaryless spin surface (M, g) of genus γ, suppose that $H : M \to (0, \infty)$ is a smooth function and $H_{max} := \max_{\xi \in M} H(\xi)$. Let ψ be a solution of the equation*

$$D_g\psi = H|\psi|^2\psi \quad \text{on } M.$$

Then the number of zeros of ψ is at most

$$\gamma - 1 + \frac{\int_M H^2|\psi|^4 d\,\mathrm{vol}_g}{4\pi}.$$

In particular, if M has $\gamma = 0$ (i.e. M is homeomorphic to the 2-sphere) and ψ satisfies

$$\int_M H|\psi|^4 d\,\mathrm{vol}_g < \frac{8\pi}{H_{max}}, \tag{7.33}$$

then ψ has no zero at all.

Proof The proof is similar to that carried out in [5, Proposition 8.1]. Here, for the sake of completeness, we will sketch the proof as follows.

On $M \setminus \psi^{-1}(0)$, let us introduce a new metric $g_1 = |\psi|^4 g$. Then the transformation formula for the Dirac operator under conformal changes yields that there is a spinor field φ_1 on $M \setminus \psi^{-1}(0)$ such that

$$D_{g_1}\varphi_1 = H\varphi_1 \quad \text{and} \quad |\varphi_1|_{g_1} = 1.$$

By Lemma 7.16, we have the scalar curvature can be estimated by $\mathrm{Scal}_{g_1} \le 2H^2$. And hence, we have the estimate for the Gauß curvature, that is $K_{g_1} \le H^2$. Moreover, we have

$$\mathrm{vol}(M \setminus \psi^{-1}(0), g_1) = \int_{M\setminus\psi^{-1}(0)} d\,\mathrm{vol}_{g_1} = \int_M |\psi|^4 d\,\mathrm{vol}_g\,. \tag{7.34}$$

Let $\psi^{-1}(0) = \{\xi_1, \ldots, \xi_l\}$ for some $l \ge 1$ if $\psi^{-1}(0) \ne \emptyset$. Let n_j be the order of the first non-vanishing term in the Taylor expansion of ψ near ξ_j, $j = 1, \ldots, l$. Then it can be calculated that the integral of the geodesic curvature κ_{g_1} over a small circle around each ξ_j is close to $-2\pi(2n_j+1)$. Now, we may remove small open disks around ξ_j from M, and we obtain a surface $\tilde{M}$ with boundary. By using the Gauss-Bonnet formula, we can derive

$$2\pi\chi(\tilde{M}) = \int_{\tilde{M}} K_{g_1} d\,\mathrm{vol}_{g_1} + \int_{\partial\tilde{M}} \kappa_{g_1} ds \le \int_M H^2|\psi|^4 d\,\mathrm{vol}_g - \sum_{j=1}^{l} 2\pi(2n_j+1).$$

And hence, return to M itself, we have

$$2\pi(2-2\gamma) = 2\pi\chi(M) \le \int_M H^2|\psi|^4 d\,\mathrm{vol}_g - 4\pi\sum_{j=1}^{l} n_j,$$

which implies l is at most $\gamma - 1 + \frac{\int_M H^2|\psi|^4 d\,\mathrm{vol}_g}{4\pi}$.

For the case $\gamma = 0$ and ψ satisfies (7.33), we assume $\psi^{-1}(0) = \{\xi_1, \ldots, \xi_l\} \ne \emptyset$. Then we have

$$1 \le l \le -1 + \frac{\int_M H^2|\psi|^4 d\,\mathrm{vol}_g}{4\pi} \le -1 + \frac{H_{max}}{4\pi}\int_M H|\psi|^4 d\,\mathrm{vol}_g < 1$$

which is impossible. Therefore ψ has no zero at all. □

7.5 Conformal Embeddings of S^2 in Euclidean 3-Space

7.5.1 The Existence Results for the PDE Problem

For notation convenience, we will write $\mathcal{L}$ instead of $\mathcal{L}_{2^*}$. The super-level sets for H will be simply denoted by

$$\{H \ge c\} = \left\{\xi \in S^m : H(\xi) \ge c\right\}$$

for any $c \in \mathbb{R}$. Recalling the hypothesis $(H*)$, it follows that H has only finitely many critical points at level d. And we may denote the corresponding critical set as

$$\mathcal{K}_d = \big\{\xi \in S^m : H(\xi) = d, \nabla H(\xi) = 0\big\} = \{\xi_1, \dots, \xi_l\}$$

for some $l \in \mathbb{N}$. We also introduce a ν-neighborhood of $\mathcal{K}_d$ as $\mathcal{O}_\nu := \cup_{k=1}^{l} B_\nu(\xi_k)$ for $\nu > 0$, where $B_r(\xi)$ is the open ball centered at $\xi \in S^m$ with radius $r > 0$ with respect to the metric g_{S^m}. Standard deformation arguments show that hypothesis $(H*)$ ensures that d is indeed a critical value for H, i.e. $\mathcal{K}_d \neq \emptyset$, and $\mathcal{H}$ is not contractible in $\{H \geq d + \sigma\}$ for any $\sigma > 0$. Moreover, for small $\nu > 0$, there exists $\sigma > 0$ such that $\mathcal{H}$ is contractible in $\{H \geq d + \sigma\} \cup \mathcal{O}_\nu$.

Now we construct the test spinors that will be helpful to characterize the level sets of the energy functional I_{p_n} on the constraint manifold $\mathscr{N}_{p_n}$ (see its definition in Sect. 7.2). Let's start with an arbitrary $\phi_0 \in \mathbb{S}_m$ (a constant spinor) such that $|\phi_0| = \frac{1}{\sqrt{2}}\big(\frac{m}{2}\big)^{\frac{m-1}{2}}$. Then, we define

$$\phi(x) = f(x)^{\frac{m}{2}}(1 - x) \cdot_{g_{\mathbb{R}^m}} \phi_0$$

where $f(x) = \frac{2}{1+|x|^2}$ for $x \in \mathbb{R}^m$. It is easy to verify that

$$D_{g_{\mathbb{R}^m}}\phi = \frac{m}{2} f\phi$$

and

$$|\phi| = \Big(\frac{m}{2}\Big)^{\frac{m-1}{2}} f^{\frac{m-1}{2}}. \tag{7.35}$$

For $\varepsilon > 0$, we set

$$\phi_\varepsilon(x) = \varepsilon^{-\frac{m-1}{2}}\phi(x/\varepsilon) \quad \text{and} \quad \phi_{y,\varepsilon}(x) = H(y)^{-\frac{m-1}{2}}\phi_\varepsilon(x)$$

where $y \in S^m$ is a fixed point. Then we have $\phi_{y,\varepsilon}$ satisfies the equation

$$D_{g_{\mathbb{R}^m}}\phi_{y,\varepsilon} = H(y)|\phi_{y,\varepsilon}|^{2^*-2}\phi_{y,\varepsilon} \quad \text{on } \mathbb{R}^m. \tag{7.36}$$

To transplant $\phi_{y,\varepsilon}$ on the sphere S^m, we need the following notation of stereographic projections. Given $y \in S^m$, we can embed S^m into $\mathbb{R}^{m+1}$ in the way that its antipodal point $-y$ is the North pole. Denoting $\mathcal{S}_y : S^m \setminus \{-y\} \to \mathbb{R}^m$ the stereographic projection, we have $\mathcal{S}_y(y) = 0$. Moreover, $S^m \setminus \{-y\}$ and $\mathbb{R}^m$ are conformally equivalent due to the fact $(\mathcal{S}_y^{-1})^* g_{S^m} = f^2 g_{\mathbb{R}^m}$ with $f(x) = \frac{2}{1+|x|^2}$.

Recall the conformal transformation property of the Dirac operator (see [23]), there is an isomorphism of spinor bundles $\iota : \mathbb{S}\big(\mathbb{R}^m, (\mathcal{S}_\xi^{-1})^* g_{S^m}\big) \to \mathbb{S}(\mathbb{R}^m, g_{\mathbb{R}^m})$ such that

$$D_{g_{\mathbb{R}^m}}\big(\iota(\varphi)\big) = \iota\big(f^{\frac{m+1}{2}} D_{(\mathcal{S}_\xi^{-1})^* g_{S^m}}(f^{-\frac{m-1}{2}}\varphi)\big),$$

where $D_{(\mathcal{S}_\xi^{-1})^* g_{S^m}}$ is the Dirac operator on $\mathbb{R}^m$ with respect to the metric $(\mathcal{S}_\xi^{-1})^* g_{S^m}$. Thus $\phi_{y,\varepsilon}$ corresponds to a spinor field $\psi_{y,\varepsilon}$ on S^m via the formula

$$\phi_{y,\varepsilon} = \iota\big(f^{\frac{m-1}{2}} \psi_{y,\varepsilon} \circ \mathcal{S}_y^{-1}\big) \tag{7.37}$$

such that $\psi_{y,\varepsilon}$ satisfies the equation

$$D\psi_{y,\varepsilon} = H(y)|\psi_{y,\varepsilon}|^{2^*-2}\psi_{y,\varepsilon} \quad \text{on } (S^m, g_{S^m})$$

Lemma 7.19 *For any $\sigma > 0$ and $\theta > 0$ there exists $\varepsilon > 0$ such that*

$$\max_{t>0} I_{p_n}(t\psi_{y,\varepsilon}^+) \le \frac{1}{2m\,(d+\sigma)^{m-1}}\Big(\frac{m}{2}\Big)^m \omega_m + 2\theta$$

uniformly for all large n and $y \in \{H \ge d + \sigma\}$.

Proof The strategy is to use Corollary 7.11. And to begin with, we first fix $\alpha > 0$ associated to θ as in Corollary 7.11.

In E^*, we have

$$\mathcal{L}'(\psi_{y,\varepsilon}) = D\psi_{y,\varepsilon} - H|\psi_{y,\varepsilon}|^{2^*-2}\psi_{y,\varepsilon} = \big(H(y) - H\big)|\psi_{y,\varepsilon}|^{2^*-2}\psi_{y,\varepsilon}.$$

Let $\psi \in E$ be such that $\|\psi\| \le 1$, we have

$$\begin{aligned}
\mathcal{L}'(\psi_{y,\varepsilon})[\psi] &= \mathrm{Re}\int_{S^m} \big(H(y) - H\big)|\psi_{y,\varepsilon}|^{2^*-2}(\psi_{y,\varepsilon}, \psi)d\,\mathrm{vol}_{g_{S^m}} \\
&= \mathrm{Re}\int_{\mathbb{R}^m} \big(H(y) - H \circ \mathcal{S}_y^{-1}\big)|\phi_{y,\varepsilon}|^{2^*-2}\big(\phi_{y,\varepsilon}, \iota(f^{\frac{m-1}{2}}\psi \circ \mathcal{S}_y^{-1})\big)d\,\mathrm{vol}_{g_{\mathbb{R}^m}} \\
&\le C\Big(\int_{\mathbb{R}^m} |H(y) - H \circ \mathcal{S}_y^{-1}|^{\frac{2^*}{2^*-1}}|\phi_{y,\varepsilon}|^{2^*} d\,\mathrm{vol}_{g_{\mathbb{R}^m}}\Big)^{\frac{2^*-1}{2^*}}.
\end{aligned} \tag{7.38}$$

Fix $\delta > 0$ arbitrarily small, we may find that $H \circ \mathcal{S}_y^{-1}(x) = H(y) + O(\delta)$ uniformly for $|x| \le \delta$. And hence, by the fact $\int_{\mathbb{R}^m} |\phi_\varepsilon|^{2^*} d\,\mathrm{vol}_{g_{\mathbb{R}^m}} \equiv \big(\frac{m}{2}\big)^m \omega_m$, we can get

$$\int_{B_\delta^0} |H(y) - H \circ \mathcal{S}_y^{-1}|^{\frac{2^*}{2^*-1}}|\phi_{y,\varepsilon}|^{2^*} d\,\mathrm{vol}_{g_{\mathbb{R}^m}} \le O(\delta^{\frac{2^*}{2^*-1}}). \tag{7.39}$$

On the other hand, we find that

$$\int_{\mathbb{R}^m\setminus B_\delta^0} \big|H(y) - H\circ \mathcal{S}_y^{-1}\big|^{\frac{2^*}{2^*-1}} |\phi_{y,\varepsilon}|^{2^*} d\,\mathrm{vol}_{g_{\mathbb{R}^m}} \le C\int_{\frac{\delta}{\varepsilon}}^{\infty} \frac{r^{m-1}}{(1+r^2)^m} dr \le C\Big(\frac{\varepsilon}{\delta}\Big)^m. \tag{7.40}$$

Therefore, combining (7.38), (7.39) and (7.40), we have

$$\|\mathcal{L}'(\psi_{y,\varepsilon})\|_{E^*} \le \frac{\alpha}{2} \quad \text{uniformly for all } y\in S^m \tag{7.41}$$

provided that $\varepsilon > 0$ is fixed small enough.

Next, we estimate the upper bound of $\mathcal{L}(\psi_{y,\varepsilon})$ in a similar way, that is

$$\begin{aligned}\mathcal{L}(\psi_{y,\varepsilon}) &= \frac{H(y)}{2m}\int_{S^m} |\psi_{y,\varepsilon}|^{2^*} d\,\mathrm{vol}_{g_{S^m}} + \frac{1}{2^*}\int_{S^m} \big(H(y)-H\big)|\psi_{y,\varepsilon}|^{2^*} d\,\mathrm{vol}_{g_{S^m}} \\ &= \frac{H(y)}{2m}\int_{\mathbb{R}^m} |\phi_{y,\varepsilon}|^{2^*} d\,\mathrm{vol}_{g_{\mathbb{R}^m}} + \frac{1}{2^*}\int_{\mathbb{R}^m} \big(H(y)-H\circ\mathcal{S}_y^{-1}\big)|\phi_{y,\varepsilon}|^{2^*} d\,\mathrm{vol}_{g_{\mathbb{R}^m}} \\ &\le \frac{H(y)}{2m}\int_{\mathbb{R}^m} |\phi_{y,\varepsilon}|^{2^*} d\,\mathrm{vol}_{g_{\mathbb{R}^m}} + \frac{\theta}{2}\end{aligned} \tag{7.42}$$

if ε is small enough. Moreover, by noticing that

$$\int_{\mathbb{R}^m} |\phi_{y,\varepsilon}|^{2^*} d\,\mathrm{vol}_{g_{\mathbb{R}^m}} = H(y)^{-m}\int_{\mathbb{R}^m} |\phi_{\varepsilon}|^{2^*} d\,\mathrm{vol}_{g_{\mathbb{R}^m}} = H(y)^{-m}\Big(\frac{m}{2}\Big)^m \omega_m, \tag{7.43}$$

we soon obtain from (7.42) and $y\in\{H\ge d+\sigma\}$ that

$$\mathcal{L}(\psi_{y,\varepsilon}) \le \frac{1}{2mH(y)^{m-1}}\Big(\frac{m}{2}\Big)^m\omega_m + \frac{\theta}{2} \le \frac{1}{2m\,(d+\sigma)^{m-1}}\Big(\frac{m}{2}\Big)^m\omega_m + \frac{\theta}{2} \tag{7.44}$$

uniformly in y.

Remark that the function $p\mapsto |\psi|_p$ is continuous provided that $\psi\in E$ is fixed. Hence if we fix ε sufficiently small, then it can be derived from (7.44) and the fact $\{H\ge d+\sigma\}$ is compact that

$$0 < c_1 \le \mathcal{L}_{p_n}(\psi_{y,\varepsilon}) \le \frac{1}{2m\,(d+\sigma)^{m-1}}\Big(\frac{m}{2}\Big)^m\omega_m + \theta \tag{7.45}$$

uniformly for all large n, where $c_1 > 0$ is some constant.

It remains to evaluate the derivatives of $\mathcal{L}_{p_n}$ for large n. Take $\psi \in E$ arbitrarily as a test spinor, then it follows that

$$\begin{aligned}
&\mathcal{L}'_{p_n}(\psi_{y,\varepsilon})[\psi] - \mathcal{L}'(\psi_{y,\varepsilon})[\psi] \\
&= \mathrm{Re}\int_{S^m} H\big(|\psi_{y,\varepsilon}|^{2^*-2} - |\psi_{y,\varepsilon}|^{p_n-2}\big)(\psi_{y,\varepsilon}, \psi) d\,\mathrm{vol}_{g_{S^m}} \\
&= \mathrm{Re}\int_{\mathbb{R}^m} (H \circ \mathcal{S}_y^{-1})\big(|\phi_{y,\varepsilon}|^{2^*-p_n} - f^{\frac{m-1}{2}(2^*-p_n)}\big)|\phi_{y,\varepsilon}|^{p_n-2} \\
&(\phi_{y,\varepsilon}, \iota(f^{\frac{m-1}{2}}\psi \circ \mathcal{S}_y^{-1})) d\,\mathrm{vol}_{g_{\mathbb{R}^m}} .
\end{aligned}$$

Remark that, by (7.35), we have

$$|\phi_{y,\varepsilon}(x)| = H(y)^{-\frac{m-1}{2}} \cdot \varepsilon^{-\frac{m-1}{2}} \cdot \Big(\frac{m}{2}\Big)^{\frac{m-1}{2}} \cdot \Big(\frac{2\varepsilon^2}{\varepsilon^2 + |x|^2}\Big)^{\frac{m-1}{2}}.$$

Thus

$$\begin{aligned}
&|\phi_{y,\varepsilon}|^{2^*-p_n} - f^{\frac{m-1}{2}(2^*-p_n)} \\
&= C_\varepsilon^{2^*-p_n}\Big(\frac{2}{1+|x|^2}\Big)^{\frac{m-1}{2}(2^*-p_n)}\Big[\Big(\frac{\varepsilon^2 + \varepsilon^2|x|^2}{\varepsilon^2 + |x|^2}\Big)^{\frac{m-1}{2}(2^*-p_n)} - 1\Big],
\end{aligned}$$

where $C_\varepsilon = H(y)^{-\frac{m-1}{2}} \cdot \varepsilon^{-\frac{m-1}{2}} \cdot \big(\frac{m}{2}\big)^{\frac{m-1}{2}} > 0$ is nothing but a constant. Since $\varepsilon > 0$ is fixed sufficiently small (say $\varepsilon < 1$), we easily get

$$\varepsilon^2 \le \frac{\varepsilon^2 + \varepsilon^2|x|^2}{\varepsilon^2 + |x|^2} \le 1 \quad \text{for all } x \in \mathbb{R}^m.$$

Therefore, due to $p_n \to 2^*$, we obtain $C_\varepsilon^{2^*-p_n} = 1 + o_n(1)$ and

$$|\phi_{y,\varepsilon}|^{2^*-p_n} - f^{\frac{m-1}{2}(2^*-p_n)} = o_n(1)\Big(\frac{2}{1+|x|^2}\Big)^{\frac{m-1}{2}(2^*-p_n)} = o_n(1) f^{\frac{m-1}{2}(2^*-p_n)}$$

uniformly in y as $n \to \infty$. This implies

$$\begin{aligned}
&\mathcal{L}'_{p_n}(\psi_{y,\varepsilon})[\psi] - \mathcal{L}'(\psi_{y,\varepsilon})[\psi] \\
&= o_n(1)\,\mathrm{Re}\int_{\mathbb{R}^m} (H \circ \mathcal{S}_y^{-1}) f^{\frac{m-1}{2}(2^*-p_n)}|\phi_{y,\varepsilon}|^{p_n-2}(\phi_{y,\varepsilon}, \iota(f^{\frac{m-1}{2}}\psi \circ \mathcal{S}_y^{-1})) d\,\mathrm{vol}_{g_{\mathbb{R}^m}} \\
&= o_n(1)\,\mathrm{Re}\int_{S^m} H|\psi_{y,\varepsilon}|^{p_n-2}(\psi_{y,\varepsilon}, \psi) d\,\mathrm{vol}_{g_{S^m}} \\
&\le o_n(1)\|\psi\|
\end{aligned}$$

uniformly for all $y \in \{H \geq d + \sigma\}$ as $n \to \infty$. Since $\psi \in E$ is arbitrary, we can derive from (7.41) that

$$\|\mathcal{L}'_{p_n} psi_{y,\varepsilon})\|_{E^*} \leq \|\mathcal{L}'(\psi_{y,\varepsilon})\|_{E^*} + \|\mathcal{L}'_{p_n}(\psi_{y,\varepsilon}) - \mathcal{L}'(\psi_{y,\varepsilon})\|_{E^*} \leq \alpha \tag{7.46}$$

uniformly for all $y \in \{H \geq d + \sigma\}$ and large n.

Combining (7.45) and (7.46), we may then apply Corollary 7.11 to get the desired assertion. □

The next result concerns with the upper bound estimate for $\max_{t>0} I_{p_n}(t\psi^+_{y,\varepsilon})$ where y locates near the critical set $\mathcal{K}_d$. Recall that, for $\nu > 0$, we denote $\mathcal{O}_\nu$ the ν-neighborhood of $\mathcal{K}_d$.

Lemma 7.20 *There exist C_0, $\nu_0 > 0$ such that*

$$\max_{t>0} I_{p_n}(t\psi^+_{y,\varepsilon}) \leq \frac{1}{2m\, d^{m-1}}\Big(\frac{m}{2}\Big)^m \omega_m - \begin{cases} C_0\varepsilon^2|\ln\varepsilon| + O(\varepsilon^2) & m = 2, \\ C_0\varepsilon^2 + O(\varepsilon^{\frac{m+3}{2}}) & m \geq 3, \end{cases}$$

uniformly for all small ε, large n and $y \in \mathcal{O}_{\nu_0}$. Particularly, for all $y \in \mathcal{O}_{\nu_0}$, the estimate

$$\max_{t>0} I_{p_n}(t\psi^+_{y,\varepsilon_0}) \leq \frac{1}{2m\, d^{m-1}}\Big(\frac{m}{2}\Big)^m \omega_m - \theta_0 \quad \textit{for some } \theta_0 > 0$$

holds as long as ε_0 is small.

Proof Since $\mathcal{K}_d = \{\xi_1, \dots, \xi_l\}$ contains only finitely many points, without loss of generality, from now on we will first restrict ourselves to a neighborhood around ξ_1. Since the hypothesis $(H*)$ ensures the Hessian of H at each $\xi \in \mathcal{K}_d$ is positive definite. Hence, there exists $R_1, \delta_0 > 0$ such that

$$H(y) \geq d + \delta_0\, \mathrm{dist}_{g_{S^m}}(y, \xi_1)^2 \tag{7.47}$$

for all $y \in B_{2R_1}(\xi_1) \subset S^m$.

Analogous to (7.42), let us estimate the following quantity

$$\mathcal{L}(\psi_{y,\varepsilon}) = \frac{H(y)}{2m}\int_{\mathbb{R}^m} |\phi_{y,\varepsilon}|^{2^*} d\,\mathrm{vol}_{g_{\mathbb{R}^m}} + \frac{1}{2^*}\int_{\mathbb{R}^m} \big(H(y) - H\circ\mathcal{S}_y^{-1}\big)|\phi_{y,\varepsilon}|^{2^*} d\,\mathrm{vol}_{g_{\mathbb{R}^m}}\,.$$

And, by (7.43) and (7.47), we find that

$$\mathcal{L}(\psi_{y,\varepsilon}) \leq \frac{1}{2m\, d^{m-1}}\Big(\frac{m}{2}\Big)^m \omega_m + \frac{1}{2^*}\int_{\mathbb{R}^m} \big(H(y) - H\circ\mathcal{S}_y^{-1}\big)|\phi_{y,\varepsilon}|^{2^*} d\,\mathrm{vol}_{g_{\mathbb{R}^m}}\,. \tag{7.48}$$

To estimate the second integral, let us recall that $H(y) = H \circ \mathcal{S}_y^{-1}(0)$ and y locates close to ξ_1, hence we have the Hessian of $H \circ \mathcal{S}_y^{-1}(\cdot)$ is positive definite in an open ball $B_{r_0}^0$, for some $r_0 < R_1$ and $y \in B_{R_1}(\xi_1)$. In particular, we deduce that

$$2H(y) - H \circ \mathcal{S}_y^{-1}(x) - H \circ \mathcal{S}_y^{-1}(-x) \leq -2\delta_0|x|^2 \quad \text{for all } x \in B_{r_0}^0. \tag{7.49}$$

At this point, let us split the last integral in (7.48) into two parts

$$\Big(\int_{B_{r_0}^0} + \int_{\mathbb{R}^m \setminus B_{r_0}^0}\Big)\big(H(y) - H \circ \mathcal{S}_y^{-1}\big)|\phi_{y,\varepsilon}|^{2^*} d\,\mathrm{vol}_{g_{\mathbb{R}^m}}.$$

And it can be seen from a similar estimate in (7.40) that

$$\int_{\mathbb{R}^m \setminus B_{r_0}^0} \big(H(y) - H \circ \mathcal{S}_y^{-1}\big)|\phi_{y,\varepsilon}|^{2^*} d\,\mathrm{vol}_{g_{\mathbb{R}^m}} = O(\varepsilon^m).$$

Notice that $|\phi_{y,\varepsilon}(x)| = |\phi_{y,\varepsilon}(-x)|$ for all $x \in \mathbb{R}^m$, it follows from (7.49) that

$$\begin{aligned}
&\int_{B_{r_0}^0} \big(H(y) - H \circ \mathcal{S}_y^{-1}\big)|\phi_{y,\varepsilon}|^{2^*} d\,\mathrm{vol}_{g_{\mathbb{R}^m}} \\
&\leq -\delta_0 \int_{B_{r_0}^0} |x|^2 |\phi_{y,\varepsilon}|^{2^*} d\,\mathrm{vol}_{g_{\mathbb{R}^m}} \\
&= -C_1\varepsilon^2 \int_0^{\frac{r_0}{\varepsilon}} \frac{r^{m+1}}{(1+r^2)^m} dr \leq \begin{cases} -C_1\varepsilon^2 |\ln\varepsilon| & m = 2, \\ -C_1\varepsilon^2 & m \geq 3, \end{cases}
\end{aligned}$$

for all ε small, where $C_1 > 0$ is a constant depending only on the dimension m and the value of $H(y)$. Therefore, by (7.48), we obtain immediately

$$\mathcal{L}(\psi_{y,\varepsilon}) \leq \frac{1}{2m\, d^{m-1}}\Big(\frac{m}{2}\Big)^m \omega_m + O(\varepsilon^m) - \begin{cases} C_1\varepsilon^2|\ln\varepsilon| & m = 2, \\ C_1\varepsilon^2 & m \geq 3, \end{cases}$$

uniformly for $y \in B_{R_1}(\xi_1)$ and for ε small.

On the other hand, for the derivative of $\mathcal{L}$, we shall use (7.38) to get

$$\|\mathcal{L}'(\psi_{y,\varepsilon})\|_{E^*} \leq C\Big(\int_{\mathbb{R}^m} \big|H(y) - H \circ \mathcal{S}_y^{-1}\big|^{\frac{2^*}{2^*-1}} |\phi_{y,\varepsilon}|^{2^*} d\,\mathrm{vol}_{g_{\mathbb{R}^m}}\Big)^{\frac{2^*-1}{2^*}}$$

for all $\varepsilon > 0$. Then, it follows that

$$\|\mathcal{L}'(\psi_{y,\varepsilon})\|_{E^*}^{\frac{2^*}{2^*-1}} \leq C|\nabla H|_{L^\infty(B_{r_0}(y))}^{\frac{2m}{m+1}} \int_{B_{r_0}^0} |x|^{\frac{2m}{m+1}} |\phi_{y,\varepsilon}|^{2^*} d\,\mathrm{vol}_{g_{\mathbb{R}^m}} + O(\varepsilon^m) \tag{7.50}$$

as $\varepsilon \to 0$. Remark that

$$\int_{\mathbb{R}^m} |x|^{\frac{2m}{m+1}} |\phi_{y,\varepsilon}|^{2^*} d\,\mathrm{vol}_{g_{\mathbb{R}^m}} = O(\varepsilon^{\frac{2m}{m+1}}) \tag{7.51}$$

as $\varepsilon \to 0$. Combining (7.50) and (7.51), we conclude that

$$\|\mathcal{L}'(\psi_{y,\varepsilon})\|_{E^*} \leq C_2|\nabla H|_{L^\infty(B_{r_0}(y))} \cdot \varepsilon + O(\varepsilon^{\frac{m+1}{2}})$$

as $\varepsilon \to 0$ with some constant $C_2 > 0$.

Now, given $\varepsilon > 0$ small enough, we see that there exists $n_\varepsilon \in \mathbb{N}$ such that for all $n \geq n_\varepsilon$

$$\mathcal{L}_{p_n}(\psi_{y,\varepsilon}) \leq \frac{1}{2m\, d^{m-1}} \Big(\frac{m}{2}\Big)^m \omega_m + O(\varepsilon^m) - \begin{cases} \frac{C_1}{2}\varepsilon^2 |\ln \varepsilon| & m = 2, \\ \frac{C_1}{2}\varepsilon^2 & m \geq 3, \end{cases} \tag{7.52}$$

and

$$\|\mathcal{L}'_{p_n}(\psi_{y,\varepsilon})\|_{E^*} \leq 2C_2|\nabla H|_{L^\infty(B_{r_0}(y))} \cdot \varepsilon + O(\varepsilon^{\frac{m+1}{2}}) \tag{7.53}$$

uniformly for $y \in B_{R_1}(\xi_1)$. Since $\nabla H(\xi_1) = 0$, we can derive that

$$|\nabla H|_{L^\infty(B_{r_0}(y))} \to 0 \quad \text{as } y \to \xi_1 \text{ and } r_0 \to 0.$$

And therefore, by combining (7.52), (7.53) and Lemma 7.10, we soon have

$$\max_{t>0} I_{p_n}(t\psi_{y,\varepsilon}^+) \leq \frac{1}{2m\, d^{m-1}} \Big(\frac{m}{2}\Big)^m \omega_m - \begin{cases} \frac{C_1}{4}\varepsilon^2 |\ln \varepsilon| + O(\varepsilon^2) & m = 2, \\ \frac{C_1}{4}\varepsilon^2 + O(\varepsilon^{\frac{m+3}{2}}) & m \geq 3, \end{cases}$$

uniformly for $y \in B_{R_1}(\xi_1)$ provided that R_1 and r_0 are fixed small enough.

If $\mathcal{K}_d = \{\xi_1, \ldots, \xi_l\}$ contains multiple points. By repeating the above argument, one shall get correspondingly $R_1, \ldots, R_l > 0$. And the proof will be completed by simply taking $\nu_0 = \min\{R_1, \ldots, R_l\}$. □

Now we are ready to proof Theorem 7.5

Proof of Theorem 7.5 The proof will be accomplished by the following four steps:

Step 1. Recall the fact that, by the hypothesis $(H*)$, for any $\nu > 0$ small there exists $\sigma > 0$ such that $\mathcal{H}$ is contractible in $\{H \geq d + \sigma\} \cup \mathcal{O}_\nu$. Now let us fix $\nu_0 > 0$ as in Lemma 7.20, then we get some $\sigma_0 > 0$ such that $\mathcal{H}$ is contractible in $\mathcal{C}_0 := \{H \geq d + \sigma_0\} \cup \mathcal{O}_{\nu_0}$. At this point, we may choose ε_0, θ_0, $\theta > 0$ small such that

$$\max\left\{\frac{1}{2m(d+\sigma_0)^{m-1}}\Big(\frac{m}{2}\Big)^m\omega_m,\ \frac{1}{2m\,d^{m-1}}\Big(\frac{m}{2}\Big)^m\omega_m - \theta_0\right\} \leq \frac{1}{2m\,d^{m-1}}\Big(\frac{m}{2}\Big)^m\omega_m - 3\theta$$

where θ_0 depends on ε_0 (by Lemma 7.20).

Recall the Nehari-Pankov manifold $\mathscr{N}_{p_n}$ defined for I_{p_n} (see (7.16)), let us introduce the sub-level sets for I_{p_n} as

$$I_{p_n}^c = \left\{u \in \mathscr{N}_{p_n} : I_{p_n}(u) \leq c\right\}$$

for $c \in \mathbb{R}$. Denoted by

$$\hat{a} = \frac{1}{2m\,d^{m-1}}\Big(\frac{m}{2}\Big)^m\omega_m - \theta, \tag{7.54}$$

then a combination of Lemmas 7.19 and 7.20 (we fix $\varepsilon > 0$ small enough such that both the lemmas hold) gives us that there exists a continuous embedding $\Phi_n : \mathcal{C}_0 \to I_{p_n}^{\hat{a}}$ for all n large. We also remark that $\hat{a} < 2\tau_{2^*} - \theta$.

Step 2. Due to the hypothesis $(H*)$, we have $H_{min} < d$. Hence if we fix $\xi_0 \in S^m$ such that $H(\xi_0) = H_{min}$, then $\xi_0 \notin \{H \geq d\}$. Let $\mathcal{S}_0 : S^m \setminus \{\xi_0\} \to \mathbb{R}^m$ be the stereographic projection from ξ_0 (that is, we treat ξ_0 as the North pole). We see that the image of $\mathcal{C}_0$ under $\mathcal{S}_0$ is a bounded set in $\mathbb{R}^m$, and thus there exists $R_0 > 0$ such that

$$\mathcal{S}_0(\mathcal{C}_0) \subset B_{R_0}^0 = \left\{x \in \mathbb{R}^m : |x| < R_0\right\}.$$

Now we can introduce a barycenter-type function $\Upsilon : E \setminus \{0\} \to \mathbb{R}^m$ as

$$\Upsilon(\psi) = \frac{\int_{S^m} \zeta \circ \mathcal{S}_0(\xi)|\psi|^{2^*} d\,\mathrm{vol}_{g_{S^m}}}{\int_{S^m} |\psi|^{2^*} d\,\mathrm{vol}_{g_{S^m}}},$$

where $\zeta : \mathbb{R}^m \to \mathbb{R}^m$ is defined by

$$\zeta(x) = \begin{cases} x & |x| < R_0, \\ \dfrac{R_0 x}{|x|} & |x| \geq R_0. \end{cases}$$

Next, we claim that $\theta > 0$ can be chosen sufficiently small such that for all large n and $u \in \mathcal{N}_{p_n}$ satisfying

$$I_{p_n}(u) < \tau_{2^*} + \theta = \frac{1}{2m(H_{max})^{m-1}}\Big(\frac{m}{2}\Big)^m \omega_m + \theta,$$

there holds

$$\mathcal{S}_0^{-1} \circ \Upsilon\big(u + h_{p_n}(u)\big) \in \mathcal{H}_\delta. \tag{7.55}$$

Suppose to the contrary that there exist $\theta_n \to 0$ and $u_n \in \mathcal{N}_{p_n}$ such that

$$I_{p_n}(u_n) < \tau_{2^*} + \theta_n \quad \text{but} \quad \mathcal{S}_0^{-1} \circ \Upsilon(\psi_n) \notin \mathcal{H}_\delta, \tag{7.56}$$

where we set $\psi_n := u_n + h_{p_n}(u_n)$. Recall the definition of τ_p for $p \in (2, 2^*]$ (see (7.18)), by Lemma 7.12 we can get

$$I_{p_n}(u_n) \ge \tau_{p_n} = \tau_{2^*} + o_n(1).$$

Thus, as $n \to \infty$, each u_n almost minimize the functional I_{p_n} on $\mathcal{N}_{p_n}$. And it follows from the well-known Ekeland Variational Principle [20, 31, 38] that for each n there exists $\tilde{u}_n \in \mathcal{N}_{p_n}$ such that

$$I_{p_n}(\tilde{u}_n) < \tau_{2^*} + 2\theta_n, \quad \big\| I'_{p_n}|_{\mathcal{N}_{p_n}}(\tilde{u}_n) \big\| \le \frac{8\theta_n}{\epsilon} \quad \text{and} \quad \|u_n - \tilde{u}_n\| \le 2\epsilon. \tag{7.57}$$

where $\epsilon > 0$ is a given small constant whose value will be fixed later. Remark that we may simply rewrite the definition of $\mathcal{N}_{p_n}$ in (7.16) as $\mathcal{N}_{p_n} = \big\{u \in E^+ : \mathcal{K}_{p_n}(u) = 0\big\}$ with $\mathcal{K}_{p_n}(u) = I'_{p_n}(u)[u]$. Then the Lagrange multiplier rule implies the existence of $t_n \in \mathbb{R}$ such that

$$I'_{p_n}|_{\mathcal{N}_{p_n}}(\tilde{u}_n) = I'_{p_n}(\tilde{u}_n) + t_n \mathcal{K}'_{p_n}(\tilde{u}_n).$$

By using the estimate $\mathcal{K}'_{p_n}(\tilde{u}_n)[\tilde{u}_n] \le -\frac{p_n-2}{p_n-1}\int_{S^m} H|\tilde{u}_n + h_{p_n}(\tilde{u}_n)|^{p_n} d\,\mathrm{vol}_{g_{S^m}}$ (a similar inequality has been shown in [13, Lemma 7.4]), we can easily conclude from (7.57) that $t_n \to 0$ and, hence, $I'_{p_n}(\tilde{u}_n) \to 0$ as $n \to \infty$. Now, set $\tilde{\psi}_n = \tilde{u}_n + h_{p_n}(\tilde{u}_n)$, it is evident that

$$\tau_{2^*} + o_n(1) \le \mathcal{L}_{p_n}(\tilde{\psi}_n) < \tau_{2^*} + 2\theta_n \quad \text{and} \quad \mathcal{L}'_{p_n}(\tilde{\psi}_n) \to 0$$

as $n \to \infty$. Then, according to (2) of Remark 7.15 and Proposition 7.13, we have the sequence $\{\tilde{\psi}_n\}$ must blow-up. Moreover, by Proposition 7.14, there exists a convergent sequence $\{a_n\} \subset S^m$, $a_n \to a \in \mathcal{H}$ as $n \to \infty$ and a sequence of positive numbers $R_n \searrow 0$ and a non-trivial solution ϕ_0 of

$$D_{g_{\mathbb{R}^m}} \phi_0 = H_{max} |\phi_0|^{2^*-2} \phi_0 \quad \text{on } \mathbb{R}^m$$

such that

$$\tilde{\psi}_n = R_n^{-\frac{m-1}{2}} \eta(\cdot) \overline{(\mu_n)_*} \circ \phi_0 \circ \mu_n^{-1} + o_n(1) \quad \text{in } E$$

as $n \to \infty$, where μ_n and η are as in Proposition 7.14. Therefore, $\int_{S^m} |\tilde{\psi}_n|^{2^*} d\,\mathrm{vol}_{g_{S^m}}$ is bounded away from 0 and particularly

$$\Upsilon(\tilde{\psi}_n) \to \mathcal{S}_0(a) \in \mathcal{S}_0(\mathcal{H}) \quad \text{as } n \to \infty.$$

Recalling the definition of Υ, we see that the derivative $\Upsilon' : E \setminus \{0\} \to E^*$ exists and the operator norm is bounded around each $\tilde{\psi}_n$. Hence, by fixing the constant ϵ small enough in (7.57), we have $|\Upsilon(\psi_n) - \Upsilon(\tilde{\psi}_n)| < \frac{\delta}{2}$, and therefore we get $\Upsilon(\psi_n) \in \mathcal{S}_0(\mathcal{H}_\delta)$ for all n large which contradicts to (7.56).

Step 3. We are going to show that I_{p_n} has a critical point in $I_{p_n}^{\hat{a}} \setminus I_{p_n}^{\hat{b}}$ where $\hat{a}$ is defined in (7.54) and

$$\hat{b} = \frac{1}{2m(H_{max})^{m-1}} \left(\frac{m}{2}\right)^m \omega_m + \theta.$$

We emphasize here that the embedding $\Phi_n : \mathcal{C}_0 \to \mathcal{N}_{p_n}$ gives us a description of low-energy levels in the sense that $\Phi_n(\mathcal{H}) \in I_{p_n}^{\hat{b}}$ and $\mathcal{S}_0^{-1} \circ \Upsilon \circ \Phi_n|_{\mathcal{H}}$ is homotopic to the inclusion $\mathcal{H} \hookrightarrow \mathcal{H}_\delta$. Fixing all these notations, we define the family

$$\Lambda_n = \Big\{ \varpi \in C\big(\mathcal{H} \times [0,1], \mathcal{N}_{p_n}\big) : \varpi(\cdot, 0) = \Phi_n \text{ and } \varpi(\cdot, 1) \text{ collapses to a single spinor} \Big\}.$$

Clearly, for all n large, $\Lambda_n \neq \emptyset$ since the contractibility of $\mathcal{H}$ in $\mathcal{C}_0$ gives us a continuous homotopy $\gamma : \mathcal{H} \times [0,1] \to \mathcal{C}_0$ such that $\gamma(\cdot, 0) = id$, $\gamma(\cdot, 1) \equiv \xi_* \in \mathcal{C}_0$ and $\Phi_n \circ \gamma \in \Lambda_n$. Since E is embedded compactly into L^p for all $p \in [1, 2^*)$, we have I_{p_n} satisfies the Palais-Smale condition for each n. And thus, by standard variational arguments, one can easily verify

$$c_n := \inf_{\varpi \in \Lambda_n} \sup_{(\xi,t) \in \mathcal{H} \times [0,1]} I_{p_n}(\varpi(\xi, t))$$

is a critical value for I_{p_n}. Particularly, $c_n \in (\hat{b}, \hat{a}]$.

Step 4. Let ψ_n^+ be a critical point of I_{p_n} at level c_n. It suffices to show that the sequence $\{\psi_n = \psi_n^+ + h_{p_n}(\psi_n^+)\}$ contains a compact subsequence.
Let us assume to the contrary that any of such sequence $\{\psi_n\}$ will blow up and let $a \in S^m$ be the associated blow-up point. According to Proposition 7.14 and the fact $\mathcal{L}_{p_n}(\psi_n) = I_{p_n}(\psi_n^+) = c_n \le \hat{a}$, the blow-up point a should locate inside $\{H \ge d + \sigma_0\}$. By hypothesis $(H*)$, there is no critical value of H in the interval (d, H_{max}) and hence we have $a \in \mathcal{H}$.
Moreover, from Proposition 7.14 again, there is a non-trivial solution ϕ_0 of

$$D_{g_{\mathbb{R}^m}}\phi = H_{max}|\phi|^{2^*-2}\phi \quad \text{on } \mathbb{R}^m$$

such that

$$\psi_n = R_n^{-\frac{m-1}{2}}\eta(\cdot)\overline{(\mu_n)_*} \circ \phi_0 \circ \mu_n^{-1} + o_n(1) \quad \text{in } E$$

as $n \to \infty$, where R_n, η, μ_n are as in Proposition 7.14. Moreover

$$\int_{S^m} H|\psi_n|^{p_n} d\,\mathrm{vol}_{g_{S^m}} = R_n^{\frac{m-1}{2}(2^*-p_n)} \int_{B^0_{2r/R_n}} H \circ \mu_n |\phi_0|^{p_n} d\,\mathrm{vol}_{g_n} + o_n(1)$$

as $n \to \infty$. Since for arbitrary $R > 0$

$$\begin{aligned} & R_n^{\frac{m-1}{2}(2^*-p_n)} \int_{B^0_{2r/R_n}\setminus B^0_R} H \circ \mu_n |\phi_0|^{p_n} d\,\mathrm{vol}_{g_n} \\ & \le C\Big(\int_{B^0_{2r/R_n}\setminus B^0_R} |\phi_0|^{2^*} d\,\mathrm{vol}_{g_{\mathbb{R}^m}}\Big)^{\frac{p_n}{2^*}} \big((2r)^m - (R_n R)^m\big)^{\frac{2^*-p_n}{2^*}} \end{aligned}$$

and

$$R_n^{\frac{m-1}{2}(2^*-p_n)} \int_{B^0_R} H \circ \mu_n |\phi_0|^{p_n} d\,\mathrm{vol}_{g_n} = H_{max} \int_{B^0_R} |\phi_0|^{2^*} d\,\mathrm{vol}_{g_n} + o_n(1),$$

we derive that

$$\int_{S^m} H|\psi_n|^{p_n} d\,\mathrm{vol}_{g_{S^m}} = H_{max} \int_{\mathbb{R}^m} |\phi_0|^{2^*} d\,\mathrm{vol}_{g_n} + o_n(1) \tag{7.58}$$

as $n \to \infty$. Notice that ϕ_0 also extends to a non-trivial solution $\bar{\phi}_0$ to the equation

$$D\bar{\phi}_0 = H_{max}|\bar{\phi}_0|^{2^*-2}\bar{\phi}_0 \quad \text{on } S^m, \tag{7.59}$$

by using the fact $\mathcal{L}_{p_n}(\psi_n) = c_n \le \hat{a} < 2\tau_{2^*}$, we have

$$H_{max} \int_{S^m} |\bar{\phi}_0|^{2^*} d\,\mathrm{vol}_{g_{S^m}} = H_{max} \int_{\mathbb{R}^m} |\phi_0|^{2^*} d\,\mathrm{vol}_{g_{\mathbb{R}^m}} < \frac{2}{(H_{max})^{m-1}} \Big(\frac{m}{2}\Big)^m \omega_m.$$

In the 2-dimensional case, we have $2^* = 4$ and

$$H_{max} \int_{S^2} |\bar{\phi}_0|^4 d\,\mathrm{vol}_{g_{S^2}} = H_{max} \int_{\mathbb{R}^2} |\phi_0|^4 d\,\mathrm{vol}_{g_{\mathbb{R}^2}} < \frac{8\pi}{H_{max}}.$$

It follows from Proposition 7.18 that $|\bar{\phi}_0|$ has no zero at all, and from the spinorial Weierstraß representation and Li-Yau's inequality that $(S^2, |\bar{\phi}_0|^4 g_{S^2})$ is embedded into $\mathbb{R}^3$ with constant mean curvature H_{max}. Now, by the Alexandrov's theorem [2], we know that $(S^2, |\bar{\phi}_0|^4 g_{S^2})$ must be a round sphere. And hence the Willmore energy satisfies

$$H_{max}^2 \int_{S^2} |\bar{\phi}_0|^4 d\,\mathrm{vol}_{g_{S^2}} = 4\pi.$$

This and (7.58) imply

$$\begin{aligned} c_n = \mathcal{L}_{p_n}(\psi_n) &= \frac{p_n - 2}{2p_n} \int_{S^2} H|\psi_n|^{p_n} d\,\mathrm{vol}_{g_{S^2}} \\ &= \frac{H_{max}}{4} \int_{\mathbb{R}^2} |\phi_0|^4 d\,\mathrm{vol}_{g_{\mathbb{R}^2}} + o_n(1) < \hat{b} \end{aligned}$$

as $n \to \infty$, which is impossible. And therefore, the compactness of the solution sequence $\{\psi_n\}$ follows.

□

Remark 7.21 The hypersurfaces with constant mean curvature are much more complicated in higher dimensions, surprising examples have been shown in [24]: there exist infinitely many distinct differentiable immersions of the 3-sphere into Euclidean 4-space having a given positive constant mean curvature, moreover, the total "area" as well as the total integral of the norm of the second fundamental form of such examples can be as large as one wants. This makes the picture of $\bar{\phi}_0$ in Step 4 unclear to us when the dimension $m \geq 3$, particularly we do not know whether or not $\bar{\phi}_0$ has the ground state energy. And it is also unclear if τ_{2^*} is an isolated critical level of the energy functional for Eq. (7.59). Our approach in this regard up to now have failed.

7.5.2 The Geometric Consequence

Thanks to the previous section, we have the existence result for the equation

$$D\psi = H|\psi|^2\psi \quad \text{on } S^2 \tag{7.60}$$

provided H is smooth and satisfy (H). Now let us give a interesting application to the problem of prescribing mean curvature on the 2-sphere.

As a direct application of Proposition 7.18, we soon have

Corollary 7.22 *On (S^2, g_{S^2}), if H is a positive smooth function satisfying criteria (H), then there exists a solution of the equation*

$$D\psi = H|\psi|^2\psi, \quad |\psi| > 0.$$

Following the spinorial Weierstrass representation [21, Theorem 13], we apply our existence results now to the problem of prescribing mean curvature on S^2:

Corollary 7.23 *On (S^2, g_{S^2}), if H is a positive smooth function satisfying criteria (H), then there is a conformal metric g_1 such that (S^2, g_1) is isometrically immersed into the Euclidean space $\mathbb{R}^3$ with mean curvature H.*

Proof By Corollary 7.22, we can introduce a conformal metric $g_1 = |\psi|^4 g_{S^2}$ on S^2. Then the conformal transformation formula implies that there exists a spinor φ_1 on (S^2, g_1) such that

$$D_{g_1}\varphi_1 = H\varphi_1 \quad \text{and} \quad |\varphi_1|_{g_1} \equiv 1.$$

Hence by the spinorial Weierstrass representation, we have there is an isometric immersion $(\widetilde{S}^2, g) \to \mathbb{R}^3$ of the universal covering $\widetilde{S}^2$ into the Euclidean space $\mathbb{R}^3$ with mean curvature H. Since S^2 is simply connected, we conclude the assertion immediately. □

Appendix

Proof of Proposition 7.13

The boundedness of $\{\psi_n\}$ follows from Lemma 7.10 (or one may follow [13, Lemma 4.1]). We then assume without loss of generality that $\psi_n \rightharpoonup \psi_0$ in E as $n \to \infty$. By (7.28), it is not difficult to check that

$$\begin{aligned} 0 &= \lim_{n\to\infty} \mathcal{L}'_{p_n}(\psi_n)[\varphi] \\ &= \lim_{n\to\infty} \left(\langle \psi_n^+, \varphi^+ \rangle - \langle \psi_n^-, \varphi^- \rangle - \mathrm{Re} \int_{S^m} H|\psi_n|^{p_n-2}(\psi_n, \varphi) d\,\mathrm{vol}_{g_{S^m}} \right) \\ &= \langle \psi_0^+, \varphi^+ \rangle - \langle \psi_0^-, \varphi^- \rangle - \mathrm{Re} \int_{S^m} H|\psi_0|^{2^*-2}(\psi_0, \varphi) d\,\mathrm{vol}_{g_{S^m}} = \mathcal{L}'_{2^*}(\psi_0)[\varphi] \end{aligned}$$

for all $\varphi \in E$. Hence ψ_0 turns out to be a solution of the critical equation (7.10).

Set $\bar{\psi}_n = \psi_n - \psi_0$. We derive that

$$
\begin{aligned}
D\bar{\psi}_n = & H|\psi_n|^{p_n-2}\psi_n - H|\bar{\psi}_n|^{p_n-2}\bar{\psi}_n - H|\psi_0|^{p_n-2}\psi_0 \\
& + H|\psi_0|^{p_n-2}\psi_0 - H|\psi_0|^{2^*-2}\psi_0 \\
& + H|\bar{\psi}_n|^{p_n-2}\bar{\psi}_n + o_n(1)
\end{aligned}
$$

where the above $o_n(1)$ term equals to $\mathcal{L}'_{p_n}(\psi_n)$ as $n \to \infty$ in E^*. Denoted by

$$
\Phi_n = H|\psi_n|^{p_n-2}\psi_n - H|\bar{\psi}_n|^{p_n-2}\bar{\psi}_n - H|\psi_0|^{p_n-2}\psi_0,
$$

since $p_n \nearrow 2^*$, it turns out that there exists $C > 0$ (independent of n) such that

$$
|\Phi_n| \le C|\bar{\psi}_n|^{p_n-2}|\psi_0| + C|\psi_0|^{p_n-2}|\bar{\psi}_n|. \tag{7.61}
$$

Thanks to the Egorov theorem, for any $\epsilon > 0$, there exists $\Omega_\epsilon \subset S^m$ such that $\text{meas}\{S^m \setminus \Omega_\epsilon\} < \epsilon$ and $\bar{\psi}_n \to 0$ uniformly on Ω_ϵ as $n \to \infty$. Therefore, by the boundedness of ψ_n, (7.61) and the Hölder inequality, we have

$$
\begin{aligned}
\text{Re}\int_{S^m}(\Phi_n, \varphi)d\,\text{vol}_{g_{S^m}} &= \text{Re}\int_{S^m\setminus\Omega_\epsilon}(\Phi_n, \varphi)d\,\text{vol}_{g_{S^m}} + \text{Re}\int_{\Omega_\epsilon}(\Phi_n, \varphi)d\,\text{vol}_{g_{S^m}} \\
&\le C\Big[\Big(\int_{S^m\setminus\Omega_\epsilon}|\psi_0|^{2^*}d\,\text{vol}_{g_{S^m}}\Big)^{\frac{1}{2^*}} \\
&\quad + \Big(\int_{S^m\setminus\Omega_\epsilon}|\psi_0|^{2^*}d\,\text{vol}_{g_{S^m}}\Big)^{\frac{p_n-2}{2^*}}\Big] \\
&\quad + \int_{\Omega_\epsilon}|\Phi_n|\cdot|\varphi|d\,\text{vol}_{g_{S^m}}\,.
\end{aligned} \tag{7.62}
$$

for arbitrary $\varphi \in E$ with $\|\varphi\| \le 1$. It is evident that the last integral in (7.62) converges to 0 as $n \to \infty$ and the remaining integrals tends to 0 uniformly in n as $\epsilon \to 0$. Thus, we get $\Phi_n = o_n(1)$ in E^*. Noting that $q \mapsto H(\cdot)|\psi_0|^{q-2}\psi_0$ is continuous in E^*, we obtain

$$
\mathcal{L}'_{p_n}(\bar{\psi}_n) = D\bar{\psi}_n - H|\bar{\psi}_n|^{p_n-2}\bar{\psi}_n = o_n(1) \quad \text{in } E^*. \tag{7.63}
$$

Now assume $\psi_0 \ne 0$ in E (otherwise we are done). Up to a subseqence, if $\mathcal{L}_{p_n}(\bar{\psi}_n) \to 0$ then, by repeating the proof in [12, Lemma 4.1], we have that $\bar{\psi}_n \to 0$ in E. This shows the compactness of $\{\psi_n\}$. In what follows, we assume that $\mathcal{L}_{p_n}(\bar{\psi}_n)$ is bounded away from 0.

Since ψ_0 is a non-trivial solution of Eq. (7.10), we can see that

$$\tau_{2^*} \leq \mathcal{L}_{2^*}(\psi_0) = \mathcal{L}_{2^*}(\psi_0) - \frac{1}{2^*}\mathcal{L}'_{2^*}(\psi_0)[\psi_0] = \frac{1}{2m}\big(\|\psi_0^+\|^2 - \|\psi_0^-\|^2\big). \qquad (7.64)$$

On the other hand, by (7.63) and $\mathcal{L}_{p_n}(\bar{\psi}_n)$ is bounded away from 0, we derive that $\{\bar{\psi}_n\}$ satisfy the condition (7.17) (where the upper bound of $\mathcal{L}_{p_n}(\bar{\psi}_n)$ follows directly from the boundedness of $\{\bar{\psi}_n\}$). And hence, by applying Lemma 7.10, we have $\tau_{p_n} \leq \max_{t>0} I_{p_n}(t\bar{\psi}_n) \leq \mathcal{L}_{p_n}(\bar{\psi}_n) + o_n(1)$. This and (7.63) imply

$$\tau_{p_n} \leq \mathcal{L}_{p_n}(\bar{\psi}_n) - \frac{1}{p_n}\mathcal{L}'_{p_n}(\bar{\psi}_n)[\bar{\psi}_n] + o_n(1) = \frac{p_n - 2}{2p_n}\big(\|\bar{\psi}_n^+\|^2 - \|\bar{\psi}_n^-\|^2\big) + o_n(1). \qquad (7.65)$$

Now, it follows from (7.64), (7.65) and Lemma 7.12 that

$$\begin{aligned}
\mathcal{L}_{p_n}(\psi_n) &= \mathcal{L}_{p_n}(\psi_n) - \frac{1}{p_n}\mathcal{L}'_{p_n}(\psi_n)[\psi_n] + o_n(1) \\
&= \frac{p_n - 2}{2p_n}\big(\|\psi_n^+\|^2 - \|\psi_n^-\|^2\big) + o_n(1) \\
&= \frac{p_n - 2}{2p_n}\big(\|\bar{\psi}_n^+\|^2 - \|\bar{\psi}_n^-\|^2\big) + \frac{p_n - 2}{2p_n}\big(\|\psi_0^+\|^2 - \|\psi_0^-\|^2\big) + o_n(1) \\
&\geq 2\tau_{2^*} + o_n(1)
\end{aligned}$$

which contradicts (7.28). Therefore, we must have $\mathcal{L}_{p_n}(\bar{\psi}_n) \to 0$ as $n \to \infty$, and the proof is hereby completed.

Proof of Proposition 7.14

We follow the steps in [25, Section 5], with necessary modifications.

To begin with, for some $\delta_0 > 0$, we introduce the singular set of $\{\psi_n\}$:

$$\Gamma := \Big\{a \in S^m : \varliminf_{r\to 0}\varliminf_{n\to\infty}\int_{B_r(a)} |\psi_n|^{p_n} d\,\mathrm{vol}_{g_{S^m}} \geq \delta_0\Big\}$$

where $B_r(a) \subset S^m$ is the distance ball of radius r with respect to the metric g_{S^m}. As was proved in [25, Lemma 5.3] (the proof can be repeated just adapting the notations in the present paper) that there exists $\delta_0 > 0$ depending only on the geometry of S^m such that $\Gamma \neq \emptyset$.

In order to have a clearer picture of ψ_n near points in Γ, for $r \geq 0$, we define

$$\Theta_n(r) = \sup_{a \in S^m} \int_{B_r(a)} |\psi_n|^{p_n} d\,\mathrm{vol}_{g_{S^m}},$$

which can be understood as a concentration function for ψ_n. Then, by choosing $\bar{\delta} > 0$ small (say $\bar{\delta} < \delta_0$), there exist a decreasing sequence $R_n \searrow 0$ as $n \to \infty$ and $\{a_n\} \subset S^m$ such that

$$\Theta_n(R_n) = \int_{B_{R_n}(a_n)} |\psi_n|^{p_n} d\,\mathrm{vol}_{g_{S^m}} = \bar{\delta}. \tag{7.66}$$

Up to a subsequence if necessary, we assume that $a_n \to a \in S^m$ as $n \to \infty$.

Recall that we have defined the rescaled geodesic normal coordinates near each a_n by $\mu_n(x) = \exp_{a_n}(R_n x)$ for $x \in \mathbb{R}^m$. Then we have a conformal equivalence $\mathbb{R}^m \supset (B_R^0, g_n) \cong (B_{R_n R}(a_n), g_{S^m}) \subset S^m$ for all large n, where $g_n = R_n^{-2} \mu_n^* g_{S^m}$. And, for any $R > 0$, g_n converges to the Euclidean metric in $C^\infty(B_R^0)$ as $n \to \infty$.

Following the idea of local trivialization introduced in [16], the coordinate map μ_n induces a spinor identification $\overline{(\mu_n)_*} : \mathbb{S}_x(B_R^0, g_n) \to \mathbb{S}_{\mu_n(x)}(B_{R_n R}(a_n), g_{S^m})$. In the sequel, we define spinors ϕ_n on B_R^0 by $\phi_n = R_n^{\frac{m-1}{2}} \overline{(\mu_n)_*}^{-1} \circ \psi_n \circ \mu_n$. By the conformal changes of the Dirac operator, a straightforward calculation shows that $D_{g_n}\phi_n = R_n^{\frac{m+1}{2}} \overline{(\mu_n)_*}^{-1} \circ (D\psi_n) \circ \mu_n$,

$$\int_{B_R^0} (D_{g_n}\phi_n, \phi_n) d\,\mathrm{vol}_{g_n} = \int_{B_{R_n R}(a_n)} (D\psi_n, \psi_n) d\,\mathrm{vol}_{g_{S^m}}, \tag{7.67}$$

$$\int_{B_R^0} |\phi_n|^{2^*} d\,\mathrm{vol}_{g_n} = \int_{B_{R_n R}(a_n)} |\psi_n|^{2^*} d\,\mathrm{vol}_{g_{S^m}}, \tag{7.68}$$

and

$$\int_{B_R^0} |\phi_n|^{p_n} d\,\mathrm{vol}_{g_n} = R_n^{-\frac{m-1}{2}(2^* - p_n)} \int_{B_{R_n R}(a_n)} |\psi_n|^{p_n} d\,\mathrm{vol}_{g_{S^m}}. \tag{7.69}$$

Moreover, since $\{\psi_n\}$ is bounded in E, we have

$$\sup_{n \geq 1} \int_{B_R^0} |\phi_n|^{2^*} d\,\mathrm{vol}_{g_n} \leq \sup_{n \geq 1} \int_{S^m} |\psi_n|^{2^*} d\,\mathrm{vol}_{g_{S^m}} < +\infty \tag{7.70}$$

for any $R > 0$.

Lemma 7.24 *There is* $\bar{\lambda} > 0$ *such that* $\bar{\lambda} \le \liminf_{n\to\infty} R_n^{\frac{m-1}{2}(2^*-p_n)} \le \limsup_{n\to\infty} R_n^{\frac{m-1}{2}(2^*-p_n)} \le 1$.

Proof It follows from (7.66), (7.69) and Hölder inequality that

$$\bar{\delta} = \int_{B_{R_n}(a_n)} |\psi_n|^{p_n} d\,\mathrm{vol}_{g_{S^m}}$$
$$\le \Big(\int_{B_1^0} |\phi_n|^{2^*} d\,\mathrm{vol}_{g_n}\Big)^{\frac{p_n}{2^*}} \Big(\int_{B_1^0} d\,\mathrm{vol}_{g_n}\Big)^{\frac{2^*-p_n}{2^*}} R_n^{\frac{m-1}{2}(2^*-p_n)}.$$

Noting that g_n converges to the Euclidean metric in the C^∞-topology on B_1^0, we can conclude immediately from $p_n \nearrow 2^*$ and (7.70) that $\bar{\delta} \le C \cdot R_n^{\frac{m-1}{2}(2^*-p_n)}$ for some constant $C > 0$.

On the other hand, suppose there exists some $\delta > 0$ such that $R_n^{\frac{m-1}{2}(2^*-p_n)} \ge 1+\delta$ for all large n. Then, we must have $\ln R_n \ge \frac{2\ln(1+\delta)}{(m-1)(2^*-p_n)} \to +\infty$ as $n \to \infty$. This implies $R_n \to +\infty$ which is absurd. □

Let ϕ_n be as above and define $\bar{L}_n = D_{g_n}\phi_n - R_n^{\frac{m-1}{2}(2^*-p_n)} H \circ \mu_n(\cdot)|\phi_n|^{p_n-2}\phi_n$. By using $\mathcal{L}'_{p_n}(\psi_n) \to 0$ in E^* and the conformal changes of the Dirac operator, it is not difficult to check that $\bar{L}_n \to 0$ in $H_{loc}^{-\frac{1}{2}}(\mathbb{R}^m, \mathbb{S}_m)$.

In what follows, by Lemma 7.24, we may assume that (up to a subsequence if necessary) $R_n^{\frac{m-1}{2}(2^*-p_n)} \to \lambda \in [\bar{\lambda}, 1]$ as $n \to \infty$. Since $\{\phi_n\}$ is bounded in $H_{loc}^{\frac{1}{2}}(\mathbb{R}^m, \mathbb{S}_m)$, we can assume that $\phi_n \rightharpoonup \phi_0$ in $H_{loc}^{\frac{1}{2}}(\mathbb{R}^m, \mathbb{S}_m)$. And, by (7.70), it is easy to see that $\phi_0 \in L^{2^*}(\mathbb{R}^m, \mathbb{S}_m)$ and satisfies the equation

$$D_{g_{\mathbb{R}^m}}\phi_0 = \lambda H(a)|\phi_0|^{2^*-2}\phi_0 \quad \text{on } \mathbb{R}^m. \tag{7.71}$$

At this point, we may apply the local analysis in [25, Lemma 5.5] to see that $\phi_n \to \phi_0$ in $H_{loc}^{\frac{1}{2}}(\mathbb{R}^m, \mathbb{S}_m)$ as $n \to \infty$, where proof is based on the fact that g_n converges to $g_{\mathbb{R}^m}$ in C^∞-topology on bounded domains in $\mathbb{R}^m$. Then, by Lemma 7.24, (7.66) and (7.69), we have

$$\lambda \int_{B_1^0} |\phi_0|^{2^*} d\,\mathrm{vol}_{g_{\mathbb{R}^m}} = \bar{\delta},$$

which implies ϕ_0 is a non-trivial solution of Eq. (7.71). And hence, by the energy gap estimate in [25, Section 4], we can derive via standard rescaling argument that

$$\int_{\mathbb{R}^m} |\phi_0|^{2^*} dx \ge \frac{1}{(\lambda H(a))^m}\Big(\frac{m}{2}\Big)^m \omega_m. \tag{7.72}$$

With these preparations in hand, we may now choose $\eta \in C^\infty(S^m)$ be such that $\eta \equiv 1$ on $B_r(a)$ and $\operatorname{supp} \eta \subset B_{2r}(a)$ for some $r > 0$ (for sure r should not be large in the sense that we assume $3r < inj_{S^m}$ where inj_{S^m} denotes the injective radius) and define a spinor field $z_n \in C^\infty(S^m, \mathbb{S}(S^m))$ by $z_n = R_n^{-\frac{m-1}{2}} \eta(\cdot)\overline{(\mu_n)_* \circ \phi_0 \circ \mu_n^{-1}}$. Setting $\varphi_n = \psi_n - z_n$, we have

Lemma 7.25

(1) $\varphi_n \rightharpoonup 0$ *in* E *as* $n \to \infty$.
(2) $\mathcal{L}'_{p_n}(z_n) \to 0$ *and* $\mathcal{L}'_{p_n} varphi_n) \to 0$ *as* $n \to \infty$.

Proof The proof of (1) is accomplished by showing $|z_n|_2 \to 0$ as $n \to \infty$ since we already have assumed $\psi_n \rightharpoonup 0$ in E. To see this, we split the L^2-integral of z_n into two parts

$$\Big(\int_{B_{R_n R}(a_n)} + \int_{S^m \setminus B_{R_n R}(a_n)}\Big)|z_n|^2 d\operatorname{vol}_{g_{S^m}}$$

where

$$\begin{aligned}\int_{B_{R_n R}(a_n)} |z_n|^2 d\operatorname{vol}_{g_{S^m}} &= R_n^{-m+1} \int_{B^0_{R_n R}} |\phi_0|^2 d\operatorname{vol}_{\mu_n^* g_{S^m}} \\ &= R_n \int_{B^0_R} |\phi_0|^2 d\operatorname{vol}_{g_n} \to 0\end{aligned}$$

and

$$\begin{aligned}\int_{S^m \setminus B_{R_n R}(a_n)} |z_n|^2 d\operatorname{vol}_{g_{S^m}} &\le C\Big(\int_{B^0_{3r/R_n} \setminus B^0_R} |\phi_0|^{2^*} d\operatorname{vol}_{g_{\mathbb{R}^m}}\Big)^{\frac{2}{2^*}} \\ &\quad \big((3r)^m - (R_n R)^m\big) \to 0\end{aligned}$$

for some constant $C > 0$. We omit the proof of (2) here, since it is standard and is similar to that of [25, Lemma 5.7]. □

Proof of Proposition 7.14: The Concentration Behavior By (2) of Lemma 7.25, we have

$$\mathcal{L}_{p_n}(z_n) + o_n(1) = \mathcal{L}_{p_n}(z_n) - \frac{1}{p_n}\mathcal{L}'_{p_n}(z_n)[z_n] = \frac{p_n - 2}{2p_n}\big(\|z_n^+\|^2 - \|z_n^-\|^2\big)$$

and

$$\mathcal{L}_{p_n}(\varphi_n) + o_n(1) = \mathcal{L}_{p_n}(\varphi_n) - \frac{1}{p_n}\mathcal{L}'_{p_n}(\varphi_n)[\varphi_n] = \frac{p_n - 2}{2p_n}\big(\|\varphi_n^+\|^2 - \|\varphi_n^-\|^2\big).$$

We claim that

Claim 7.1 $\mathcal{L}_{p_n}(\psi_n) = \mathcal{L}_{p_n}(z_n) + \mathcal{L}_{p_n}(\varphi_n) + o_n(1)$ as $n \to \infty$.

Assuming Claim 7.1 for the moment, then we are going to show that $\mathcal{L}_{p_n}(\varphi_n) \to 0$ as $n \to \infty$. Indeed, suppose to the contrary that (up to a subsequence) $\mathcal{L}_{p_n}(\varphi_n) \geq c > 0$, it follows from the boundedness of $\{\varphi_n\}$ in E, Lemma 7.10 and 7.12 that

$$\tau_{2^*} = \tau_{p_n} + o_n(1) \leq \max_{t>0} I_{p_n}(t\varphi_n^+) + o_n(1) \leq \mathcal{L}_{p_n}(\varphi_n) + o_n(1). \tag{7.73}$$

On the other hand, we have

$$\begin{aligned}
\mathcal{L}_{p_n}(z_n) &= \mathcal{L}_{p_n}(z_n) - \frac{1}{2}\mathcal{L}'_{p_n}(z_n)[z_n] = \frac{p_n - 2}{2p_n}\int_{S^m} H|z_n|^{p_n} d\,\mathrm{vol}_{g_{S^m}} + o_n(1) \\
&= \frac{p_n - 2}{2p_n} R_n^{\frac{m-1}{2}(2^* - p_n)} \int_{B_R^0} (H \circ \mu_n)|\phi_0|^{p_n} d\,\mathrm{vol}_{g_n} + o_n(1) + o_R(1) \\
&= \frac{1}{2m}\lambda H(a) \int_{B_R^0} |\phi_0|^{2^*} d\,\mathrm{vol}_{g_{\mathbb{R}^m}} + o_n(1) + o_R(1)
\end{aligned}$$

for $R > 0$ large. Thus, by (7.21), (7.72), $\lambda \leq 1$ and $H(a) \leq H_{max}$, we obtain

$$\mathcal{L}_{p_n}(z_n) \geq \frac{1}{2m(\lambda H(a))^{m-1}}\Big(\frac{m}{2}\Big)^m \omega_m + o_n(1) \geq \tau_{2^*} + o_n(1). \tag{7.74}$$

Combining Claim 7.1, (7.73) and (7.74), we obtain $\mathcal{L}_{p_n}(\psi_n) \geq 2\tau_{2^*} + o_n(1)$ which contradicts to (7.28). Therefore, we have $\mathcal{L}_{p_n}(\varphi_n) \to 0$ as $n \to \infty$ and this, together with $\mathcal{L}'_{p_n}(\varphi_n) \to 0$, implies $\varphi_n \to 0$ in E as $n \to \infty$ (follow the proof of [12, Lemma 4.1]). Moreover, we can get a lower bound for λ, that is $\lambda > 2^{-\frac{1}{m-1}}$, since $H(a) \leq H_{max}$ and $\mathcal{L}_{p_n}(\psi_n) < 2\tau_{2^*}$.

Now it remains to prove Claim 7.1. We point out here that this is equivalent to show

$$\int_{S^m} (D\psi_n, \psi_n) d\,\mathrm{vol}_{g_{S^m}} = \int_{S^m} (Dz_n, z_n) d\,\mathrm{vol}_{g_{S^m}} + \int_{S^m} (D\varphi_n, \varphi_n) d\,\mathrm{vol}_{g_{S^m}} + o_n(1). \tag{7.75}$$

And since $\varphi_n = \psi_n - z_n$, it suffices to prove $\int_{S^m} (Dz_n, \varphi_n) d\,\mathrm{vol}_{g_{S^m}} = o_n(1)$ as $n \to \infty$. In fact, for arbitrary $R > 0$, we have

$$\begin{aligned}
\int_{S^m} (Dz_n, \varphi_n) d\,\mathrm{vol}_{g_{S^m}} &= \Big(\int_{B_{R_n R}(a_n)} + \int_{B_{3r}(a_n)\setminus B_{R_n R}(a_n)}\Big)(Dz_n, \varphi_n) d\,\mathrm{vol}_{g_{S^m}} \\
&= \Big(\int_{B_R^0} + \int_{B_{3r/R_n}^0 \setminus B_R^0}\Big)(D_{g_n}\phi_0, \phi_n - \phi_0) d\,\mathrm{vol}_{g_n},
\end{aligned}$$

where the first integral goes to 0 as $n \to \infty$ since $\phi_n \to \phi_0$ in $H^{\frac{1}{2}}_{loc}(\mathbb{R}^m, \mathbb{S}_m)$. Meanwhile to estimate the second integral, we first observe that (through the conformal transformation)

$$\sup_n \int_{B^0_{3r/R_n}} |\phi_n - \phi_0|^{2^*} d\,\mathrm{vol}_{g_{\mathbb{R}^m}} \le \sup_n \int_{B_{3r}(a_n)} |\psi_n - z_n|^{2^*} d\,\mathrm{vol}_{g_{S^m}} < +\infty$$

Thus, by $d\,\mathrm{vol}_{g_n} \le C d\,\mathrm{vol}_{g_{\mathbb{R}^m}}$, we have

$$\Big| \int_{B^0_{3r/R_n} \setminus B^0_R} (D_{g_n}\phi_0, \phi_n - \phi_0) d\,\mathrm{vol}_{g_n} \Big| \le C\Big(\int_{B^0_{3r/R_n} \setminus B^0_R} |\nabla \phi_0|^{\frac{2m}{m+1}} d\,\mathrm{vol}_{g_{\mathbb{R}^m}} \Big)^{\frac{m+1}{2m}} \to 0$$

as $R \to \infty$. Therefore we obtain (7.75) is valid. □

In the above, we proved the concentration behavior $\psi_n = R_n^{-\frac{m-1}{2}} \eta(\cdot)\overline{(\mu_n)_* \circ \phi_0 \circ \mu_n^{-1}} + o_n(1)$ in E and the lower estimate for $\mathcal{L}_{p_n}(\psi_n)$. It remains to locate the blow-up point $a \in S^m$ and to evaluate the parameter $\lambda \in [\bar{\lambda}, 1]$ that are involved in the analysis.

For arbitrary $\xi \in S^m$, we embed S^m into $\mathbb{R}^{m+1}$ in the way that its antipodal point $-\xi$ is the North pole. Denoting $\mathcal{S}_\xi$ the stereographic projection from $S^m \setminus \{-\xi\}$ to $\mathbb{R}^m$, we have $\mathcal{S}_\xi(\xi) = 0$. Particularly, $S^m \setminus \{-\xi\}$ and $\mathbb{R}^m$ are conformally equivalent, where $(\mathcal{S}_\xi^{-1})^* g_{S^m} = f^2 g_{\mathbb{R}^m}$ with $f(x) = \frac{2}{1+|x|^2}$.

Let $\iota : \mathbb{S}(\mathbb{R}^m, (\mathcal{S}_\xi^{-1})^* g_{S^m}) \to \mathbb{S}(\mathbb{R}^m, g_{\mathbb{R}^m})$ denote the isomorphism of spinor bundles induced by the conformal transformation of Dirac operator [23]. If $\psi \in E$ is a solution to Eq. (7.13) for some $p \in (2, 2^*]$, then $\phi := \iota(f^{\frac{m-1}{2}} \psi \circ \mathcal{S}_\xi^{-1})$ will satisfy the transformed equation $D_{g_{\mathbb{R}^m}} \phi = f^{\frac{m-1}{2}(2^*-p)}(H \circ \mathcal{S}_\xi^{-1})|\phi|^{p-2}\phi$ on $(\mathbb{R}^m, g_{\mathbb{R}^m})$.

Let ψ_n be as before, and now we require further that $\mathcal{L}'_{p_n}(\psi_n) \equiv 0$. By the regularity results proved in [3], we have ψ_n is of $C^{1,\alpha}$ for some $\alpha \in (0, 1)$ and are classical solutions to (7.27). The proof of Proposition 7.14 will be accomplished by collecting following two results.

Lemma 7.26 *Let $a \in S^m$ be the associate blow-up point. Then $\nabla H(a) = 0$.*

Lemma 7.27 *Let $\lambda \in [\bar{\lambda}, 1]$ be the parameter involved in* (7.71)*. Then $\lambda = 1$.*

Proof of Lemma 7.26 Let us consider the stereographic projection $\mathcal{S}_a : S^m \setminus \{-a\} \to \mathbb{R}^m$ and the associated bundle isomorphism $\iota : \mathbb{S}(\mathbb{R}^m, (\mathcal{S}_a^{-1})^* g_{S^m}) \to \mathbb{S}(\mathbb{R}^m, g_{\mathbb{R}^m})$. Denoted by $\tilde{\phi}_n = \iota(f^{\frac{m-1}{2}} \psi_n \circ \mathcal{S}_a^{-1})$, we have that $\tilde{\phi}_n$ satisfies

$$D_{g_{\mathbb{R}^m}} \tilde{\phi}_n = H_n(x)|\tilde{\phi}_n|^{p_n-2}\tilde{\phi}_n \quad \text{on } (\mathbb{R}^m, g_{\mathbb{R}^m}). \tag{7.76}$$

where, for ease of notation, we denote $H_n(x) := f(x)^{\frac{m-1}{2}(2^*-p_n)}(H \circ \mathcal{S}_a^{-1})(x)$.

Take $\beta \in C_c^\infty(S^m)$ be a cut-off function on S^m such that $\beta \equiv 1$ on $B_{2r}(a)$ and $\operatorname{supp}\beta \subset B_{3r}(a)$ where $r > 0$ is the fixed radius in the definition of z_n. Then we are allowed to multiply (7.76) by $\phi_{n,k} = \partial_k\big((\beta \circ \mathcal{S}_a^{-1})\tilde{\phi}_n\big)$ as a test spinor for each $k = 1, 2, \ldots, m$, and consequently

$$\operatorname{Re}\int_{\mathbb{R}^m}(D_{g_{\mathbb{R}^m}}\tilde{\phi}_n, \phi_{n,k})d\,\mathrm{vol}_{g_{\mathbb{R}^m}} = \operatorname{Re}\int_{\mathbb{R}^m} H_n|\tilde{\phi}_n|^{p_n-2}(\tilde{\phi}_n, \phi_{n,k})d\,\mathrm{vol}_{g_{\mathbb{R}^m}}\,. \tag{7.77}$$

Note that $(\beta \circ \mathcal{S}_a^{-1})\tilde{\phi}_n$ has a compact support, we may integrate by parts to get

$$\begin{aligned} 0 &= \operatorname{Re}\int_{\mathbb{R}^m}\partial_k\big(D_{g_{\mathbb{R}^m}}\tilde{\phi}_n, (\beta \circ \mathcal{S}_a^{-1})\tilde{\phi}_n\big)d\,\mathrm{vol}_{g_{\mathbb{R}^m}} \\ &= 2\operatorname{Re}\int_{\mathbb{R}^m}(D_{g_{\mathbb{R}^m}}\tilde{\phi}_n, \phi_{n,k})d\,\mathrm{vol}_{g_{\mathbb{R}^m}} \\ &\quad + \operatorname{Re}\int_{\mathbb{R}^m}\big(\partial_k\tilde{\phi}_n, \nabla(\beta \circ \mathcal{S}_a^{-1}) \cdot_{g_{\mathbb{R}^m}} \tilde{\phi}_n\big)d\,\mathrm{vol}_{g_{\mathbb{R}^m}} \\ &\quad - \operatorname{Re}\int_{\mathbb{R}^m}\big(D_{g_{\mathbb{R}^m}}\tilde{\phi}_n, \partial_k(\beta \circ \mathcal{S}_a^{-1})\tilde{\phi}_n\big)d\,\mathrm{vol}_{g_{\mathbb{R}^m}}, \end{aligned} \tag{7.78}$$

where $\cdot_{g_{\mathbb{R}^m}}$ denotes the Clifford multiplication with respect to $g_{\mathbb{R}^m}$. Now let us evaluate the last two integrals of the previous equality. First of all we see from the conformal transformation and the regularity results (see [3]) that $\{\nabla\tilde{\phi}_n\}$ is uniformly bounded in $L^{\frac{2m}{m+1}}(\mathbb{R}^m, \mathbb{S}_m)$. Hence, by the concentration behavior of ψ_n, we derive

$$\begin{aligned} &\Big|\int_{\mathbb{R}^m}\big(\partial_k\tilde{\phi}_n, \nabla(\beta \circ \mathcal{S}_a^{-1}) \cdot_{g_{\mathbb{R}^m}} \tilde{\phi}_n\big)d\,\mathrm{vol}_{g_{\mathbb{R}^m}}\Big| \\ &\quad \le C\Big(\int_{B_{3r}(a)\setminus B_{2r}(a)}|\psi_n|^{2^*}d\,\mathrm{vol}_{g_{S^m}}\Big)^{\frac{1}{2^*}} \to 0 \end{aligned}$$

as $n \to \infty$. And analogously, we have $\big|\int_{\mathbb{R}^m}\big(D_{g_{\mathbb{R}^m}}\tilde{\phi}_n, \partial_k(\beta\circ\mathcal{S}_a^{-1})\tilde{\phi}_n\big)d\,\mathrm{vol}_{g_{\mathbb{R}^m}}\big| \to 0$ as $n \to \infty$. Thus, we conclude from (7.78) that

$$\operatorname{Re}\int_{\mathbb{R}^m}\big(D_{g_{\mathbb{R}^m}}\tilde{\phi}_n, \phi_{n,k}\big)d\,\mathrm{vol}_{g_{\mathbb{R}^m}} = o_n(1) \quad \text{as } n \to \infty. \tag{7.79}$$

On the other hand, to evaluate the integral in the right side of (7.77), we first observe that

$$\begin{aligned} 0 &= \int_{\mathbb{R}^m}\partial_k\big[H_n \cdot (\beta \circ \mathcal{S}_a^{-1})|\tilde{\phi}_n|^{p_n}\big]d\,\mathrm{vol}_{g_{\mathbb{R}^m}} \\ &= \int_{\mathbb{R}^m}\partial_k H_n \cdot (\beta \circ \mathcal{S}_a^{-1})|\tilde{\phi}_n|^{p_n}d\,\mathrm{vol}_{g_{\mathbb{R}^m}} \end{aligned}$$

$$+p_n \operatorname{Re} \int_{\mathbb{R}^m} H_n |\tilde{\phi}_n|^{p_n-2} (\tilde{\phi}_n, \phi_{n,k}) d\, \mathrm{vol}_{g_{\mathbb{R}^m}}$$

$$-(p_n - 1) \int_{\mathbb{R}^m} H_n \cdot \partial_k(\beta \circ \mathcal{S}_a^{-1}) |\tilde{\phi}_n|^{p_n} d\, \mathrm{vol}_{g_{\mathbb{R}^m}} . \tag{7.80}$$

It is evident that the last integral converges to 0 as $n \to \infty$, since $\partial_k(\beta \circ \mathcal{S}_a^{-1})$ vanishes on B_{2r}^0. We only need to estimate the first integral. Notice that $f(x) = \frac{2}{1+|x|^2}$ and $\beta \circ \mathcal{S}_a^{-1}$ has a compact support on $\mathbb{R}^m$, we see that H_n, H_n^{-1} and ∇H_n are bounded uniformly on $\mathrm{supp}(\beta \circ \mathcal{S}_a^{-1})$. Hence, due to the concentration behavior of $\tilde{\phi}_n$, we obtain

$$\int_{\mathbb{R}^m} \partial_k H_n \cdot (\beta \circ \mathcal{S}_a^{-1}) |\tilde{\phi}_n|^{p_n} d\, \mathrm{vol}_{g_{\mathbb{R}^m}} = \partial_k H_n(0) \int_{\mathbb{R}^m} |\tilde{\phi}_n|^{p_n} d\, \mathrm{vol}_{g_{\mathbb{R}^m}} + o_n(1)$$

as $n \to \infty$. This and (7.80) imply

$$\operatorname{Re} \int_{\mathbb{R}^m} H_n |\tilde{\phi}_n|^{p_n-2} (\tilde{\phi}_n, \phi_{n,k}) d\, \mathrm{vol}_{g_{\mathbb{R}^m}}$$

$$= -\frac{1}{p_n} \partial_k H_n(0) \int_{\mathbb{R}^m} |\tilde{\phi}_n|^{p_n} d\, \mathrm{vol}_{g_{\mathbb{R}^m}} + o_n(1). \tag{7.81}$$

Combining (7.77), (7.79) and (7.81), we conclude that

$$\partial_k H_n(0) \int_{\mathbb{R}^m} |\tilde{\phi}_n|^{p_n} d\, \mathrm{vol}_{g_{\mathbb{R}^m}} = o_n(1) \tag{7.82}$$

as $n \to \infty$. Since we already know from the blow-up analysis that

$$\lim_{n\to\infty} \int_{\mathbb{R}^m} f^{\frac{m-1}{2}(2^*-p_n)} |\tilde{\phi}_n|^{p_n} d\, \mathrm{vol}_{g_{\mathbb{R}^m}} = \lim_{n\to\infty} \int_{S^m} |\psi_n|^{p_n} d\, \mathrm{vol}_{g_{S^m}} > 0,$$

(7.82) tells us nothing but $\partial_k H_n(0) \equiv 0$. Notice that $\nabla f(0) = 0$ and k can be varying from 1 to m, we have $\nabla(H \circ \mathcal{S}_a^{-1})(0) = 0$, i.e. $\nabla H(a) = 0$ which completes the proof. □

Proof of Lemma 7.27 The proof is very similar to that of Lemma 7.26. Let us recall the equation under stereographic projection (7.76) and consider the conformal transformation of $\tilde{\phi}_n$ by $\tilde{\phi}_{n,R}(x) = R^{\frac{m-1}{2}} \tilde{\phi}_n(Rx)$ for $R > 0$. Then we have

$$D_{g_{\mathbb{R}^m}} \tilde{\phi}_{n,R} = R^{\frac{m-1}{2}(2^*-p_n)} \widehat{H}_{n,R} |\tilde{\phi}_{n,R}|^{p_n-2} \tilde{\phi}_{n,R} \quad \text{on } \mathbb{R}^m \tag{7.83}$$

where, for ease of notations, we have denoted $\widehat{H}_{n,R}(x) = f^{\frac{m-1}{2}(2^*-p_n)}(Rx) \cdot H \circ \mathcal{S}_a^{-1}(Rx)$.

Let $\beta \in C_c^\infty(S^m)$ be the same cut-off function as in (7.77), and set $\hat{\phi}_{n,R}(x) = \beta \circ \mathcal{S}_a^{-1}(Rx) \cdot \tilde{\phi}_{n,R}(x)$. Then we can multiply (7.83) by $\hat{\phi}_n^* = \frac{\partial}{\partial R}\big|_{R=R_n} \hat{\phi}_{n,R}$ (where $R_n > 0$ was determined in (7.66)) to get

$$\begin{aligned}\mathrm{Re} \int_{\mathbb{R}^m} &(D_{g_{\mathbb{R}^m}} \tilde{\phi}_{n,R_n}, \hat{\phi}_n^*) d\,\mathrm{vol}_{g_{\mathbb{R}^m}} \\ &= R_n^{\frac{m-1}{2}(2^*-p_n)} \mathrm{Re} \int_{\mathbb{R}^m} \widehat{H}_{n,R_n} |\tilde{\phi}_{n,R_n}|^{p_n-2} (\tilde{\phi}_{n,R_n}, \hat{\phi}_n^*) d\,\mathrm{vol}_{g_{\mathbb{R}^m}} .\end{aligned}$$

In what follows, we are going to estimate the above two integrals.

Via the conformal transformation property of the Dirac operator, we see that

$$\int_{\mathbb{R}^m} (D_{g_{\mathbb{R}^m}} \tilde{\phi}_{n,R}, \hat{\phi}_{n,R}) d\,\mathrm{vol}_{g_{\mathbb{R}^m}} \equiv \int_{S^m} (D\psi_n, \beta\psi_n) d\,\mathrm{vol}_{g_{S^m}} \tag{7.84}$$

and

$$\int_{\mathbb{R}^m} (\beta \circ \mathcal{S}_a^{-1})(Rx) \cdot \widehat{H}_{n,R} |\tilde{\phi}_{n,R}|^{p_n} d\,\mathrm{vol}_{g_{\mathbb{R}^m}} \equiv R^{\frac{m-1}{2}(p_n - 2^*)} \int_{S^m} \beta H |\psi_n|^{p_n} d\,\mathrm{vol}_{g_{S^m}} \tag{7.85}$$

for all $R > 0$. Hence, by taking derivative with respect to R in (7.84) and (7.85), and estimate similar as in the proof of Lemma 7.26, we can obtain

$$(2^* - p_n) R_n^{-1} \int_{S^m} \beta H |\psi_n|^{p_n} d\,\mathrm{vol}_{g_{S^m}} = O_n(1) \quad \text{as } n \to \infty.$$

Since the concentration behavior suggests that $\lim_{n\to\infty} \int_{S^m} \beta H |\psi_n|^{p_n} d\,\mathrm{vol}_{g_{S^m}} > 0$, we find $2^* - p_n = O(R_n)$ as $n \to \infty$. Therefore we derive

$$\lambda = \lim_{n\to\infty} R_n^{\frac{m-1}{2}(2^*-p_n)} = \lim_{n\to\infty} e^{O(1) R_n \ln R_n} = 1.$$

This completes the proof. □

References

1. Adams, R.: Sobolev Spaces. Academic Press, New York (1975)
2. Alexandrov, A.D.: Uniqueness theorems for surfaces in the large. Am. Math. Soc. Transl. Ser. **21**, 441–519 (1962)
3. Ammann, B.: A Variational Problem in Conformal Spin Geometry. Habilitationsschrift, Universität Hamburg, Hamburg (2003)
4. Ammann, B.: A spin-conformal lower bound of the first positive Dirac eigenvalue. Differ. Geom. Appl. **18**(1), 21–32 (2003)
5. Ammann, B.: The smallest Dirac eigenvalue in a spin-conformal class and cmc immersions. Commun. Anal. Geom. **17**(3), 429–479 (2009)

6. Ammann, B., Humbert, E., Morel, B.: Mass endomorphism and spinorial Yamabe type problems on conformally flat manifolds. Commun. Anal. Geom. **14**(1), 163–182 (2006)
7. Ammann, B., Humbert, E., Ould Ahmedou, M.: An obstruction for the mean curvature of a conformal immersion $S^n \to \mathbb{R}^{n+1}$. Proc. Am. Math. Soc. **135**(2), 489-493 (2007)
8. Ammann, B., Grossjean, J.F., Humbert, E., Morel, B.: A spinorial analogue of Aubin's inequality. Math. Z. **260**, 127–151 (2008)
9. Anderson, M.: Conformal immersions of prescribed mean curvature in $\mathbb{R}^3$. Nonlinear Anal. **114**, 142–157 (2015)
10. Aubin, T.: Équations différentielles non linéaires et problème de Yamabe concernant la courbure scalaire. J. Math. Pures Appl. **55**, 269–296 (1976)
11. Bär, C.: On nodal sets for Dirac and Laplace operators. Commun. Math. Phys. **188**(3), 709–721 (1997)
12. Bartsch, T., Xu, T.: Strongly localized semiclassical states for nonlinear Dirac equations. Discrete Contin. Dyn. Syst. **41**(1), 29–60 (2021). Special issue celebrating 25 years of excellence
13. Bartsch, T., Xu, T.: A spinorial analogue of the Brezis-Nirenberg theorem involving the critical Sobolev exponent. J. Funct. Anal. **280**, 108991 (2021)
14. Baum, H., Friedrich, T., Grunewald, R., Kath, I.: Twistors and Killing Spinors on Riemannian Manifolds. Teubner, Leipzig (1991)
15. Booss-Bavnbek, B., Marcolli, M., Wang, B.L.: Weak UCP and perturbed monopole equations. Int. J. Math. **13**, 987 (2002)
16. Bourguignon, J.P., Gauduchon, P.: Spineurs, opérateurs de Dirac et variations de métriques. Commun. Math. Phys. **144**(3), 581–599 (1992)
17. Buffoni, B., Jeanjean, L., Stuart, C.A.: Existence of a nontrivial solution to a strongly indefinite semilinear equation. Proc. Am. Math. Soc. **119**(1), 179–186 (1993)
18. Chen, Q., Jost, J., Wang, G.: Nonlinear Dirac equations on Riemann surfaces. Ann. Global Anal. Geom. **33**, 253–270 (2008)
19. Eisenhart, L.P.: A treatise on the differential geometry of curves and surfaces. Ginn and company, Boston (1909)
20. Ekeland, I.: On the variational principle. J. Math. Anal. Appl. **47**, 324–353 (1974)
21. Friedrich, T.: On the spinor representation of surfaces in Euclidean 3-space. J. Geom. Phys. **28**(1–2), 143–157 (1998)
22. Friedrich, T.: Dirac operators in Riemannian Geometry. Graduate Studies in Mathematics, vol. 25. American Mathematical Society, Providence (2000)
23. Ginoux, N.: The Dirac Spectrum. Lecture Notes in Mathematics, vol. 1976. Springer, Berlin (2009)
24. Hsiang, W.Y., Teng, Z.H., Yu, W.C.: New examples of constant mean curvature immersions of $(2k-1)$-spheres into Euclidean $2k$-Space. Ann. Math. **117**, 609–625 (1983)
25. Isobe, T.: Nonlinear Dirac equations with critical nonlinearities on compact spin manifolds. J. Funct. Anal. **260**, 253–307 (2011)
26. Isobe, T.: A perturbation method for spinorial Yamabe type equations on S^m and its application. Math. Ann. **355**(4), 1255–1299 (2013)
27. Kazdan, J., Warner, F.: Existence and conformal deformation of metrics with prescribed Gaussian and scalar curvature. Ann. Math. **101**, 317–331 (1975)
28. Kusner, R., Schmitt, N.: Representation of surfaces in space (1996). arXiv:dg-ga/9610005
29. Lawson, H.B., Michelson, M.L.: Spin Geometry. Princeton University Press, Princeton (1989)
30. Li, P., Yau, S.T.: A new conformal invariant and its applications to the Willmore conjecture and the first eigenvalue on compact surfaces. Invent. Math. **69**, 269–291 (1982)
31. Mawhin, J., Willem, M.: Critical Point Theory and Hamiltonian Systems. Applied Mathematical Sciences, vol. 74. Springer, New York (1989)
32. Raulot, S.: A Sobolev-like inequality for the Dirac operator. J. Funct. Anal. **256**, 1588–1617 (2009)
33. Sulanke, S.: Die Berechnung des Spektrums des Quadrates des Dirac-Operators auf der Sphäre, Doktorarbeit. Humboldt-Universität zu Berlin, Berlin (1979)

34. Taimanov, I.A.: Surfaces of revolution in terms of solitons. Ann. Global Anal. Geom. **15**(5), 419–435 (1997)
35. Taimanov, I.A.: Modified Novikov-Veselov equation and differential geometry of surfaces. Am. Math. Soc. Transl. Ser. **179**, 133–151 (1997)
36. Taimanov, I.A.: The Weierstrass representation of closed surfaces in $\mathbb{R}^3$. Funct. Anal. Appl. **32**(4), 258–267 (1998)
37. Wang, C.Y.: A remark on nonlinear Dirac equations. Proc. Am. Math. Soc. **138**(10), 3753–3758 (2010)
38. Willem, M.: Minimax Theorems. Birkhäuser, Basel (1996)
39. Xu, T.: Conformal embeddings of $S^2 \to \mathbb{R}^3$ with prescribed mean curvature: a variational approach. Commun. Anal. Geom. **32**(1), 203–254 (2024)
40. Yamabe, H.: On the deformation of Riemannian structures on compact manifolds. Osaka Math. J. **12**, 21–37 (1960)

Chapter 8
Further Problems with Strongly Indefinite Structures

In this chapter, we shall survey two further problems arising from the study of strongly indefinite structures. First we will focus on the role of geometric datum of a Riemannian manifold in Hamiltonian elliptic systems, see Sect. 8.1. Next, in Sect. 8.2, we will deal with some coupled-Dirac systems.

8.1 Singularly Perturbed Elliptic System on Riemannian Manifolds

Let (M, g) be a smooth closed Riemannian m-manifold, $m \geq 3$. We consider the singularly perturbed nonlinear elliptic problem

$$\begin{cases} -\varepsilon^2 \Delta_g v + v = f_1(u), & \text{on } M \\ -\varepsilon^2 \Delta_g u + u = f_2(v), & \\ u, v > 0, & \end{cases} \tag{8.1}$$

where $\Delta_g = \mathrm{div}_g \nabla_g$ is the Laplace-Beltrami operator on (M, g) and f_1, f_2 fall into a typical class of superlinear functions. Beside the existence of a solution for a fixed value of ε, we are interested in the blow-up profile (the asymptotic behavior) of these solutions as ε goes to zero. Particularly, we are looking for a local geometric quantity of (M, g) which affects the location of a global maximum point of a solution of (8.1) as $\varepsilon \searrow 0$.

And in order to describe our results, it is useful to recall some notations and definitions. For a closed Riemannian m-manifold (M, g), let $\exp : TM \to M$ be the exponential map defined on the tangent bundle TM of M. Since we assume M to be closed, there exists $r > 0$ such that $\exp_\xi : B_r(0) \supset \mathbb{R}^m \cong T_\xi M \to B_r(\xi) \subset M$ is a diffeomorphism for any $\xi \in M$. In what follows, $B_r(0)$ will denote the ball in

© The Author(s), under exclusive license to Springer Nature Singapore Pte Ltd. 2025
Y. Ding, T. Xu, *Localization Approaches in Strongly Indefinite Problems*, Series in Contemporary Mathematics 6, https://doi.org/10.1007/978-981-97-9539-0_8

$\mathbb{R}^m$ centered at 0 with radius r and, for $\xi \in M$, $B_r(\xi)$ will denote the ball in M centered at ξ with respect to the metric g. For vector fields X, Y, W, Z on M, the Riemannian curvature tensor $\mathcal{R}$ is given by

$$\mathcal{R}(X, Y, W, Z) = g(\nabla_Y \nabla_X W, Z) - g(\nabla_X \nabla_Y W, Z) - g(\nabla_{[Y,X]} W, Z) \tag{8.2}$$

where ∇ is the Riemannian connection. The Ricci tensor Ric : $TM \times TM \to \mathbb{R}$ is defined by the trace of $\mathcal{R}$, that is,

$$\mathrm{Ric}_\xi(X, Y) = \sum_{j=1}^{m} \mathcal{R}_\xi(X, e_j, Y, e_j) \tag{8.3}$$

where $\{e_1, \ldots, e_m\}$ is a basis of the (Riemann) normal coordinates in $T_\xi M$. The scalar curvature will be denoted by Scal_g.

8.1.1 The Functional Settings

Let (M, g) be a closed Riemannian m-manifold, $m \geq 3$, we set

$$H_g^1(M) = \left\{ u : M \to \mathbb{R} : \int_M |\nabla_g u|^2 + |u|^2 d\,\mathrm{vol}_g < +\infty \right\}$$

where $d\,\mathrm{vol}_g$ is the volume form on M associated to the metric g. We also set $L_g^p(M)$, $p \geq 1$, for the space of functions on (M, g) for which the p-th power of their absolute value is integrable, and is equipped with its canonical norm. It is clear that $L_g^2(M)$ is a Hilbert space.

Let us take the index $\alpha = \varepsilon \in (0, +\infty)$, and introduce the self-adjoint operator

$$S_\varepsilon : \mathcal{D}(S_\varepsilon) \subset L_g^2(M) \times L_g^2(M) \to L_g^2(M) \times L_g^2(M)$$

defined by

$$S_\varepsilon \begin{pmatrix} u \\ v \end{pmatrix} = \begin{pmatrix} -\varepsilon^2 \Delta_g v + v \\ -\varepsilon^2 \Delta_g u + u \end{pmatrix}.$$

Then (8.1) can be equivalently rewritten as

$$S_\varepsilon z = \mathcal{F}(z) \tag{8.4}$$

where the nonlinear part is defined by

$$\mathcal{F}(z) = \begin{pmatrix} f_1(u) \\ f_2(v) \end{pmatrix} \quad \text{for } z = \begin{pmatrix} u \\ v \end{pmatrix} \in \mathcal{D}(S_\varepsilon).$$

Associated with (8.4), we choose the Hilbert space $\mathcal{H} = H_g^1(M) \times H_g^1(M)$, and equip it with a family of inner products

$$\begin{aligned} \langle z_1, z_2 \rangle_\varepsilon &= \frac{1}{\varepsilon^m} \int_M \big(\varepsilon^2 \nabla_g u_1 \cdot \nabla_g u_2 + u_1 u_2\big) d\,\mathrm{vol}_g \\ &\quad + \frac{1}{\varepsilon^m} \int_M \big(\varepsilon^2 \nabla_g v_1 \cdot \nabla_g v_2 + v_1 v_2\big) d\,\mathrm{vol}_g \end{aligned}$$

for $z_i = \begin{pmatrix} u_i \\ v_i \end{pmatrix} \in \mathcal{H}$, $i = 1, 2$, and the induced norms $\|\cdot\|_\varepsilon$.

We now take the operator $L_\varepsilon : (\mathcal{H}, \langle \cdot, \cdot \rangle_\varepsilon) \to (\mathcal{H}, \langle \cdot, \cdot \rangle_\varepsilon)$ by

$$L_\varepsilon \begin{pmatrix} u \\ v \end{pmatrix} = \begin{pmatrix} v \\ u \end{pmatrix}.$$

Then we have L_ε is self-adjoint, continuous, invertible and $\sigma(L_\varepsilon) = \{1, -1\}$. Particularly, we can split the space $\mathcal{H}$ with respect to L_ε as

$$\mathcal{H} = \mathcal{H}_\varepsilon^+ \oplus \mathcal{H}_\varepsilon^-, \quad \text{where } \mathcal{H}_\varepsilon^\pm \equiv \big\{(u, \pm u)^T : u \in H_g^1(M)\big\}.$$

It is clear that (2.16) and (2.17) are satisfied on $\mathcal{H}_\varepsilon^+$ and $\mathcal{H}_\varepsilon^-$, respectively, with the constant $\delta = 1$. Moreover, (8.4) (or equivalently (8.1)) reads as

$$L_\varepsilon z = N_\varepsilon(z) \quad \text{for } z \in \mathcal{H}$$

where the relation between S_ε and L_ε is given by

$$L_\varepsilon = T_\varepsilon^{-1} \circ S_\varepsilon \quad \text{with } T_\varepsilon = -\varepsilon^2 \Delta_g + 1 : \mathcal{H} \to \mathcal{H}^*$$

and, similarly, $N_\varepsilon = T_\varepsilon^{-1} \circ \mathcal{F}$.

In order to study this nonlinear problem, we impose the following conditions on the nonlinear functions f_i, $i = 1, 2$, involved in (8.1):

(F) $f_i \in C^1(\mathbb{R})$, $f_i(s) = 0$ for $s \le 0$, and there exist $q_1, q_2 > 2$ such that $\frac{1}{q_1} + \frac{1}{q_2} > \frac{m-2}{m}$ and

$$\lim_{s \to +\infty} \frac{f_1'(s)}{s^{q_1 - 2}} = \lim_{s \to +\infty} \frac{f_2'(s)}{s^{q_2 - 2}} = 0. \tag{8.5}$$

Moreover, for some $\theta > 1$ and every $s > 0$,

$$0 < \theta f_i(s) \le f_i'(s)s \quad i = 1, 2. \tag{8.6}$$

As was explained in [13, 15], we can assume without loss of generality that the numbers q_1, q_2 in (F) are such that $2 < q_1 = q_2 < \frac{2m}{m-2}$. In fact, since the solutions we work with are of ground-state type, they are a priori bounded in L^∞-norm and this bound keeps invariant even if we truncate f_1 and f_2 at infinity so that (F) is satisfied for the modified functions with $q_1 = q_2$ in (8.5).

Now, we denote $F_i(s) = \int_0^s f_i(t)dt$, $i = 1, 2$, the primitive functions of f_i, and $\Psi_\varepsilon : (\mathcal{H}, \langle\cdot,\cdot\rangle_\varepsilon) \to \mathbb{R}$ with

$$\Psi_\varepsilon(z) = \frac{1}{\varepsilon^m}\int_M F_1(u)d\,\mathrm{vol}_g + \frac{1}{\varepsilon^m}\int_M F_2(v)d\,\mathrm{vol}_g \quad \text{for } z = \begin{pmatrix} u \\ v \end{pmatrix} \in \mathcal{H}.$$

We also equip $L_g^p(M) \times L_g^p(M)$, $p \ge 1$, with the norm

$$|z|_{p,\varepsilon} = \Big(\frac{1}{\varepsilon^m}\int_M |z|^p d\,\mathrm{vol}_g\Big)^{\frac{1}{p}} \quad \text{for } z = \begin{pmatrix} u \\ v \end{pmatrix} \in L_g^p(M) \times L_g^p(M).$$

By Sobolev's embedding theorem, we have that there exists $c_p > 0$, independent of ε, such that $|z|_{p,\varepsilon} \le c_p\|z\|_\varepsilon$ for each $p \in [2, \frac{2m}{m-2}]$. Then we verify the hypotheses (H1) to (H6) of Theorem 3.34 as follows.

Lemma 8.1 *Under condition (F), we have $\Psi_\varepsilon \in C^2(\mathcal{H}, \mathbb{R})$ satisfies (H1)-(H6) of Theorem 3.34 for every $\varepsilon > 0$.*

Proof Since the proof is quite standard, we will be sketchy.

The C^2-smoothness comes directly from (8.5) and our choice of $q := q_1 = q_2 \in (2, \frac{2m}{m-2})$. Hence (H1) is verified with $N_\varepsilon = \nabla\Psi_\varepsilon$. To proceed, we mention that (8.6) implies

$$\theta F_i(s) = \theta\int_0^s f_i(t)dt \le \int_0^s f_i'(t)tdt = f_i(t)t\Big|_{t=0}^{t=s} - \int_0^s f_i(t)dt = f_i(s)s - F_i(s).$$

Thus $(\theta + 1)F_i(s) \le f_i(s)s$, $i = 1, 2$, and this gives (H2). Using the positiveness of $f_i'(s)$ in (8.6), the convexity follows immediately from the computation: for $z = \begin{pmatrix} u \\ v \end{pmatrix}$, $w = \begin{pmatrix} \tilde{u} \\ \tilde{v} \end{pmatrix} \in \mathcal{H}$

$$\langle\nabla N_\varepsilon(z)[w], w\rangle_\varepsilon = \frac{1}{\varepsilon^m}\int_M f_1'(u)\tilde{u}^2 d\,\mathrm{vol}_g + \frac{1}{\varepsilon^m}\int_M f_2'(v)\tilde{v}^2 d\,\mathrm{vol}_g \ge 0.$$

Moreover, since $f_i(0) = f_i'(0) = 0$, we can see from (8.5) that for arbitrary $\sigma > 0$ there exists $c_\sigma > 0$ such that $f_i(s) \le \sigma s + c_\sigma \big(f_i(s)s\big)^{\frac{q-1}{q}}$, for all $s \ge 0$. And hence,

$$
\begin{aligned}
\langle N_\varepsilon(z), w\rangle_\varepsilon &= \frac{1}{\varepsilon^m}\int_M f_1(u)\tilde{u}d\,\mathrm{vol}_g + \frac{1}{\varepsilon^m}\int_M f_2(v)\tilde{v}d\,\mathrm{vol}_g \\
&\le \frac{\sigma}{\varepsilon^m}\int_M \big(|u||\tilde{u}| + |v||\tilde{v}|\big)d\,\mathrm{vol}_g \\
&\quad + \frac{c_\sigma}{\varepsilon^m}\int_M \Big(\big(f_1(u)u\big)^{\frac{q-1}{q}}|\tilde{u}| + \big(f_2(v)v\big)^{\frac{q-1}{q}}|\tilde{v}|\Big)d\,\mathrm{vol}_g \\
&\le \sigma|z|_{2,\varepsilon}|w|_{2,\varepsilon} + c_\sigma\,\langle N_\varepsilon(z), z\rangle_\varepsilon^{\frac{q-1}{q}}\,|w|_{q,\varepsilon} \quad \text{by Hölder inequality}
\end{aligned}
$$

which verifies (H4) by choosing σ small. To see (H5), we first remark that, for $s > 0$ and $t \in \mathbb{R}$,

$$
\begin{aligned}
f_i'(s)(s+t)^2 - 2f_i(s)t &= f_i'(s)\Big(s^2 + 2\Big(1 - \frac{f_i(s)}{f_i'(s)s}\Big)st + t^2\Big) \\
&= f_i'(s)\Big(\Big(1 - \frac{f_i(s)}{f_i'(s)s}\Big)s + t\Big)^2 \\
&\quad + f_i'(s)\Big(1 - \Big(1 - \frac{f_i(s)}{f_i'(s)s}\Big)^2\Big)s^2 \\
&\ge \Big(2 - \frac{f_i(s)}{f_i'(s)s}\Big)f_i(s)s \quad \text{by the positiveness of } f_i'(s).
\end{aligned}
$$

And since, by (8.6), we have $0 < \frac{f_i(s)}{f_i'(s)s} \le \frac{1}{\theta} < 1$. This, together with above estimate, implies (H5). Finally, the hypothesis (H6) follows from the compact embedding

$$
(\mathcal{H}, \|\cdot\|_\varepsilon) \hookrightarrow \big(L_g^q(M) \times L_g^q(M), |\cdot|_{q,\varepsilon}\big),
$$

for $q \in (2, \frac{2m}{m-2})$. □

By virtue of the above lemma, we see that solutions of (8.1) correspond to critical points of the functional

$$
\Phi_\varepsilon(z) = \frac{1}{2}\langle L_\varepsilon z, z\rangle_\varepsilon - \Psi_\varepsilon(z) \quad \text{for } z \in \mathcal{H} \text{ and } \varepsilon \in (0, +\infty).
$$

And, in order to apply Theorem 3.34 (with the index $\alpha = \varepsilon$), one needs to find a sequence of test functions satisfy (3.42) for the functionals Φ_ε. In the sequel, the following problem on the Euclidean space $\mathbb{R}^m$ will play a key role:

$$\begin{cases} -\Delta v + v = f_1(u), & \\ -\Delta u + u = f_2(v), & \text{on } \mathbb{R}^m \\ u, v > 0 & \end{cases} \tag{8.7}$$

whose energy functional is given by

$$\Phi_e(z) = \int_{\mathbb{R}^m} \big(\nabla u \cdot \nabla v + uv - F_1(u) - F_2(v)\big)dx, \text{ for } z = \begin{pmatrix} u \\ v \end{pmatrix} \in H^1(\mathbb{R}^m) \times H^1(\mathbb{R}^m).$$

Let us denote γ_e the ground-state critical value of Φ_e, that is,

$$\gamma_e = \inf\big\{\Phi_e(z) : \nabla\Phi_e(z) = 0,\ z \neq 0\big\}.$$

Under the same condition (F), it follows from [14, 16] that $\gamma_e > 0$ is attained at some positive functions U, V which satisfy the system (8.7). Moreover, by elliptic estimates, U and V decay exponentially and they are radially symmetric with respect to a common point of $\mathbb{R}^m$ (see for instance [2]). For later use, we denote $\mathcal{L}_e$ the set of all ground-state solutions (U, V) of Eq. (8.7) satisfying $U(0) = \max_{x\in\mathbb{R}^m} U(x)$ and $V(0) = \max_{x\in\mathbb{R}^m} V(x)$. Then it follows that any $(U, V) \in \mathcal{L}_e$ is radially symmetric with respect to the origin and $\frac{\partial U}{\partial\rho} < 0$, $\frac{\partial V}{\partial\rho} < 0$ for $\rho = |x| > 0$ (see [2, 4]).

Lemma 8.2 *Let $(U, V) \in \mathcal{L}_e$. Then, for any $k \geq 0$*

$$\int_{\mathbb{R}^m} \Big(\nabla U \cdot \nabla V + UV - F_1(U) - F_2(V)\Big)|x|^k dx = \frac{2(k+1)}{m+k}\int_{\mathbb{R}^m} \big(\nabla U \cdot \nabla V\big)|x|^k dx \tag{8.8}$$

Proof Since U, V are decreasing exponentially at infinity, multiplying both sides of system (8.7) by the couple $(|x|^k x \cdot \nabla U, |x|^k x \cdot \nabla V)$, integrating by parts on $B_R(0)$ and taking $R \to \infty$, we get the identity. □

Next, we define a map $\mathcal{S} : M \times \mathcal{L}_e \to \mathbb{R}$ by

$$\begin{aligned} \mathcal{S}(\xi, Z) := &\sum_{i,j=1}^{m} \int_{\mathbb{R}^m} \mathcal{R}_\xi(X, e_i, X, e_j)\partial_i U \partial_j V dx - \frac{1}{2}\int_{\mathbb{R}^m} \mathrm{Ric}_\xi(X, X)\nabla U \cdot \nabla V dx \\ &- \frac{1}{2}\int_{\mathbb{R}^m} \mathrm{Ric}_\xi(X, X) UV dx + \frac{1}{2}\int_{\mathbb{R}^m} \mathrm{Ric}_\xi(X, X)\big(F_1(U) + F_2(V)\big)dx \end{aligned} \tag{8.9}$$

where $X \in T_\xi M$, $X = \sum_{i=1}^m x_i e_i$. Then we have the following

Proposition 8.3 *Let $Z = (U, V) \in \mathcal{L}_e$. Then, there holds that*

$$\mathcal{S}(\xi, Z) = -\frac{3\mathrm{Scal}_g(\xi)}{m(m+2)} \int_{\mathbb{R}^m} \nabla U \cdot \nabla V |x|^2 dx,$$

where $\mathrm{Scal}_g(\xi)$ *is the scalar curvature of M at ξ.*

Proof Note that $X = \sum_{i=1}^m x_i e_i$ and $Z = (U, V)$ is radially symmetric, it follows from (8.2) and (8.3) that

$$\sum_{i,j=1}^m \int_{\mathbb{R}^m} \mathcal{R}_\xi(X, e_i, X, e_j) \partial_i U \partial_j V dx$$
$$= \sum_{i,j=1}^m \sum_{k,l=1}^m \int_{\mathbb{R}^m} \mathcal{R}_\xi(e_k, e_i, e_l, e_j) \frac{x_k x_l x_i x_j}{|x|^2} \frac{\partial U}{\partial \rho} \frac{\partial V}{\partial \rho} dx = 0$$

and

$$\int_{\mathbb{R}^m} \mathrm{Ric}_\xi(X, X) \nabla U \cdot \nabla V dx = \sum_{i,j=1}^m \int_{\mathbb{R}^m} \mathrm{Ric}_\xi(e_i, e_j) x_i x_j \nabla U \cdot \nabla V dx$$
$$= \sum_{i=1}^m \mathrm{Ric}_\xi(e_i, e_i) \int_{\mathbb{R}^m} (x_i)^2 \nabla U \cdot \nabla V dx$$
$$= \frac{\mathrm{Scal}_g(\xi)}{m} \int_{\mathbb{R}^m} \nabla U \cdot \nabla V |x|^2 dx.$$

where in the last equality, we have used $\mathrm{Scal}_g(\xi) = \sum_{i=1}^m \mathrm{Ric}_\xi(e_i, e_i)$. Analogously, we have

$$\int_{\mathbb{R}^m} \mathrm{Ric}_\xi(X, X) U V dx = \frac{\mathrm{Scal}_g(\xi)}{m} \int_{\mathbb{R}^m} U V |x|^2 dx$$

and

$$\int_{\mathbb{R}^m} \mathrm{Ric}_\xi(X, X) \big(F_1(U) + F_2(V)\big) dx = \frac{\mathrm{Scal}_g(\xi)}{m} \int_{\mathbb{R}^m} \big(F_1(U) + F_2(V)\big) |x|^2 dx.$$

Finally, the assertion can be straightforwardly verified via Lemma 8.2. □

8.1.2 Scalar Curvature and the Localization

For the given Riemannian metric g, we write it locally as $g = \sum_{i,j=1}^{m} g_{ij} dx_i \otimes dx_j$ on $B_r(\xi)$, $\xi \in M$. We also let $G(\xi) = (g_{ij}(\xi))_{ij}$ and $G^{-1}(\xi) = (g^{ij}(\xi))_{ij}$. Henceforth, det stands for the determinant. We now consider the Riemann normal coordinates on $B_r(\xi)$. For ease of notation, we identify $x = (x_1, \ldots, x_m) \in \mathbb{R}^m$ with $X = \sum_i x_i e_i \in T_\xi M$, then it is well known (see [3]) that

$$g_{ij}(\exp_\xi x) = \delta_{ij} - \frac{1}{3}\mathcal{R}_\xi(x, e_i, x, e_j) + O(|x|^3), \tag{8.10}$$

$$g^{ij}(\exp_\xi x) = \delta_{ij} + \frac{1}{3}\mathcal{R}_\xi(x, e_i, x, e_j) + O(|x|^3) \tag{8.11}$$

and

$$\sqrt{\det G}(\exp_\xi x) = 1 - \frac{1}{6}\,\mathrm{Ric}_\xi(x, x) + O(|x|^3), \tag{8.12}$$

where δ_{ij} is the Kronecker delta.

Let $0 < r < inj(M)/2$ where $inj(M) > 0$ is the injectivity radius of M, and let $\eta \in C_c^\infty(\mathbb{R}^m, [0, 1])$ be such that $|\nabla\eta| \le 2/r$, $\eta(x) = 1$ for $|x| \le r$ and $\eta(x) = 0$ for $|x| \ge 2r$. For a ground-state solution $Z = (U, V) \in \mathcal{L}_e$, we set $U_\varepsilon(x) = U(\frac{x}{\varepsilon})$ and $V_\varepsilon(x) = V(\frac{x}{\varepsilon})$. Define $W_\varepsilon = (\eta U_\varepsilon, \eta V_\varepsilon)$ and $z_\varepsilon = (u_\varepsilon, v_\varepsilon) = W_\varepsilon \circ \exp_\xi^{-1}$. Then, via the expression

$$\Delta_g u = \frac{1}{\sqrt{\det G}} \sum_{i,j} \partial_i\big(g^{ij}\sqrt{\det G}\,\partial_j u\big),$$

we have

Lemma 8.4 *As* $\varepsilon \to 0$,

$$\Phi_\varepsilon(z_\varepsilon) = \gamma_e + \frac{\varepsilon^2}{3}\mathcal{S}(\xi, Z) + O(\varepsilon^3) \quad \text{and} \quad \|\nabla\Phi_\varepsilon(z_\varepsilon)\|_\varepsilon = O(\varepsilon^2),$$

i.e. $\{z_\varepsilon\}$ *verifies the condition* (3.42).

Proof Since $Z = (U, V) \in \mathcal{L}_e$ decays exponentially at infinity, by the change of variables and (8.10)-(8.12), we see immediately that

$$\begin{aligned}
\Phi_\varepsilon(z_\varepsilon) &= \frac{1}{\varepsilon^m}\int_M \Big(\varepsilon^2 \sum_{i,j=1}^m g^{ij}\partial_i u_\varepsilon \partial_j v_\varepsilon + u_\varepsilon v_\varepsilon\Big) d\,\mathrm{vol}_g \\
&\quad - \frac{1}{\varepsilon^m}\int_M \big(F_1(u_\varepsilon) + F_2(v_\varepsilon)\big) d\,\mathrm{vol}_g \\
&= \frac{1}{\varepsilon^m}\int_{B_{2r}(0)} \Big(\varepsilon^2 \sum_{i,j=1}^m g^{ij}(\exp_\xi x)\partial_i(\eta U_\varepsilon)\partial_j(\eta V_\varepsilon) + \eta^2 U_\varepsilon V_\varepsilon\Big) \\
&\qquad \sqrt{\det G}(\exp_\xi x)dx \\
&\quad - \frac{1}{\varepsilon^m}\int_{B_{2r}(0)} \big(F_1(\eta U_\varepsilon) + F_2(\eta V_\varepsilon)\big)\sqrt{\det G}(\exp_\xi x)dx \\
&= \Phi_e(Z) + \frac{\varepsilon^2}{3}\mathcal{S}(\xi, Z) + O(\varepsilon^3) \qquad \text{use the formulation of } \mathcal{S} \text{ in (8.9)}
\end{aligned}$$

and analogously, for arbitrary $z = \binom{u}{v} \in \mathcal{H}$,

$$\begin{aligned}
\langle \nabla\Phi_\varepsilon(z_\varepsilon), z\rangle_\varepsilon &= \frac{1}{\varepsilon^m}\int_{B_{2r}(0)} \Big(\varepsilon^2 \sum_{i,j=1}^m g^{ij}(\exp_\xi x)\partial_i u_\varepsilon \partial_j v + u_\varepsilon v\Big) \\
&\qquad \sqrt{\det G}(\exp_\xi x)dx \\
&\quad - \frac{1}{\varepsilon^m}\int_{B_{2r}(0)} f_2(\eta V_\varepsilon) v \sqrt{\det G}(\exp_\xi x)dx \\
&\quad + \frac{1}{\varepsilon^m}\int_{B_{2r}(0)} \Big(\varepsilon^2 \sum_{i,j=1}^m g^{ij}(\exp_\xi x)\partial_i u \partial_j v_\varepsilon + u v_\varepsilon\Big) \\
&\qquad \sqrt{\det G}(\exp_\xi x)dx \\
&\quad - \frac{1}{\varepsilon^m}\int_{B_{2r}(0)} f_1(\eta U_\varepsilon) u \sqrt{\det G}(\exp_\xi x)dx \\
&= O(\varepsilon^2)\|z\|_\varepsilon, \qquad \text{use the exponential decay of } U \text{ and } V
\end{aligned}$$

hence the lemma is proved. □

Since we have checked that all the hypotheses in Theorem 3.34 hold true, we are now ready to state

Proposition 8.5 *For all small $\varepsilon > 0$, there exists a ground-state solution $\bar{z}_\varepsilon = (\bar{u}_\varepsilon, \bar{v}_\varepsilon)$ of* (8.1) *with energy evaluation*

$$\gamma_\varepsilon := \Phi_\varepsilon(\bar{z}_\varepsilon) \leq \gamma_e + \frac{\varepsilon^2}{3}\mathcal{S}(\xi, Z) + O(\varepsilon^3), \tag{8.13}$$

where $\xi \in M$ and $Z \in \mathcal{L}_e$ are given arbitrarily. Moreover, for each ε, there exists a unique maximum point $\xi_\varepsilon \in M$ of $\bar{u}_\varepsilon + \bar{v}_\varepsilon$ such that for some constants $C, c > 0$,

$$\bar{u}_\varepsilon(\xi) + \bar{v}_\varepsilon(\xi) + |\nabla \bar{u}_\varepsilon(\xi)| + |\nabla \bar{v}_\varepsilon(\xi)| \leq Ce^{-\frac{c}{\varepsilon}\mathrm{dist}_g(\xi, \xi_\varepsilon)}, \quad \xi \in M$$

where dist_g is the distance induced by the metric g. And, up to a subsequence, the transformed functions $\bar{U}_\varepsilon(x) = \bar{u}_\varepsilon \circ \exp_{\xi_\varepsilon}(\varepsilon x)$ and $\bar{V}_\varepsilon(x) = \bar{v}_\varepsilon \circ \exp_{\xi_\varepsilon}(\varepsilon x)$ converge to a ground-state solution $\bar{Z} = (\bar{U}, \bar{V}) \in \mathcal{L}_e$ of (8.7) *in C^2_{loc}-sense.*

Proof Since the existence and the energy evaluation are direct consequences of Lemma 8.1, 8.4 and Theorem 3.34, we only need to describe the asymptotic behavior of the solutions $\bar{z}_\varepsilon = (\bar{u}_\varepsilon, \bar{v}_\varepsilon)$ as $\varepsilon \to 0$.

By the Sobolev embedding and the arguments in Lemma 3.28, we deduce that $\|\bar{z}_\varepsilon\|_\varepsilon$ is bounded. And by bootstrap arguments, we see also that $\|\bar{u}_\varepsilon\|_\infty$ and $\|\bar{v}_\varepsilon\|_\infty$ are bounded. Furthermore, via the maximum principle, we have $\|\bar{u}_\varepsilon + \bar{v}_\varepsilon\|_\infty$ is bounded away from 0.

Let $\xi_\varepsilon \in M$ be such that $\liminf_{\varepsilon\to 0} \bar{u}_\varepsilon(\xi_\varepsilon) + \bar{v}_\varepsilon(\xi_\varepsilon) > 0$. Consider the Riemann normal coordinates on $B_r(\xi_\varepsilon) \subset M$, for $r > 0$ fixed small. Denoted by $\bar{U}_\varepsilon(x) = \bar{u}_\varepsilon \circ \exp_{\xi_\varepsilon}(\varepsilon x)$ and $\bar{V}_\varepsilon(x) = \bar{v}_\varepsilon \circ \exp_{\xi_\varepsilon}(\varepsilon x)$, we find

$$\begin{cases} -\dfrac{1}{\sqrt{\det G}(\exp_{\xi_\varepsilon}(\varepsilon x))}\displaystyle\sum_{i,j}\partial_i\big(g^{ij}(\exp_{\xi_\varepsilon}(\varepsilon x))\sqrt{\det G}(\exp_{\xi_\varepsilon}(\varepsilon x))\partial_j\bar{U}_\varepsilon\big) \\ \quad + \bar{U}_\varepsilon = f_2(\bar{V}_\varepsilon), \\ -\dfrac{1}{\sqrt{\det G}(\exp_{\xi_\varepsilon}(\varepsilon x))}\displaystyle\sum_{i,j}\partial_i\big(g^{ij}(\exp_{\xi_\varepsilon}(\varepsilon x))\sqrt{\det G}(\exp_{\xi_\varepsilon}(\varepsilon x))\partial_j\bar{V}_\varepsilon\big) \\ \quad + \bar{V}_\varepsilon = f_1(\bar{U}_\varepsilon), \end{cases}$$

on $B_{r/\varepsilon}(0)$. Then, from elliptic estimates, we deduce that $\bar{U}_\varepsilon$ and $\bar{V}_\varepsilon$ converge locally uniformly in C^2-topology to a finite energy solution $\bar{Z} = (\bar{U}, \bar{V})$ of (8.7). Note that $\Phi_e(\bar{Z}) \geq \gamma_e$, we can conclude that there exist a non-negative integer l and $\xi_\varepsilon^i \in M$ for $1 \leq i \leq l$ satisfying

$$\lim_{\varepsilon\to 0}\frac{\min_{i\neq j}\mathrm{dist}(\xi_\varepsilon^i, \xi_\varepsilon^j)}{\varepsilon} = \infty$$

and $\liminf_{\varepsilon\to 0}\bar{u}_\varepsilon(\xi_\varepsilon^i)+\bar{v}_\varepsilon(\xi_\varepsilon^i)>0$, $i=1,\ldots,l$. It also follows from the standard comparison principle that for some constants $C, c>0$

$$\bar{u}_\varepsilon(\xi)+\bar{v}_\varepsilon(\xi)+|\nabla\bar{u}_\varepsilon(\xi)|+|\nabla\bar{v}_\varepsilon(\xi)|\le C\exp\Big(-\frac{c}{\varepsilon}\operatorname{dist}_g\big(\xi,\{\xi_\varepsilon^1,\ldots,\xi_\varepsilon^l\}\big)\Big),\ \xi\in M.$$

This implies that

$$\liminf_{\varepsilon\to 0}\Phi_\varepsilon(\bar{z}_\varepsilon)\ge l\gamma_e.$$

Since we have already known from Theorem 3.34 the upper energy estimate (8.13), we see that $l=1$. And thus, for a unique maximum point $\xi_\varepsilon\in M$ of $\bar{u}_\varepsilon+\bar{v}_\varepsilon$, we have

$$\bar{u}_\varepsilon(\xi)+\bar{v}_\varepsilon(\xi)+|\nabla\bar{u}_\varepsilon(\xi)|+|\nabla\bar{v}_\varepsilon(\xi)|\le C\exp\Big(-\frac{c}{\varepsilon}\operatorname{dist}_g(\xi,\xi_\varepsilon)\Big),\qquad \xi\in M.$$

and, in this way, the transformed functions $\bar{U}_\varepsilon$ and $\bar{V}_\varepsilon$ converge locally in C^2-topology to a ground-state solution $\bar{Z}=(\bar{U},\bar{V})\in\mathcal{L}_e$ as $\varepsilon\to 0$. □

The next result gives a refined lower energy estimate.

Lemma 8.6 *Let $\bar{z}_\varepsilon=(\bar{u}_\varepsilon,\bar{v}_\varepsilon)$, $\bar{Z}=(\bar{U},\bar{V})\in\mathcal{L}_e$ and $\xi_\varepsilon\in M$ be as in Proposition 8.5. Suppose that, up to a subsequence, $\xi_\varepsilon\to\xi$ in M (with respect to the Riemannian metric g) as $\varepsilon\to 0$, then*

$$\gamma_\varepsilon\ge\gamma_e+\frac{\varepsilon^2}{3}\mathcal{S}(\xi,\bar{Z})+o(\varepsilon^2).$$

Proof For $\varepsilon>0$ small, let us denote $\bar{z}_\varepsilon=(\bar{u}_\varepsilon,\bar{v}_\varepsilon)$ the ground-state solution of (8.1) as in the previous proposition and $\xi_\varepsilon\in M$ be the maximum point of $\bar{u}_\varepsilon+\bar{v}_\varepsilon$. Take $\beta_\varepsilon\in C^\infty(M,[0,1])$ be such that $|\nabla_g\beta_\varepsilon|\le 4/r$, $\beta_\varepsilon\equiv 1$ on $B_r(\xi_\varepsilon)$ and $\operatorname{supp}\beta_\varepsilon\subset B_{2r}(\xi_\varepsilon)$, for some $r<inj(M)/2$.

Since $\bar{u}_\varepsilon$ and $\bar{v}_\varepsilon$ solves (8.1), we have

$$\begin{aligned}\Phi_\varepsilon(\bar{z}_\varepsilon)&=\frac{1}{\varepsilon^m}\int_M\Big(\frac12 f_1(\bar{u}_\varepsilon)\bar{u}_\varepsilon-F_1(\bar{u}_\varepsilon)\Big)d\,\mathrm{vol}_g\\&\quad+\frac{1}{\varepsilon^m}\int_M\Big(\frac12 f_2(\bar{v}_\varepsilon)\bar{v}_\varepsilon-F_2(\bar{v}_\varepsilon)\Big)d\,\mathrm{vol}_g\end{aligned}$$

and, multiplying both sides of (8.1) by $(\beta_\varepsilon^2\bar{u}_\varepsilon, \beta_\varepsilon^2\bar{v}_\varepsilon)$ and integrating by parts, we find

$$\begin{aligned}\Phi_\varepsilon(\beta_\varepsilon\bar{z}_\varepsilon) &= \frac{1}{\varepsilon^m}\int_M\Big(\frac{\beta_\varepsilon^2}{2}f_1(\bar{u}_\varepsilon)\bar{u}_\varepsilon - F_1(\beta_\varepsilon\bar{u}_\varepsilon)\Big)d\,\mathrm{vol}_g\\ &+\frac{1}{\varepsilon^m}\int_M\Big(\frac{\beta_\varepsilon^2}{2}f_2(\bar{v}_\varepsilon)\bar{v}_\varepsilon - F_2(\beta_\varepsilon\bar{v}_\varepsilon)\Big)d\,\mathrm{vol}_g\\ &+\frac{1}{\varepsilon^m}\int_M \varepsilon^2\bar{u}_\varepsilon\bar{v}_\varepsilon|\nabla_g\beta_\varepsilon|^2 d\,\mathrm{vol}_g\,.\end{aligned}$$

For $i = 1, 2$ and $s > 0$, let us consider the function $h_i(t) = \frac{t^2}{2}f_i(s)s - F_i(ts)$. It is all clear that $h_i(0) = 0$ and $h_i'(0) = h_i'(1) = 0$. Assume $t_0 > 0$ satisfies $h_i'(t_0) = 0$, then it follows from $0 = h_i'(t_0) = t_0 f_i(s)s - f_i(t_0 s)s$ and (8.6) that

$$\begin{aligned}h_i''(t_0) &= f_i(s)s - f_i'(t_0 s)s^2 = \frac{f_i(t_0 s)}{t_0}s - f_i'(t_0 s)s^2\\ &\le \frac{1}{\theta}f_i'(t_0 s)s^2 - f_i'(t_0 s)s^2 < 0 \qquad \text{by (8.6) and } \theta > 1\end{aligned}$$

i.e., $t_0 > 0$ is a strict maximum point of h_i. Hence, we must have $t_0 = 1$, $h_i(t)$ is increasing for $t \in [0, 1]$ and $h_i(1)$ is the unique global maximum. Together with the exponential decay of $\bar{u}_\varepsilon$ and $\bar{v}_\varepsilon$ on M, this implies

$$\Phi_\varepsilon(\beta_\varepsilon\bar{z}_\varepsilon) \le \Phi_\varepsilon(\bar{z}_\varepsilon) + \frac{1}{\varepsilon^m}\int_M \varepsilon^2\bar{u}_\varepsilon\bar{v}_\varepsilon|\nabla_g\beta_\varepsilon|^2 d\,\mathrm{vol}_g = \gamma_\varepsilon + o(\varepsilon^2). \tag{8.14}$$

On the other hand, use the local coordinates around ξ_ε and argue similarly as in Lemma 8.4, we can get

$$\begin{aligned}\Phi_\varepsilon(\beta_\varepsilon\bar{z}_\varepsilon) &= \int_{B_{2r/\varepsilon}(0)}\Big(\sum_{i,j=1}^m g^{ij}(\exp_{\xi_\varepsilon}\varepsilon x)\partial_i(\bar{\beta}_\varepsilon\bar{U}_\varepsilon)\partial_j(\bar{\beta}_\varepsilon\bar{V}_\varepsilon) + \bar{\beta}_\varepsilon^2\bar{U}_\varepsilon\bar{V}_\varepsilon\Big)\\ &\sqrt{\det G}(\exp_{\xi_\varepsilon}\varepsilon x)dx\\ &-\int_{B_{2r/\varepsilon}(0)}\big(F_1(\bar{\beta}_\varepsilon\bar{U}_\varepsilon) + F_2(\bar{\beta}_\varepsilon\bar{V}_\varepsilon)\big)\sqrt{\det G}(\exp_{\xi_\varepsilon}\varepsilon x)dx\\ &= \Phi_e(\bar{W}_\varepsilon) + \frac{\varepsilon^2}{3}\mathcal{S}(\xi_\varepsilon, \bar{W}_\varepsilon) + o(\varepsilon^2)\end{aligned}$$

use the exponential decay of $\bar{U}_\varepsilon$ and $\bar{V}_\varepsilon$

where, for simplicity, we denote $\bar{U}_\varepsilon(x) = \bar{u}_\varepsilon \circ \exp_{\xi_\varepsilon}(\varepsilon x)$, $\bar{V}_\varepsilon(x) = \bar{v}_\varepsilon \circ \exp_{\xi_\varepsilon}(\varepsilon x)$, $\bar{\beta}_\varepsilon(x) = \beta_\varepsilon \circ \exp_{\xi_\varepsilon}(\varepsilon x)$ and $\bar{W}_\varepsilon = (\bar{\beta}_\varepsilon \bar{U}_\varepsilon, \bar{\beta}_\varepsilon \bar{V}_\varepsilon)$. Note that, by Proposition 8.5, $\bar{W}_\varepsilon$ converges in C^2_{loc}-topology to $\bar{Z} \in \mathcal{L}_e$. Hence, together with $\xi_\varepsilon \to \xi$ in M, we have

$$\mathcal{S}(\xi_\varepsilon, \bar{W}_\varepsilon) = \mathcal{S}(\xi, \bar{Z}) + o_\varepsilon(1).$$

where $o_\varepsilon(1) \to 0$ as $\varepsilon \to 0$. Note that the energy estimate in Theorem 3.34 also applies to the functional Φ_e (since the condition (F) guarantees the validity of (H1) to (H5)), we have

$$\Phi_e(\bar{Z}) = \gamma_e \le \Phi_e(\bar{W}_\varepsilon) + O(\|\nabla \Phi_e(\bar{W}_\varepsilon)\|^2). \tag{8.15}$$

where $\|\cdot\|$ is the usual norm on $H^1(\mathbb{R}^m) \times H^1(\mathbb{R}^m)$. Since, for arbitrary $W = (U, V)$,

$$\begin{aligned}
\big\langle \nabla \Phi_e(\bar{W}_\varepsilon), W \big\rangle &= \int_{\mathbb{R}^m} \big(\nabla(\bar{\beta}_\varepsilon \bar{U}_\varepsilon) \cdot \nabla V + \bar{\beta}_\varepsilon \bar{U}_\varepsilon V - f_2(\bar{\beta}_\varepsilon \bar{V}_\varepsilon) V\big) dx \\
&\quad + \int_{\mathbb{R}^m} \big(\nabla U \cdot \nabla(\bar{\beta}_\varepsilon \bar{V}_\varepsilon) + \bar{\beta}_\varepsilon U \bar{V}_\varepsilon - f_1(\bar{\beta}_\varepsilon \bar{U}_\varepsilon) U\big) dx \\
&= \langle \nabla \Phi_\varepsilon(\beta_\varepsilon \bar{z}_\varepsilon), w_\varepsilon \rangle_\varepsilon + O(\varepsilon^2)\|W\| \quad \text{with } w_\varepsilon(\xi) = W\Big(\frac{\exp_{\xi_\varepsilon}^{-1} \xi}{\varepsilon}\Big) \\
&= \langle \nabla \Phi_\varepsilon(\bar{z}_\varepsilon), w_\varepsilon \rangle_\varepsilon + o(\varepsilon^2)\|w_\varepsilon\|_\varepsilon + O(\varepsilon^2)\|W\|,
\end{aligned}$$

and $\|w_\varepsilon\|_\varepsilon = O(\|W\|)$, we have $\|\nabla \Phi_e(\bar{W}_\varepsilon)\| = O(\varepsilon^2)$. Therefore, by (8.14) and (8.15), we obtain

$$\gamma_\varepsilon \ge \Phi_e(\bar{W}_\varepsilon) + \frac{\varepsilon^2}{3}\mathcal{S}(\xi_\varepsilon, \bar{W}_\varepsilon) + o(\varepsilon^2) \ge \gamma_e + \frac{\varepsilon^2}{3}\mathcal{S}(\xi, \bar{Z}) + o(\varepsilon^2)$$

as desired. □

Combining Proposition 8.3, 8.5 and Lemma 8.6, we are now ready to state

Theorem 8.7 (Xu [19]) *Under the condition (F), for small $\varepsilon > 0$, there exists a ground-state solution $\bar{z}_\varepsilon = (\bar{u}_\varepsilon, \bar{v}_\varepsilon)$ of* (8.1) *and a unique maximum point $\xi_\varepsilon \in M$ of $\bar{u}_\varepsilon + \bar{v}_\varepsilon$ such that*

(1) *for some constants $C, c > 0$,*

$$\bar{u}_\varepsilon(\xi) + \bar{v}_\varepsilon(\xi) \le C \exp\Big(-\frac{c}{\varepsilon}\,\mathrm{dist}_g(\xi, \xi_\varepsilon)\Big);$$

(2) $\lim_{\varepsilon \to 0} \mathrm{Scal}_g(\xi_\varepsilon) = \max_{\xi \in M} \mathrm{Scal}_g(\xi)$;

(3) *up to a subsequence, the transformed functions* $\bar{U}_\varepsilon(x) = \bar{u}_\varepsilon \circ \exp_{\xi_\varepsilon}(\varepsilon x)$ *and* $\bar{V}_\varepsilon(x) = \bar{v}_\varepsilon \circ \exp_{\xi_\varepsilon}(\varepsilon x)$ *converge to a ground-state solution* $\bar{Z} = (\bar{U}, \bar{V})$ *of* (8.7)*, where* $\bar{Z}$ *is selected by*

$$\int_{\mathbb{R}^m} \nabla\bar{U} \cdot \nabla\bar{V}|x|^2 dx$$
$$= \begin{cases} \max\limits_{Z=(U,V)\in\mathcal{L}_e} \displaystyle\int_{\mathbb{R}^m} \nabla U \cdot \nabla V|x|^2 dx, & \text{if } \max\limits_{\xi\in M} \mathrm{Scal}_g(\xi) > 0, \\ \min\limits_{Z=(U,V)\in\mathcal{L}_e} \displaystyle\int_{\mathbb{R}^m} \nabla U \cdot \nabla V|x|^2 dx, & \text{if } \max\limits_{\xi\in M} \mathrm{Scal}_g(\xi) < 0. \end{cases}$$

8.2 Semiclassical Standing Waves of Maxwell-Dirac Systems on $\mathbb{R}^3$

In this section, we consider the following specified coupled Dirac system which has been already introduced in Chap. 2 (see (2.14)),

$$\begin{cases} \alpha \cdot (i\varepsilon\nabla + Q(x)\mathbf{A})\, w - a\beta w - \omega w - Q(x)A_0 w = P(x)g(|w|)w, \\ -\Delta A_k = 4\pi Q(x)(\alpha_k w)\bar{w} \quad k = 0, 1, 2, 3. \end{cases} \tag{8.16}$$

Writing $G(|w|) := \int_0^{|w|} g(s)s ds$, we make the following hypotheses:

(g_1) $g(0) = 0$, $g \in C^1(0, \infty)$, $g'(s) > 0$ *for* $s > 0$, *and there exist* $p \in (2, 3)$, $c_1 > 0$ *such that* $g(s) \le c_1(1 + s^{p-2})$ *for* $s \ge 0$;
(g_2) *there exist* $\sigma > 2$, $\theta > 2$ *and* $c_0 > 0$ *such that* $c_0 s^\sigma \le G(s) \le \frac{1}{\theta} g(s)s^2$ *for all* $s > 0$.

A typical example is the power function $g(s) = s^{\sigma-2}$. For describing the charge distribution and external fields we always assume that $Q(x)$ and $P(x)$ verify, respectively

(Q_0) $Q \in C^{0,1}(\mathbb{R}^3) \cap L^\infty(\mathbb{R}^3)$ with $Q(x) \ge 0$ a.e. on $\mathbb{R}^3$;
(P_0) $P \in C^{0,1}(\mathbb{R}^3) \cap L^\infty(\mathbb{R}^3)$ with $\inf P > 0$ and $\limsup\limits_{|x|\to\infty} P(x) < \max P(x)$.

For showing the concentration phenomena, we set $m := \max_{x\in\mathbb{R}^3} P(x)$ and

$$\mathscr{P} := \{x \in \mathbb{R}^3 : P(x) = m\}.$$

Our result reads as

Theorem 8.8 (Ding and Xu [7]) *Assume that* $\omega \in (-a, a)$, (g_1)-(g_2), (Q_0) *and* (P_0) *are satisfied. Then for all* $\varepsilon > 0$ *small,*

(i) *The system (8.16) has at least one least energy solution* $w_\varepsilon \in W^{1,q}$ *for all* $q \geq 2$*. In addition, if* $P,\ Q \in C^{1,1}(\mathbb{R}^3)$ *the solutions will be in* C^1 *class.*

(ii) *The set of all least energy solutions is compact in* $W^{1,q}$ *for all* $q \geq 2$*;*

(iii) *There is a maximum point* x_ε *of* $|w_\varepsilon|$ *with* $\lim_{\varepsilon\to 0} \mathrm{dist}(x_\varepsilon, \mathscr{P}) = 0$ *such that* $u_\varepsilon(x) := w_\varepsilon(\varepsilon x + x_\varepsilon)$ *converges uniformly to a least energy solution of (the limit equation)*

$$i\alpha \cdot \nabla u - a\beta u - \omega u = mg(|u|)u. \tag{8.17}$$

(iv) $|w_\varepsilon(x)| \leq C \exp\left(-\frac{c}{\varepsilon}|x - x_\varepsilon|\right)$ *for some* $C, c > 0$.

Remark 8.9 This localized existence theorem is comparable with the results obtained in Chap. 4. In fact, a new feature of the Maxwell-Dirac system is that the linear external potentials A_0, $\mathbf{A}$ and Q do not effect the localization of the least energy solution. In other words, the localized behavior of the solution is governed by the spinor part of the system.

8.2.1 The Variational Framework

In this section we discuss the variational setting for the equivalent system

$$\begin{cases} \alpha \cdot (i\nabla + Q_\varepsilon \mathbf{A}_\varepsilon)\, u - a\beta u - \omega u - Q_\varepsilon A_{\varepsilon,0} u = P_\varepsilon\, g(|u|)u, \\ -\Delta A_{\varepsilon,k} = \varepsilon^2 4\pi\, Q_\varepsilon J_k \quad k = 0, 1, 2, 3, \end{cases} \tag{8.18}$$

where $Q_\varepsilon(x) = Q(\varepsilon x)$, $P_\varepsilon(x) = P(\varepsilon x)$, $\mathbf{A}_\varepsilon(x) = \mathbf{A}(\varepsilon x)$, $A_{\varepsilon,k}(x) = A_k(\varepsilon x)$, $k = 0, 1, 2, 3$, and

$$J_k = (\alpha_k u)\bar{u} \ \text{ for } k = 0, 1, 2, 3.$$

And without loss of generality, we assume $0 \in \mathscr{P}$ and the conditions (g_1)–(g_2), (P_0) and (Q_0) are satisfied.

In the sequel, by $|\cdot|_q$ we denote the usual L^q-norm, and $(\cdot,\cdot)_2$ the usual L^2-inner product. Let $H_0 = i\alpha \cdot \nabla - a\beta$ denote the self-adjoint operator on $L^2 \equiv L^2(\mathbb{R}^3, \mathbb{C}^4)$ with domain $\mathcal{D}(H_0) = H^1 \equiv H^1(\mathbb{R}^3, \mathbb{C}^4)$. It is well known that $\sigma(H_0) = \sigma_c(H_0) = \mathbb{R} \setminus (-a, a)$ where $\sigma(\cdot)$ and $\sigma_c(\cdot)$ denote the spectrum and the continuous spectrum. Thus the space L^2 possesses the orthogonal decomposition:

$$L^2 = L^+ \oplus L^-, \quad u = u^+ + u^-$$

so that H_0 is positive definite (resp. negative definite) in L^+ (resp. L^-). Let $E := \mathcal{D}(|H_0|^{1/2}) = H^{1/2}$ be equipped with the inner product

$$\langle u, v\rangle = \mathrm{Re}(|H_0|^{1/2} u, |H_0|^{1/2} v)_2$$

and the induced norm $\|u\| = \langle u, u\rangle^{1/2}$, where $|H_0|$ and $|H_0|^{1/2}$ denote respectively the absolute value of H_0 and the square root of $|H_0|$. Since $\sigma(H_0) = \mathbb{R} \setminus (-a, a)$, one has

$$a|u|_2^2 \le \|u\|^2 \quad \text{for all } u \in E.$$

Note that this norm is equivalent to the usual $H^{1/2}$-norm, hence E embeds continuously into L^q for all $q \in [2, 3]$ and compactly into L^q_{loc} for all $q \in [1, 3)$. It is clear that E possesses the following decomposition

$$E = E^+ \oplus E^- \ \text{ with } \ E^\pm = E \cap L^\pm,$$

orthogonal with respect to both $(\cdot,\cdot)_2$ and $\langle\cdot,\cdot\rangle$ inner products. This decomposition induces also a natural decomposition of L^p, hence there is $d_p > 0$ such that

$$d_p \left|u^\pm\right|_p^p \le |u|_p^p \ \text{ for all } u \in E.$$

Let $\mathcal{D}^{1,2} \equiv \mathcal{D}^{1,2}(\mathbb{R}^3, \mathbb{R})$ be the completion of $C_c^\infty(\mathbb{R}^3, \mathbb{R})$ with respect the Dirichlet norm

$$\|u\|_{\mathcal{D}}^2 = \int |\nabla u|^2 \, dx.$$

Then (8.18) can be reduced to a single equation with a non-local term. Actually, since Q is bounded and $u \in L^q$ for all $q \in [2, 3]$, one has $Q_\varepsilon |u|^2 \in L^{6/5}$ for all $u \in E$, and there holds, for all $v \in \mathcal{D}^{1,2}$,

$$\begin{aligned} \left|\int Q_\varepsilon(x) J_k \cdot v dx\right| &\le \left(\int \left|Q_\varepsilon(x)\,|u|^2\right|^{6/5} dx\right)^{5/6} \left(\int |v|^6\right)^{1/6} \\ &\le S^{-1/2} \left|Q_\varepsilon\,|u|^2\right|_{6/5} \|v\|_{\mathcal{D}}, \end{aligned} \tag{8.19}$$

where S is the Sobolev embedding constant: $S|u|_6^2 \le \|u\|_{\mathcal{D}}^2$ for all $u \in \mathcal{D}^{1,2}$. Hence there exists a unique $A^k_{\varepsilon,u} \in \mathcal{D}^{1,2}$ for $k = 0, 1, 2, 3$ such that

$$\int \nabla A^k_{\varepsilon,u} \nabla v dx = \varepsilon^2 4\pi \int Q_\varepsilon(x) J_k v \, dx \tag{8.20}$$

for all $v \in \mathcal{D}^{1,2}$. It follows that $A^k_{\varepsilon,u}$ satisfies the Poisson equation

$$-\Delta A^k_{\varepsilon,u} = \varepsilon^2 4\pi Q_\varepsilon(x) J_k$$

and there holds

$$A^k_{\varepsilon,u}(x) = \varepsilon^2 \int \frac{Q_\varepsilon(y) J_k(y)}{|x-y|} dy = \frac{\varepsilon^2}{|x|} * (Q_\varepsilon J_k). \tag{8.21}$$

Substituting $A^k_{\varepsilon,u}$, $k = 0, 1, 2, 3$, in (8.18), we are led to the equation

$$H_0 u - \omega u - Q_\varepsilon(x) A^0_{\varepsilon,u} u + \sum_{k=1}^{3} Q_\varepsilon(x) \alpha_k A^k_{\varepsilon,u} u = P_\varepsilon(x) g(|u|) u. \tag{8.22}$$

On E we define the functional

$$\Phi_\varepsilon(u) = \frac{1}{2}\left(\|u^+\|^2 - \|u^-\|^2 - \omega\, |u|_2^2\right) - \Gamma_\varepsilon(u) - \Psi_\varepsilon(u)$$

for $u = u^+ + u^-$, where

$$\Gamma_\varepsilon(u) = \frac{1}{4}\int Q_\varepsilon(x) A^0_{\varepsilon,u}(x) J_0 dx - \frac{1}{4}\sum_{k=1}^{3}\int Q_\varepsilon(x) A^k_{\varepsilon,u} J_k\, dx$$

and

$$\Psi_\varepsilon(u) = \int P_\varepsilon(x) G(|u|) dx.$$

Lemma 8.10 *Under the hypotheses* (g_1)-(g_2), *one has* $\Phi_\varepsilon \in C^2(E, \mathbb{R})$ *and any critical point of* Φ_ε *is a solution of* (8.18).

Proof Clearly, $\Psi_\varepsilon \in C^2(E, \mathbb{R})$. It remains to check that $\Gamma_\varepsilon \in C^2(E, \mathbb{R})$. It suffices to show that, for any $u, v \in E$,

$$|\Gamma_\varepsilon(u)| \le \varepsilon^2 C_1 \, |Q|^2_\infty \, \|u\|^4, \tag{8.23}$$

$$\left|\Gamma'_\varepsilon(u) v\right| \le \varepsilon^2 C_2 \, |Q|^2_\infty \, \|u\|^3 \, \|v\|, \tag{8.24}$$

$$\left|\Gamma''_\varepsilon(u)[v, v]\right| \le \varepsilon^2 C_3 \, |Q|^2_\infty \, \|u\|^2 \, \|v\|^2. \tag{8.25}$$

Observe that one has, by (8.19) and (8.20) with $v = A^k_{\varepsilon,u}$,

$$|A^k_{\varepsilon,u}|_6 \le S^{-1/2} \|A^k_{\varepsilon,u}\|_{\mathcal{D}} \le \varepsilon^2 C_1 \, |Q|_\infty \, \|u\|^2. \tag{8.26}$$

This, together with the Hölder inequality (with $r = 6, r' = 6/5$), implies (8.23). Note that $\Gamma_\varepsilon'(u)v = \frac{d}{dt}\Gamma_\varepsilon(u+tv)\big|_{t=0}$, so

$$\begin{aligned}\Gamma_\varepsilon'(u)v =& \frac{\varepsilon^2}{2}\iint \frac{Q_\varepsilon(x)Q_\varepsilon(y)}{|x-y|}\Big(J_0(x)\operatorname{Re}[\alpha_0 u\overline{v}(y)] + J_0(y)\operatorname{Re}[\alpha_0 u\overline{v}(x)]\\ &- \sum_{k=1}^{3}\big(J_k(x)\operatorname{Re}[\alpha_k u\overline{v}(y)] + J_k(y)\operatorname{Re}[\alpha_k u\overline{v}(x)]\big)\Big)dydx\\ =& \int\Big(Q_\varepsilon A_{\varepsilon,u}^0\operatorname{Re}[\alpha_0 u\overline{v}] - \sum_{k=1}^{3} Q_\varepsilon A_{\varepsilon,u}^k\operatorname{Re}[\alpha_k u\overline{v}]\Big)dx\end{aligned} \tag{8.27}$$

which, together with the Hölder's inequality and (8.26), shows (8.24). Similarly,

$$\begin{aligned}\Gamma_\varepsilon''(u)[v,v] =& \int Q_\varepsilon\Big(A_{\varepsilon,u}^0 J_k^v - \sum_{k=1}^{3} A_{\varepsilon,u}^k J_k^v\Big)dx\\ &+ 2\varepsilon^2\iint \frac{Q_\varepsilon(x)Q_\varepsilon(y)}{|x-y|}\Big[\big(\operatorname{Re}[\alpha_0 u\overline{v}(x)]\big)\big(\operatorname{Re}[\alpha_0 u\overline{v}(y)]\big)\\ &- \sum_{k=1}^{3}\big(\operatorname{Re}[\alpha_k u\overline{v}(x)]\big)\big(\operatorname{Re}[\alpha_k u\overline{v}(y)]\big)\Big]dxdy\end{aligned}$$

where $J_k^u = \alpha_k u\overline{u}$ and $J_k^v = \alpha_k v\overline{v}$, and one gets (8.25).

Now it is a standard to verify that critical points of Φ_ε are solutions of (8.18). □

We show further the following:

Lemma 8.11 *For every $\varepsilon > 0$, Γ_ε is nonnegative and weakly sequentially lower semi-continuous.*

Proof Firstly, let us recall some technical results in [9]: For any $\xi = (\xi_0, \xi_1, \xi_2, \xi_3) \in \mathbb{R}^4$ and $u \in \mathbb{C}^4$, we have

$$\begin{aligned}&\Big|\xi_0(\beta u, u) + \sum_{k=1}^{3}\xi_k(\alpha_k u, u)\Big|^2\\ =&\Big|\Big(\beta u, \Big[\xi_0 + \sum_{k=1}^{3}\xi_k\pi_k\Big]u\Big)\Big|^2\\ \le&|\beta u|_{\mathbb{C}^4}^2\Big(u, \Big(\xi_0 - \sum_{k=1}^{3}\xi_k\pi_k\Big)\Big(\xi_0 + \sum_{k=1}^{3}\xi_k\pi_k\Big)u\Big)\\ =&|\xi|_{\mathbb{R}^4}^2|u|_{\mathbb{C}^4}^4.\end{aligned} \tag{8.28}$$

Here, we have used the formulas $(u, v) = u\bar{v}$ for all $u, v \in \mathbb{C}^4$, $\pi_k = \beta \cdot \alpha_k$, $\beta^* = \beta$, $\pi_k^* = -\pi_k$ and $\pi_i\pi_j + \pi_j\pi_j = -2\delta_{ij}$, $1 \le i, j \le 3$. As a consequence, we find

$$(\beta u, u)^2 + \sum_{k=1}^{3} (\beta u, \pi_k u)^2 \le |u|^4_{\mathbb{C}^4}. \tag{8.29}$$

So, taking $u(x) \in E$, $x, y \in \mathbb{R}^3$, $\xi_0 = \pm(\beta u, u)(y)$, $\xi_k = (\beta u, \pi_k u)(y)$, we get from (8.28) and (8.29) that

$$\begin{aligned} &\pm (\beta u, u)(y)(\beta u, u)(x) + \sum_{k=1}^{3} (\beta u, \pi_k u)(y)(\beta u, \pi_k u)(x) \\ &= \pm (\beta u, u)(y)(\beta u, u)(x) + \sum_{k=1}^{3} (\alpha_k u, u)(y)(\alpha_k u, u)(x) \\ &\le |\xi|_{\mathbb{R}^4} |u(x)|^2_{\mathbb{C}^4} \le |u(y)|^2_{\mathbb{C}^4} |u(x)|^2_{\mathbb{C}^4}. \end{aligned} \tag{8.30}$$

It is not difficult to see from (8.30) that

$$J_0(x)J_0(y) - \sum_{k=1}^{3} J_k(x)J_k(y) \ge 0. \tag{8.31}$$

And hence (see (8.21))

$$\Gamma_\varepsilon(u) = \frac{\varepsilon^2}{4} \iint \frac{Q_\varepsilon(x)Q_\varepsilon(y)\left(J_0(x)J_0(y) - \sum_{k=1}^{3} J_k(x)J_k(y)\right)}{|x-y|} dxdy \ge 0.$$

And if $u_n \rightharpoonup u$ in E, then $u_n \to u$ a.e.. Therefore (8.31) and Fatou's lemma yield

$$\Gamma_\varepsilon(u) \le \liminf_{n\to\infty} \Gamma_\varepsilon(u_n)$$

as claimed. □

Set, for $r > 0$, $B_r = \{u \in E : \|u\| \le r\}$, and for $e \in E^+$

$$E_e := E^- \oplus \mathbb{R}^+ e$$

with $\mathbb{R}^+ = [0, +\infty)$. In virtue of the assumptions (g_1)-(g_2), for any $\delta > 0$, there exist $r_\delta > 0$, $c_\delta > 0$ and $c'_\delta > 0$ such that

$$\begin{cases} g(s) < \delta \ \text{ for all } 0 \le s \le r_\delta; \\ G(s) \ge c_\delta\, s^\theta - \delta\, s^2 \ \text{ for all } s \ge 0; \\ G(s) \le \delta\, s^2 + c'_\delta\, s^p \ \text{ for all } s \ge 0 \end{cases} \tag{8.32}$$

and

$$\widehat{G}(s) := \frac{1}{2}g(s)s^2 - G(s) \ge \frac{\theta-2}{2\theta}g(s)s^2 \ge \frac{\theta-2}{2}G(s) \ge c_\theta s^\sigma \tag{8.33}$$

for all $s \ge 0$, where $c_\theta = c_0(\theta-2)/2$.

Lemma 8.12 *For all $\varepsilon \in (0, 1]$, Φ_ε possess the linking structure:*

(1) *There are $r > 0$ and $\tau > 0$, both independent of ε, such that $\Phi_\varepsilon|_{B_r^+} \ge 0$ and $\Phi_\varepsilon|_{S_r^+} \ge \tau$, where*

$$B_r^+ = B_r \cap E^+ = \{u \in E^+ : \|u\| \le r\},$$

$$S_r^+ = \partial B_r^+ = \{u \in E^+ : \|u\| = r\}.$$

(2) *For any $e \in E^+ \setminus \{0\}$, there exist $R = R_e > 0$ and $C = C_e > 0$, both independent of ε, such that, for all $\varepsilon > 0$, there hold $\Phi_\varepsilon(u) < 0$ for all $u \in E_e \setminus B_R$ and $\max \Phi_\varepsilon(E_e) \le C$.*

Proof Recall that $|u|_p^p \le C_p \|u\|^p$ for all $u \in E$ by Sobolev's embedding theorem. Assertion (1) follows easily because, for $u \in E^+$ and $\delta > 0$ small enough

$$\begin{aligned} \Phi_\varepsilon(u) &= \frac{1}{2}\|u\|^2 - \frac{\omega}{2}|u|_2^2 - \Gamma_\varepsilon(u) - \Psi_\varepsilon(u) \\ &\ge \frac{1}{2}\|u\|^2 - \frac{\omega}{2}|u|_2^2 - \varepsilon^2 C_1 |Q|_\infty^2 \|u\|^4 - |P|_\infty(\delta|u|_2^2 + c'_\delta|u|_p^p) \end{aligned}$$

with C_1, C_p independent of u and $p > 2$ (see (8.23) and (8.32)).

For checking (2), take $e \in E^+ \setminus \{0\}$. In virtue of (8.32), one gets, for $u = se + v \in E_e$,

$$\begin{aligned} \Phi_\varepsilon(u) &= \frac{1}{2}\|se\|^2 - \frac{1}{2}\|v\|^2 - \frac{\omega}{2}|u|_2^2 - \Gamma_\varepsilon(u) - \Psi_\varepsilon(u) \\ &\le \frac{1}{2}s^2\|e\|^2 - \frac{1}{2}\|v\|^2 - c_\delta d_\theta \inf P \cdot s^\theta |e|_\theta^\theta \end{aligned} \tag{8.34}$$

proving the conclusion. □

Recall that a sequence $\{u_n\} \subset E$ is called to be a $(PS)_c$-sequence for functional $\Phi \in C^1(E, \mathbb{R})$ if $\Phi(u_n) \to c$ and $\Phi'(u_n) \to 0$, and is called to be $(C)_c$-sequence for Φ if $\Phi(u_n) \to c$ and $(1 + \|u_n\|)\Phi'(u_n) \to 0$. It is clear that if $\{u_n\}$ is a $(PS)_c$-sequence with $\{\|u_n\|\}$ bounded then it is also a $(C)_c$-sequence. Below we are going to study $(C)_c$-sequences for Φ_ε but firstly we observe the following

Lemma 8.13 *Let* $\{u_n\} \subset E \setminus \{0\}$ *be bounded in* $L^\sigma(\mathbb{R}^3)$*, where* $\sigma > 0$ *is the constant in* (g_2)*. Then* $\left\{\frac{A^k_{\varepsilon,u_n}}{\|u_n\|}\right\}$ *is bounded in* $L^6(\mathbb{R}^3)$ *uniformly in* $\varepsilon \in (0, 1]$*, for* $k = 0, 1, 2, 3$.

Proof Set $v_n = \frac{u_n}{\|u_n\|}$. Notice that A^k_{ε,u_n} satisfies the equation

$$-\Delta A^k_{\varepsilon,u_n} = \varepsilon^2 4\pi Q_\varepsilon(x)(\alpha_k u_n)\overline{u_n},$$

hence,

$$-\Delta \frac{A^k_{\varepsilon,u_n}}{\|u_n\|} = \varepsilon^2 4\pi Q_\varepsilon(x)(\alpha_k u_n)\overline{v_n}.$$

Observe that $\|v_n\| = 1$, E embeds continuously into L^q for $q \in [2, 3]$, and

$$\begin{aligned}\left|\int Q_\varepsilon(x)(\alpha_k u_n)\overline{v_n} \cdot \psi dx\right| &\le |Q|_\infty |u_n|_\sigma |v_n|_q |\psi|_6 \\ &\le S^{-1/2} |Q|_\infty |u_n|_\sigma |v_n|_q \|\psi\|_{\mathcal{D}}\end{aligned}$$

for any $\psi \in \mathcal{D}^{1,2}(\mathbb{R}^3, \mathbb{C}^4)$ and $\frac{1}{\sigma} + \frac{1}{q} + \frac{1}{6} = 1$. We infer

$$\left\|\frac{A^k_{\varepsilon,u_n}}{\|u_n\|}\right\|_{\mathcal{D}} \le \varepsilon^2 \tilde{C} |Q|_\infty |u_n|_\sigma ,$$

which yields the conclusion. □

We now turn to an estimate on boundness of $(C)_c$-sequences which is the key ingredient in the sequel. Recall that, by (g_1), there exist $r_1 > 0$ and $a_1 > 0$ such that

$$g(s) \le \frac{a - |\omega|}{2|P|_\infty} \quad \text{for all } s \le r_1, \tag{8.35}$$

and, for $s \ge r_1$, $g(s) \le a_1 s^{p-2}$, so $g(s)^{\sigma_0 - 1} \le a_2 s^2$ with

$$\sigma_0 := \frac{p}{p-2} > 3$$

which, jointly with (g_2), yields (see (8.33))

$$g(s)^{\sigma_0} \le a_2 g(s)s^2 \le a_3\widehat{G}(s) \quad \text{for all } s \ge r_1. \tag{8.36}$$

Lemma 8.14 *For any $\lambda > 0$, denoting $I = [0, \lambda]$, there is $\Lambda > 0$ independent of ε such that, for all $\varepsilon \in (0, 1]$, any $(C)_c$-sequence $\{u_n^\varepsilon\}$ of Φ_ε with $c \in I$, there holds (up to a subsequence)*

$$\|u_n^\varepsilon\| \le \Lambda$$

for all $n \in \mathbb{N}$.

Proof Let $\{u_n^\varepsilon\}$ be a $(C)_c$-sequence of Φ_ε with $c \in I$: $\Phi_\varepsilon(u_n^\varepsilon) \to c$ and $(1 + \|u_n^\varepsilon\|)\|\Phi_\varepsilon'(u_n^\varepsilon)\| \to 0$. Without loss of generality we may assume that $\|u_n^\varepsilon\| \ge 1$. The form of Φ_ε and the representation (8.27) $(\Gamma_\varepsilon'(u)u = 4\Gamma_\varepsilon(u))$ implies that

$$2\lambda > c + o(1) = \Phi_\varepsilon(u_n^\varepsilon) - \frac{1}{2}\Phi_\varepsilon'(u_n^\varepsilon)u_n^\varepsilon = \Gamma_\varepsilon(u_n^\varepsilon) + \int P_\varepsilon(x)\widehat{G}(|u_n^\varepsilon|) \tag{8.37}$$

and

$$\begin{aligned} o(1) &= \Phi_\varepsilon'(u_n^\varphi)(u_n^{\varepsilon+} - u_n^{\varepsilon-}) \\ &= \|u_n^\varepsilon\|^2 - \omega\big(|u_n^{\varepsilon+}|_2^2 - |u_n^{\varepsilon-}|_2^2\big) - \Gamma_\varepsilon'(u_n^\varepsilon)(u_n^{\varepsilon+} - u_n^{\varepsilon-}) \\ &\quad - \operatorname{Re}\int P_\varepsilon(x)g(|u_n^\varepsilon|)u_n^\varepsilon \cdot \overline{u_n^{\varepsilon+} - u_n^{\varepsilon-}}. \end{aligned} \tag{8.38}$$

By Lemma 8.11, (8.33) and (8.37), $\{u_n^\varepsilon\}$ is bounded in L^σ uniformly in ε with the upper bound, denoted by C_1, depending on λ, σ, θ and $\inf P$. It follows from (8.38) that

$$\begin{aligned} &o(1) + \frac{a - |\omega|}{a}\|u_n^\varepsilon\|^2 \\ &\le \Gamma_\varepsilon'(u_n^\varepsilon)(u_n^{\varepsilon+} - u_n^{\varepsilon-}) + \operatorname{Re}\int P_\varepsilon(x)g(|u_n^\varepsilon|)u_n^\varepsilon \cdot \overline{u_n^{\varepsilon+} - u_n^{\varepsilon-}}. \end{aligned}$$

This, together with (8.35), shows

$$\begin{aligned} &o(1) + \frac{a - |\omega|}{2a}\|u_n^\varepsilon\|^2 \\ &\le \Gamma_\varepsilon'(u_n^\varepsilon)(u_n^{\varepsilon+} - u_n^{\varepsilon-}) + \operatorname{Re}\int_{|u_n^\varepsilon| \ge r_1} P_\varepsilon(x)g(|u_n^\varepsilon|)u_n^\varepsilon \cdot \overline{u_n^{\varepsilon+} - u_n^{\varepsilon-}}. \end{aligned} \tag{8.39}$$

Recall that (g_1) and (g_2) imply $2 < \sigma \le p$. Setting $t = \frac{p\sigma}{2\sigma - p}$, one sees

$$2 < t < p, \quad \frac{1}{\sigma_0} + \frac{1}{\sigma} + \frac{1}{t} = 1.$$

By Hölder's inequality, the fact $\Gamma_\varepsilon(u_n^\varepsilon) \ge 0$, (8.36) and (8.37), the boundedness of $\{|u_n^\varepsilon|_\sigma\}$ uniformly in ε, and the embedding of E into L^t, we have

$$\begin{aligned}&\int_{|u_n^\varepsilon| \ge r_1} P_\varepsilon(x)\, g(|u_n^\varepsilon|)\, |u_n^\varepsilon|\, |u_n^{\varepsilon+} - u_n^{\varepsilon-}| \\ &\le |P|_\infty \Big(\int_{|u_n^\varepsilon| \ge r_1} g(|u_n^\varepsilon|)^{\sigma_0} \Big)^{1/\sigma_0} \Big(\int |u_n^\varepsilon|^\sigma \Big)^{1/\sigma} \Big(|u_n^{\varepsilon+} - u_n^{\varepsilon-}|^t \Big)^{1/t} \\ &\le C_2 \|u_n^\varepsilon\|\end{aligned} \tag{8.40}$$

with C_2 independent of ε.

Let $q = \frac{6\sigma}{5\sigma - 6}$. Then $2 < q < 3$ and $\frac{1}{\sigma} + \frac{1}{q} + \frac{1}{6} = 1$. Set

$$\zeta = \begin{cases} 0 & \text{if } q = \sigma; \\ \dfrac{2(\sigma - q)}{q(\sigma - 2)} & \text{if } q < \sigma; \\ \dfrac{3(q - \sigma)}{q(3 - \sigma)} & \text{if } q > \sigma \end{cases}$$

and note that

$$|u|_q \le \begin{cases} |u|_2^\zeta \cdot |u|_\sigma^{1-\zeta} & \text{if } 2 < q \le \sigma \\ |u|_3^\zeta \cdot |u|_\sigma^{1-\zeta} & \text{if } \sigma < q < 3. \end{cases}$$

By virtue of the Hölder inequality, Lemma 8.11, the boundedness of $\{|u_n^\varepsilon|_\sigma\}$, and the embedding of E to L^2 and L^3, we obtain that

$$\begin{aligned}&\Big| \mathrm{Re} \int Q_\varepsilon(x) A^k_{\varepsilon, u_n^\varepsilon}(\alpha_k u_n^\varepsilon) \cdot \overline{u_n^{\varepsilon+} - u_n^{\varepsilon-}} \Big| \\ &= \Big| \|u_n^\varepsilon\|\, \mathrm{Re} \int Q_\varepsilon(x) \frac{A^k_{\varepsilon, u_n^\varepsilon}}{\|u_n^\varepsilon\|}(\alpha_k u_n^\varepsilon) \cdot \overline{u_n^{\varepsilon+} - u_n^{\varepsilon-}} \Big| \\ &\le |Q|_\infty\, \|u_n^\varepsilon\| \left| \frac{A^k_{\varepsilon, u_n^\varepsilon}}{\|u_n^\varepsilon\|} \right|_6 |u_n^\varepsilon|_\sigma\, |u_n^{\varepsilon+} - u_n^{\varepsilon-}|_q \\ &\le \varepsilon^2 C_3 \|u_n^\varepsilon\|\, |u_n^\varepsilon|_q \le \varepsilon^2 C_4 \|u_n^\varepsilon\|^{1+\zeta}\end{aligned}$$

with C_4 independent of ε. This, together with the representation of (8.27), implies that

$$|\Gamma'_\varepsilon(u_n^\varepsilon)(u_n^{\varepsilon+} - u_n^{\varepsilon-})| \le C_5 \|u_n^\varepsilon\|^{1+\zeta} \tag{8.41}$$

with C_5 independent of ε.

Now the combination of (8.39), (8.40) and (8.41) shows that

$$\|u_n^\varepsilon\|^2 \le M_1 \|u_n^\varepsilon\| + M_2 \big\|u_n^\varepsilon\big\|^{1+\zeta} \tag{8.42}$$

with M_1 and M_2 being independent of $\varepsilon \le 1$. Therefore, either $\|u_n^\varepsilon\| \le 1$ or there is $\Lambda \ge 1$ independent of ε such that

$$\|u_n^\varepsilon\| \le \Lambda$$

as desired. □

Finally, for the later aim we define the operator $\mathcal{A}_{\varepsilon,k} : E \to \mathcal{D}^{1,2}(\mathbb{R}^3)$ by $\mathcal{A}_{\varepsilon,k}(u) = A^k_{\varepsilon,u}$. We have

Lemma 8.15 *For* $k = 0, 1, 2, 3$,

(1) $\mathcal{A}_{\varepsilon,k}$ *maps bounded sets into bounded sets;*
(2) $\mathcal{A}_{\varepsilon,k}$ *is continuous;*

Proof Clearly, (1) is a straight consequence of (8.26). (2) follows easily because, for $u, v \in E$, one sees that $A^j_{\varepsilon,u} - A^j_{\varepsilon,v}$ satisfies

$$-\Delta(A^j_{\varepsilon,u} - A^j_{\varepsilon,v}) = \varepsilon^2 4\pi Q_\varepsilon(x)\big[(\alpha_j u)\bar{u} - (\alpha_j v)\bar{v}\big].$$

Hence

$$\begin{aligned}
\Big\|A^j_{\varepsilon,u} - A^j_{\varepsilon,v}\Big\|_{\mathcal{D}^{1,2}} &\le \varepsilon^2 C\,|Q|_\infty \left|(\alpha_j u)\bar{u} - (\alpha_j v)\bar{v}\right|_{6/5} \\
&\le \varepsilon^2 C\,|Q|_\infty \Big(|u - v|_{12/5}\,|u|_{12/5} + |u - v|_{12/5}\,|v|_{12/5} \Big) \\
&\le \varepsilon^2 C_1\,|Q|_\infty \left(\|u - v\| \cdot \|u\| + \|u - v\| \cdot \|v\|\right),
\end{aligned}$$

and this implies the desired conclusion. □

8.2.2 *Auxiliary Functionals: Annihilation of the Non-convex Term*

Observe that the non-local term Γ_ε is rather complex. The main purpose of this section is, by cut-off arguments, to introduce an auxiliary functional which will simplify our proofs.

In order to prove our main result, we will make use of the limit equation:

$$i\alpha \cdot \nabla u - a\beta u - \omega u = \mu g(|u|)u.$$

where $\mu > 0$. Its solutions are critical points of the functional

$$\begin{aligned}\mathscr{T}_\mu(u) &:= \frac{1}{2}\left(\|u^+\|^2 - \|u^-\|^2 - \omega\,|u|_2^2\right) - \mu \int G(|u|)\\ &= \frac{1}{2}\left(\|u^+\|^2 - \|u^-\|^2 - \omega\,|u|_2^2\right) - \mathscr{G}_\mu(u).\end{aligned}$$

defined for $u = u^+ + u^- \in E = E^+ \oplus E^-$. Denote the critical set, the least energy and the set of least energy solutions of $\mathscr{T}_\mu$ as follows

$$\begin{aligned}\mathscr{K}_\mu &:= \{u \in E : \ \mathscr{T}_\mu'(u) = 0\},\\ \gamma_\mu &:= \inf\{\mathscr{T}_\mu(u) : \ u \in \mathscr{K}_\mu \setminus \{0\}\},\\ \mathscr{R}_\mu &:= \{u \in \mathscr{K}_\mu : \ \mathscr{T}_\mu(u) = \gamma_\mu, \ |u(0)| = |u|_\infty\}.\end{aligned}$$

The following lemma is from [5] (see also [6])

Lemma 8.16 *There hold the following:*

(*i*) $\mathscr{K}_\mu \neq \emptyset$, $\gamma_\mu > 0$ *and* $\mathscr{K}_\mu \subset \cap_{q\geq 2} W^{1,q}$,
(*ii*) γ_μ *is attained and* $\mathscr{R}_\mu$ *is compact in* $H^1(\mathbb{R}^3, \mathbb{C}^4)$,
(*iii*) *there exist* $C, c > 0$ *such that*

$$|u(x)| \leq C \exp(-c\,|x|)$$

for all $x \in \mathbb{R}^3$ *and* $u \in \mathscr{R}_\mu$.

Recalling the infinite-dimensional reduction introduced in Proposition 3.13 and 3.29, we denote

$$J_\mu : E^+ \to \mathbb{R}, \ \ J_\mu(u) = \mathscr{T}_\mu(u + \mathscr{J}_\mu(u)),$$

$$\mathscr{M}_\mu := \{u \in E^+ \setminus \{0\} : \ J_\mu'(u)u = 0\},$$

where $(J_\mu, \mathscr{J}_\mu)$ is the reduction couple for $\mathscr{T}_\mu$. Then, it follows that, for each $u \in E^+ \setminus \{0\}$ there is a unique $t = t(u) > 0$ such that $tu \in \mathscr{M}_\mu$ and

$$\gamma_\mu = \inf\{J_\mu(u) : u \in \mathscr{M}_\mu\} = \inf_{e \in E^+ \setminus \{0\}} \max_{u \in E_e} \mathscr{T}_\mu(u).$$

The following lemma is can be derived from Chap. 4.

Lemma 8.17 *There hold:*

(1) *Let $u \in \mathscr{M}_\mu$ be such that $J_\mu(u) = \gamma_\mu$ and set $E_u = E^- \oplus \mathbb{R}^+ u$. Then*

$$\max_{w \in E_u} \mathscr{T}_\mu(w) = J_\mu(u).$$

(2) *If $\mu_1 < \mu_2$, then $\gamma_{\mu_1} > \gamma_{\mu_2}$.*

In order to make the reduction method work for Φ_ε as ε small, we circumvent by cutting off the nonlocal terms. We find our current framework is more delicate, since the solutions we look for are at the least energy level and Γ_ε is not convex (even for u with $\|u\|$ large). By cutting off the nonlocal terms, and using the reduction method, we are able to find a critical point via an appropriate min-max scheme. The critical point will eventually be shown to be a least energy solution of the original equation when ε is sufficiently small.

By virtue of (P_0), set $\mu = b := \inf P(x) > 0$, take $e_0 \in \mathscr{M}_b$ such that $J_b(e_0) = \gamma_b$, and set $E_{e_0} = E^- \oplus \mathbb{R}^+ e_0$. One has

Lemma 8.18 *For all $\varepsilon > 0$, $\max_{v \in E_{e_0}} \Phi_\varepsilon(v) \le \gamma_b$.*

Proof It is clear that $\Phi_\varepsilon(u) \le \mathscr{T}_b(u)$ for all $u \in E$, hence, by Lemma 8.17

$$\max_{v \in E_{e_0}} \Phi_\varepsilon(v) \le \max_{v \in E_{e_0}} \mathscr{T}_b(v) = J_b(e_0) = \gamma_b$$

as claimed. □

To introduce the modified functional, by virtue of Lemma 8.14, for $\lambda = \gamma_b$ and $I = [0, \gamma_b]$, let $\Lambda \ge 1$ be the associated constant (independent of ε). Denote $T := (\Lambda + 1)^2$ and let $\eta : [0, \infty) \to [0, 1]$ be a smooth function with $\eta(t) = 1$ if $0 \le t \le T$, $\eta(t) = 0$ if $t \ge T + 1$, $\max |\eta'(t)| \le c_1$ and $\max |\eta''(t)| \le c_2$. Define

$$\begin{aligned}
\widetilde{\Phi}_\varepsilon(u) &= \frac{1}{2}\left(\|u^+\|^2 - \|u^-\|^2 - \omega |u|_2^2\right) - \eta(\|u\|^2)\Gamma_\varepsilon(u) - \Psi_\varepsilon(u) \\
&= \frac{1}{2}\left(\|u^+\|^2 - \|u^-\|^2 - \omega |u|_2^2\right) - \mathscr{F}_\varepsilon(u) - \Psi_\varepsilon(u).
\end{aligned}$$

By definition, $\Phi_\varepsilon|_{B_T} = \widetilde{\Phi}_\varepsilon|_{B_T}$. It is easy to see that $0 \le \mathscr{F}_\varepsilon(u) \le \Gamma_\varepsilon(u)$ and

$$\left|\mathscr{F}'_\varepsilon(u)v\right| \le \left|2\eta'(\|u\|^2)\Gamma_\varepsilon(u)\,\langle u, v\rangle\right| + \left|\Gamma'_\varepsilon(u)v\right|$$

for $u, v \in E$.

Lemma 8.19 *There exists $\varepsilon_1 > 0$ such that, for any $\varepsilon \le \varepsilon_1$, if $\{u_n^\varepsilon\}$ is a $(C)_c$ sequence of $\widetilde{\Phi}_\varepsilon$ with $c \in I$ then $\|u_n^\varepsilon\| \le \Lambda + \frac{1}{2}$, and consequently $\widetilde{\Phi}_\varepsilon(u_n^\varepsilon) = \Phi_\varepsilon(u_n^\varepsilon)$.*

Proof We repeat the arguments of Lemma 8.14. Let $\{u_n^\varepsilon\}$ be a $(C)_c$-sequence of $\widetilde{\Phi}_\varepsilon$ with $c \in I$. If $\|u_n^\varepsilon\|^2 \ge T + 1$ then $\widetilde{\Phi}_\varepsilon(u_n^\varepsilon) = \Phi_\varepsilon(u_n^\varepsilon)$ so, by Lemma 8.14, one has $\|u_n^\varepsilon\| \le \Lambda$, a contradiction. Thus we assume that $\|u_n^\varepsilon\|^2 \le T + 1$. Then, using (8.23), $|\eta'(\|u_n^\varepsilon\|^2)\|u_n^\varepsilon\|^2\Gamma_\varepsilon(u_n^\varepsilon)| \le \varepsilon^2 d_1$ (here and in the following, by d_j we denote positive constants independent of ε). Similarly to (8.37),

$$2\gamma_b > c + o(1) \ge \left(\eta(\|u_n^\varepsilon\|^2) + 2\eta'(\|u_n^\varepsilon\|^2)\|u_n^\varepsilon\|^2\right)\Gamma_\varepsilon(u_n^\varepsilon) + \int P_\varepsilon(x)\widehat{G}(|u_n^\varepsilon|)$$

which yields

$$2\gamma_b + \varepsilon^2 d_1 > \eta(\|u_n^\varepsilon\|^2)\Gamma_\varepsilon(u_n^\varepsilon) + \int P_\varepsilon(x)\widehat{G}(|u_n^\varepsilon|),$$

consequently $|u_n^\varepsilon|_\sigma \le d_2$. Similarly to (8.39) we get that

$$\begin{aligned}\frac{a - |\omega|}{2a}\|u_n^\varepsilon\|^2 \le{}& \varepsilon^2 d_3 + \eta(\|u_n^\varepsilon\|^2)\Gamma'_\varepsilon(u_n^\varepsilon)(u_n^{\varepsilon+} - u_n^{\varepsilon-}) \\ &+ \operatorname{Re}\int_{|u_n^\varepsilon| \ge r_1} P_\varepsilon(x) g(|u_n^\varepsilon|) u_n^\varepsilon \cdot \overline{u_n^{\varepsilon+} - u_n^{\varepsilon-}}\end{aligned}$$

which, together with (8.40) and (8.41), implies either $\|u_n^\varepsilon\| \le 1$ or as (8.42)

$$\|u_n^\varepsilon\|^2 \le \varepsilon^2 d_4 + M_1\|u_n^\varepsilon\| + M_2\|u_n^\varepsilon\|^{1+\zeta},$$

thus

$$\|u_n^\varepsilon\| \le \varepsilon^2 d_5 + \Lambda.$$

The proof is complete. □

Based on this lemma, to prove Theorem 8.8 it suffices to study $\widetilde{\Phi}_\varepsilon$ and get its critical points with critical values in $[0, \gamma_b]$. This will be done via a series of arguments. The first is to introduce the minimax values of $\widetilde{\Phi}_\varepsilon$. It is easy to verify the following lemma.

Lemma 8.20 *$\widetilde{\Phi}_\varepsilon$ possesses a linking structure and we can replace Φ_ε by $\widetilde{\Phi}_\varepsilon$ in Lemma 8.12. In addition,*

$$\max_{v \in E_{e_0}} \widetilde{\Phi}_\varepsilon(v) \le \gamma_b,$$

where $e_0 \in \mathscr{M}_b$ such that $J_b(e_0) = \gamma_b$ and $E_{e_0} = E^- \oplus \mathbb{R}^+ e_0$

Proof One can follow the proofs of Lemmas 8.12 and Lemma 8.18 with minor changes. □

Define

$$c_\varepsilon := \inf_{e \in E^+ \setminus \{0\}} \max_{u \in E_e} \widetilde{\Phi}_\varepsilon(u).$$

As a consequence of Lemma 8.20 we have

Lemma 8.21 $\tau \le c_\varepsilon \le \gamma_b$.

We now describe further the minimax value c_ε. As before, for a fixed $u \in E^+$ we define $\phi_u : E^- \to \mathbb{R}$ by

$$\phi_u(v) = \widetilde{\Phi}_\varepsilon(u + v).$$

A direct computation gives, for any $v, z \in E^-$,

$$\begin{aligned}\phi_u''(v)[z, z] &= -\|z\|^2 - \omega\,|z|_2^2 - \mathscr{F}_\varepsilon''(u+v)[z,z] - \Psi_\varepsilon''(u+v)[z,z],\\ &\le -\frac{(a - |\omega|)}{a}\,\|z\|^2 - \mathscr{F}_\varepsilon''(u+v)[z,z],\end{aligned}$$

and

$$\begin{aligned}&\mathscr{F}_\varepsilon''(u+v)[z,z]\\ &= \Big(4\eta''(\|u+v\|^2)\,|\langle u+v, z\rangle|^2 + 2\eta'(\|u+v\|^2)\,\|z\|^2\Big)\,\Gamma_\varepsilon(u+v)\\ &\quad + 4\eta'(\|u+v\|^2)\,\langle u+v, z\rangle\,\Gamma_\varepsilon'(u+v)z\\ &\quad + \eta(\|u+v\|^2)\Gamma_\varepsilon''(u+v)[z,z].\end{aligned}$$

Combining (8.23)–(8.25) yields that there exists $\varepsilon_0 \in (0, \varepsilon_1]$ such that

$$\phi_u''(v)[z,z] \le -\frac{a - |\omega|}{2a}\,\|z\|^2 \quad \text{if } 0 < \varepsilon \le \varepsilon_0.$$

Since

$$\phi_u(v) \le \frac{a+|\omega|}{2a}\,\|u\|^2 - \frac{a-|\omega|}{2a}\,\|v\|^2,$$

there is $h_\varepsilon : E^+ \to E^-$, uniquely defined, such that

$$\phi_u(h_\varepsilon(u)) = \max_{v\in E^-} \phi_u(v)$$

and

$$v \ne h_\varepsilon(u) \Leftrightarrow \widetilde{\Phi}_\varepsilon(u+v) < \widetilde{\Phi}_\varepsilon(u+h_\varepsilon(u)).$$

It is clear that, for all $v \in E^-$, $0 = \phi_u'(h_\varepsilon(u))v$. Observe that we have for $u \in E^+$ and $v \in E^-$

$$\begin{aligned}
&\widetilde{\Phi}_\varepsilon(u+h_\varepsilon(u)) - \widetilde{\Phi}_\varepsilon(u+v)\\
&= \int_0^1 (1-t)\Big[\mathscr{F}_\varepsilon''(u+h_\varepsilon(u)+t(v-h_\varepsilon(u)))[v-h_\varepsilon(u), v-h_\varepsilon(u)]\\
&+ \Psi_\varepsilon''(u+h_\varepsilon(u)+t(v-h_\varepsilon(u)))[v-h_\varepsilon(u), v-h_\varepsilon(u)]\Big]dt\\
&+ \frac{1}{2}\,\|v-h_\varepsilon(u)\|^2 + \frac{\omega}{2}\,|v-h_\varepsilon(u)|_2^2 .
\end{aligned} \tag{8.43}$$

Define $I_\varepsilon : E^+ \to \mathbb{R}$ by

$$I_\varepsilon(u) = \widetilde{\Phi}_\varepsilon(u+h_\varepsilon(u)),$$

and set

$$\mathscr{N}_\varepsilon := \{u \in E^+ \setminus \{0\} : \ I_\varepsilon'(u)[u] = 0\}.$$

Lemma 8.22 *For any $u \in E^+ \setminus \{0\}$, there is a unique $t = t(u) > 0$ such that $tu \in \mathscr{N}_\varepsilon$.*

Proof This proof is quite technical, for details we refer [1, 6]. We only give a sketch of the proof. Firstly, we observe that for any $u \in E \setminus \{0\}$ and $v \in E$,

$$
\begin{aligned}
&\big(\Gamma''_\varepsilon(u)[u,u] - \Gamma'_\varepsilon(u)u\big) + 2\big(\Gamma''_\varepsilon(u)[u,v] - \Gamma'_\varepsilon(u)v\big) + \Gamma''_\varepsilon(u)[v,v] \\
&= 2\varepsilon^2 \iint \frac{Q_\varepsilon(x)Q_\varepsilon(y)}{|x-y|}\Big[J_0(x)[\alpha_0(u+v)\overline{(u+v)}](y) \\
&\quad - \sum_{k=1}^{3} J_k(x)[\alpha_k(u+v)\overline{(u+v)}](y)\Big]dxdy \\
&\quad + 2\varepsilon^2 \iint \frac{Q_\varepsilon(x)Q_\varepsilon(y)}{|x-y|}\Big[\big(\mathrm{Re}[\alpha_0 u\bar{v}(x)]\big)\big(\mathrm{Re}[\alpha_0 u\bar{v}(y)]\big) \\
&\quad - \sum_{k=1}^{3}\big(\mathrm{Re}[\alpha_k u\bar{v}(x)]\big)\big(\mathrm{Re}[\alpha_k u\bar{v}(y)]\big)\Big]dxdy \\
&\geq O(\varepsilon^2)\|u\|^2\|v\|^2.
\end{aligned}
$$

Here we used the formula

$$
\begin{aligned}
&\pm(\beta z, z)(y)(\beta u, u)(x) + \sum_{k=1}^{3}(\beta z, \pi_k z)(y)(\beta u, \pi_k u)(x) \\
&= \pm(\beta z, z)(y)(\beta u, u)(x) + \sum_{k=1}^{3}(\alpha_k z, z)(y)(\alpha_k u, u)(x) \\
&\leq |z(y)|^2_{\mathbb{C}^4}|u(x)|^2_{\mathbb{C}^4}
\end{aligned}
$$

which follows from (8.28) and (8.29) with $z = u + v \in E$. Consequently, we deduce that

$$
\begin{aligned}
&\big(\mathscr{F}''_\varepsilon(u)[u,u] - \mathscr{F}'_\varepsilon(u)u\big) + 2\big(\mathscr{F}''_\varepsilon(u)[u,v] - \mathscr{F}'_\varepsilon(u)v\big) \\
&+ \mathscr{F}''_\varepsilon(u)[v,v] \geq o(\varepsilon)\|v\|^2 + o(\varepsilon)\,.
\end{aligned}
$$

Invoking the arguments in [1], if $z \in E^+ \setminus \{0\}$ with $I'_\varepsilon(z)z = 0$, we see by a delicate calculation that, for ε sufficiently small,

$$
I''_\varepsilon(z)[z,z] < 0. \tag{8.44}
$$

Now for a fixed $u \in E^+ \setminus \{0\}$, we set $f(t) = I_\varepsilon(tu)$. From Lemma 8.20, we see that $f(0) = 0$, $f(t) > 0$ for $t > 0$ sufficiently small, and $f(t) \to -\infty$ as $t \to \infty$. Thus there exists $t(u) > 0$ such that

$$
I_\varepsilon(t(u)u) = \sup_{t \geq 0} I_\varepsilon(tu).
$$

It is clear that

$$\left.\frac{d\, I_\varepsilon(tu)}{dt}\right|_{t=t(u)} = I_\varepsilon'(t(u)u)u = \frac{1}{t(u)} I_\varepsilon'(t(u)u)t(u)u = 0,$$

and consequently by (8.44)

$$I_\varepsilon''(t(u)u)[t(u)u, t(u)u] < 0.$$

Therefore, one sees that such $t(u) > 0$ is unique. □

Lemma 8.23 $c_\varepsilon = \inf_{u\in\mathscr{N}_\varepsilon} I_\varepsilon(u)$.

Proof Indeed, denoting $\hat{c}_\varepsilon = \inf_{u\in\mathscr{N}_\varepsilon} I_\varepsilon(u)$, given $e \in E^+$, if $u = v + se \in E_e$ with $\widetilde{\Phi}_\varepsilon(u) = \max_{z\in E_e} \widetilde{\Phi}_\varepsilon(z)$ then the restriction $\widetilde{\Phi}_\varepsilon|_{E_e}$ of $\widetilde{\Phi}_\varepsilon$ on E_e satisfies $(\widetilde{\Phi}_\varepsilon|_{E_e})'(u) = 0$ which implies $v = h_\varepsilon(se)$ and $I_\varepsilon'(se)(se) = 0$, i.e. $se \in \mathscr{N}_\varepsilon$. Thus $\hat{c}_\varepsilon \le c_\varepsilon$. On the other hand, if $w \in \mathscr{N}_\varepsilon$ then $(\widetilde{\Phi}_\varepsilon|_{E_w})'(w + h_\varepsilon(w)) = 0$, hence, $c_\varepsilon \le \max_{u\in E_w} \widetilde{\Phi}_\varepsilon(u) = I_\varepsilon(w)$. Thus $\hat{c}_\varepsilon \ge c_\varepsilon$. □

Lemma 8.24 *For any $e \in E^+ \setminus \{0\}$, there is $T_e > 0$ independent of ε such that $t_\varepsilon \le T_e$ for $t_\varepsilon > 0$ satisfying $t_\varepsilon e \in \mathscr{N}_\varepsilon$.*

Proof Since $I_\varepsilon'(t_\varepsilon e)(t_\varepsilon e) = 0$, one gets

$$\widetilde{\Phi}_\varepsilon(t_\varepsilon e + h_\varepsilon(t_\varepsilon e)) = \max_{w\in E_e} \widetilde{\Phi}_\varepsilon(w) \ge \tau.$$

This, together with Lemma 8.20, shows the assertion. □

Let $\mathscr{K}_\varepsilon := \{u \in E : \widetilde{\Phi}_\varepsilon'(u) = 0\}$ be the critical set of $\widetilde{\Phi}_\varepsilon$. Since critical points of I_ε and $\widetilde{\Phi}_\varepsilon$ are in one-to-one correspondence via the injective map $u \mapsto u + h_\varepsilon(u)$ from E^+ into E, let us denoted by

$$\mathscr{C}_\varepsilon := \{u \in \mathscr{K}_\varepsilon : \widetilde{\Phi}_\varepsilon(u) = c_\varepsilon\},$$

from Lemma 8.23, one easily sees that if $\mathscr{C}_\varepsilon \ne \emptyset$ then $c_\varepsilon = \inf\{\widetilde{\Phi}_\varepsilon(u) : u \in \mathscr{K}_\varepsilon \setminus \{0\}\}$. Next we estimate the regularity of critical points of $\widetilde{\Phi}_\varepsilon$. By using the same iterative argument of [8] one obtains easily the following

Lemma 8.25 *If $u \in \mathscr{K}_\varepsilon$ with $|\widetilde{\Phi}_\varepsilon(u)| \le C_1$, then, for any $q \in [2, +\infty)$, $u \in W^{1,q}(\mathbb{R}^3, \mathbb{C}^4)$ with $\|u\|_{W^{1,q}} \le \Lambda_q$ where Λ_q depends only on C_1 and q.*

Proof See [8]. We outline the proof as follows. From (8.22), we write

$$u = H_0^{-1}\Big(\omega u + Q_\varepsilon(x)A^0_{\varepsilon,u}u - \sum_{k=1}^{3} Q_\varepsilon(x)\alpha_k A^k_{\varepsilon,u}u + P_\varepsilon(x)g(|u|)u\Big).$$

For the later use, let $\rho : [0, \infty) \to [0, 1]$ be a smooth function satisfying $\rho(s) = 1$ if $s \in [0, 1]$ and $\rho(s) = 0$ if $s \in [2, \infty)$. Then we have

$$\begin{aligned} g(s) &:= g_1(s) + g_2(s) \\ &= \rho(s)g(s) + (1 - \rho(s))g(s). \end{aligned}$$

Consequently, $u = u_1 + u_2 + u_3$ with

$$\begin{aligned} u_1 &= H_0^{-1} \left(\omega u + P_\varepsilon \cdot g_1(|u|)u\right), \\ u_2 &= H_0^{-1} \left(Q_\varepsilon \cdot A^0_{\varepsilon,u} u - \sum Q_\varepsilon \cdot \alpha_k A^k_{\varepsilon,u} u\right), \\ u_3 &= H_0^{-1} \left(P_\varepsilon \cdot g_2(|u|)u\right). \end{aligned}$$

Noting that, by Hölder's inequality, for $q \geq 2$

$$\left|Q_\varepsilon \alpha_k A^k_{\varepsilon,u} u\right|_s \leq |Q|_\infty \left|A^k_{\varepsilon,u}\right|_6 |u|_q$$

with $\frac{1}{s} = \frac{1}{6} + \frac{1}{q}$ and since

$$\left|P_\varepsilon(x) \, |u|^{p-2} u\right|_t \leq |P|_\infty \, |u|^{p-1}_{t(p-1)},$$

one has

$$u_1 \in W^{1,2} \cap W^{1,3}, \; u_2 \in W^{1,s}, \; u_3 \in W^{1,t}.$$

Then, denoting $s^* = \frac{3s}{3-s}$ and $t^* = \frac{3t}{3-t}$, $u \in W^{1,q}$ with $q = \min\{s^*, t^*\}$. A standard bootstrap argument shows that $u \in \cap_{q\geq 2} L^q$, $u_1 \in \cap_{q\geq 2} W^{1,q}$, $u_2 \in \cap_{6>q\geq 2} W^{1,q}$ and $u_3 \in \cap_{q\geq 2} W^{1,q}$.

By the Sobolev's embedding theorems, $u \in C^{0,\gamma}$ for some $\gamma \in (0, 1)$. This, together with elliptic regularity (see [10]), shows $A^k_{\varepsilon,u} \in W^{2,6}_{loc} \cap L^6$ for $k = 0, 1, 2, 3$ and

$$\left\|A^k_{\varepsilon,u}\right\|_{W^{2,6}(B_1(x))} \leq C_2\left(\varepsilon^2 \, |Q|_\infty \, |u|^2_{L^{12}(B_2(x))} + \left|A^k_{\varepsilon,u}\right|_{L^6(B_2(x))}\right)$$

for all $x \in \mathbb{R}^3$, with C_2 independent of x and ε, where $B_r(x) = \{y \in \mathbb{R}^3 : |y - x| < r\}$ for $r > 0$. Since $W^{2,6}(B_1(x)) \hookrightarrow C^1(B_1(x))$, we have

$$\left\|A^k_{\varepsilon,u}\right\|_{C^1(B_1(x))} \leq C_3\left(\varepsilon^2 \, |Q|_\infty \, |u|^2_{L^{12}(B_2(x))} + \left|A^k_{\varepsilon,u}\right|_{L^6(B_2(x))}\right) \tag{8.45}$$

for all $x \in \mathbb{R}^3$ with C_3 independent of x and ε. Consequently $A^k_{\varepsilon,u} \in L^\infty$, and that yields

$$\left|Q_\varepsilon \alpha_k A^k_{\varepsilon,u} u\right|_s \le |Q|_\infty \left|A^k_{\varepsilon,u}\right|_\infty |u|_s .$$

Thus $u_2 \in \bigcap_{q\ge 2} W^{1,q}$, and combining with $u_1, u_3 \in \bigcap_{q\ge 2} W^{1,q}$ the conclusion is obtained. □

Remark 8.26 Let $\mathscr{L}_\varepsilon$ denote the set of all least energy solutions of $\widetilde{\Phi}_\varepsilon$. If $u \in \mathscr{L}_\varepsilon$, $\widetilde{\Phi}_\varepsilon(u) = c_\varepsilon \le \gamma_b$. Recall that $\mathscr{L}_\varepsilon$ is bounded in E with an upper bound Λ independent of ε. Therefore, as a consequence of Lemma 8.25 we see that, for each $q \in [2, +\infty)$ there is $C_q >$ independent of ε such that

$$\|u\|_{W^{1,q}} \le C_q \quad \text{for all } u \in \mathscr{L}_\varepsilon. \tag{8.46}$$

This, together with the Sobolev embedding theorem, implies that there is $C_\infty > 0$ independent of ε with

$$\|u\|_\infty \le C_\infty \quad \text{for all } u \in \mathscr{L}_\varepsilon. \tag{8.47}$$

8.2.3 *The Localization Result*

Recall that we always assume $0 \in \mathscr{P}$, the main theorem will be carried in three parts: *Existence, Concentration, and Exponential decay.*

Part 1: Existence

Keeping the notation of Sect. 8.2.1 we now turn to the existence result of the main theorem. The proof is carried out in three lemmas. The modified problem gives us an access to Lemma 8.27, which is the key ingredient for Lemma 8.28.

Recall that γ_m denotes the least energy of $\mathscr{T}_m$, where $\mu = m := \max_{x\in\mathbb{R}^3} P(x)$, and J_m denotes the associated reduction functional on E^+. We have

Lemma 8.27 *$c_\varepsilon \to \gamma_m$ as $\varepsilon \to 0$.*

Lemma 8.28 *c_ε is attained for all small $\varepsilon > 0$.*

Lemma 8.29 *$\mathscr{L}_\varepsilon$ is compact in $W^{1,q}$ for each $q \ge 2$, for all small $\varepsilon > 0$.*

Proof of Lemma 8.27 Firstly we show that

$$\liminf_{\varepsilon\to 0} c_\varepsilon \ge \gamma_m. \tag{8.48}$$

Arguing indirectly, assume that $\liminf_{\varepsilon\to 0} c_\varepsilon < \gamma_m$. By the definition of c_ε and Lemma 8.23 we can choose an $e_j \in \mathcal{N}_\varepsilon$ and $\delta > 0$ such that

$$\max_{u\in E_{e_j}} \widetilde{\Phi}_{\varepsilon_j}(u) \le \gamma_m - \delta$$

as $\varepsilon_j \to 0$. Since $P_\varepsilon(x) \le m$ and $\mathscr{F}(u) = o(1)$ as $\varepsilon \to 0$ uniformly in u (by (8.23) and the definition of η), the representations of $\widetilde{\Phi}_\varepsilon$ and $\mathscr{T}_m$ imply that $\widetilde{\Phi}_\varepsilon(u) \ge \mathscr{T}_m(u) - \delta/2$ for all $u \in E$ and ε small. Note also that $\gamma_m \le J_m(e_j) \le \max_{u\in E_{e_j}} \mathscr{T}_m(u)$. Therefore we get, for all ε_j small,

$$\gamma_m - \delta \ge \max_{u\in E_{e_j}} \widetilde{\Phi}_{\varepsilon_j}(u) \ge \max_{u\in E_{e_j}} \mathscr{T}_m(u) - \frac{\delta}{2} \ge \gamma_m - \frac{\delta}{2},$$

a contradiction.

We now turn to prove the desired conclusion. Set $P^0(x) = m - P(x)$ and $P^0_\varepsilon(x) = P^0(\varepsilon x)$. Then

$$\widetilde{\Phi}_\varepsilon(u) = \mathscr{T}_m(u) - \mathscr{F}_\varepsilon(u) + \int P^0_\varepsilon(x)G(|u|). \tag{8.49}$$

In virtue of Lemma 8.16, let $u = u^+ + u^- \in \mathscr{R}_m$, be a least energy solution of the limit equation with $\mu = m$, and set $e = u^+$. Clearly, $e \in \mathscr{M}_m$, $\mathscr{J}_m(e) = u^-$ and $J_m(e) = \gamma_m$. There is a unique $t_\varepsilon > 0$ such that $t_\varepsilon e \in \mathcal{N}_\varepsilon$ and one has

$$c_\varepsilon \le I_\varepsilon(t_\varepsilon e). \tag{8.50}$$

By Lemma 8.24 t_ε is bounded. Hence, without loss of generality we can assume $t_\varepsilon \to t_0$ as $\varepsilon \to 0$. Using (8.43), we infer

$$\begin{aligned}\frac{1}{2}\|v_\varepsilon\|^2 + (I) &= \widetilde{\Phi}_\varepsilon(w_\varepsilon) - \widetilde{\Phi}_\varepsilon(u_\varepsilon)\\ &= \mathscr{T}_m(w_\varepsilon) - \mathscr{T}_m(u_\varepsilon) - \mathscr{F}_\varepsilon(w_\varepsilon) + \mathscr{F}_\varepsilon(u_\varepsilon)\\ &\quad + \int P^0_\varepsilon(x)G(|w_\varepsilon|) - \int P^0_\varepsilon(x)G(|u_\varepsilon|)\end{aligned}$$

where, setting

$$u_\varepsilon = t_\varepsilon e + \mathscr{J}_m(t_\varepsilon e),\ w_\varepsilon = t_\varepsilon e + h_\varepsilon(t_\varepsilon e),\ v_\varepsilon = u_\varepsilon - w_\varepsilon,$$

$$(I) := \frac{\omega}{2}|v_\varepsilon|_2^2 + \int_0^1 (1-s)\big(\mathscr{F}''_\varepsilon(w_\varepsilon + sv_\varepsilon)[v_\varepsilon, v_\varepsilon] + \Psi''_\varepsilon(w_\varepsilon + sv_\varepsilon)[v_\varepsilon, v_\varepsilon]\big)dt.$$

Taking into account that

$$\mathscr{F}_\varepsilon(u_\varepsilon) - \mathscr{F}_\varepsilon(w_\varepsilon) = \mathscr{F}_\varepsilon'(w_\varepsilon)v_\varepsilon + \int_0^1 (1-s)\mathscr{F}_\varepsilon''(w_\varepsilon + sv_\varepsilon)[v_\varepsilon, v_\varepsilon]dt$$

and

$$\begin{aligned}&\int P_\varepsilon^0(x)\big(G(|w_\varepsilon|) - G(|u_\varepsilon|)\big)\\ &= -\int P_\varepsilon^0(x)g(|u_\varepsilon|)u_\varepsilon \cdot \overline{v_\varepsilon} + \int_0^1 (1-s)\mathscr{G}_m''(u_\varepsilon - sv_\varepsilon)[v_\varepsilon, v_\varepsilon]dt\\ &\quad - \int_0^1 (1-s)\Psi_\varepsilon''(u_\varepsilon - sv_\varepsilon)[v_\varepsilon, v_\varepsilon]dt,\end{aligned}$$

setting

$$(II) := \int_0^1 (1-s)\Psi_\varepsilon''(u_\varepsilon - sv_\varepsilon)[v_\varepsilon, v_\varepsilon]dt,$$

one has

$$\begin{aligned}&\frac{1}{2}\|v_\varepsilon\|^2 + (I) + (II)\\ &\le \mathscr{F}_\varepsilon'(w_\varepsilon)v_\varepsilon + \int_0^1 (1-s)\mathscr{F}_\varepsilon''(w_\varepsilon + sv_\varepsilon)[v_\varepsilon, v_\varepsilon] - \int P_\varepsilon^0(x)g(|u_\varepsilon|)u_\varepsilon \cdot \overline{v_\varepsilon}.\end{aligned}$$

So we deduce, noticing that $0 \le P_\varepsilon^0(x) \le m$,

$$\begin{aligned}&\frac{1}{2}\|v_\varepsilon\|^2 + \frac{\omega}{2}|v_\varepsilon|_2^2 + \int_0^1 (1-s)\Psi_\varepsilon''(w_\varepsilon + sv_\varepsilon)[v_\varepsilon, v_\varepsilon]\\ &\le \left|\mathscr{F}_\varepsilon'(w_\varepsilon)v_\varepsilon\right| + \int P_\varepsilon^0(x)g(|u_\varepsilon|)\,|u_\varepsilon|\cdot|v_\varepsilon|.\end{aligned} \tag{8.51}$$

Since $t_\varepsilon \to t_0$, it is clear that $\{u_\varepsilon\}, \{w_\varepsilon\}$ and $\{v_\varepsilon\}$ are bounded, hence, by the definitions and (8.23), (8.24),

$$\mathscr{F}_\varepsilon(z_\varepsilon) = o(1), \quad \|\mathscr{F}_\varepsilon'(z_\varepsilon)\| = o(1)$$

as $\varepsilon \to 0$ for $z_\varepsilon = u_\varepsilon, w_\varepsilon, v_\varepsilon$. In addition, by noting that for $q \in [2, 3]$

$$\limsup_{r\to\infty} \int_{|x|>r} |u_\varepsilon|^q = 0,$$

using the assumption $0 \in \mathscr{P}$ one deduces

$$\begin{aligned}
&\int \left(P_\varepsilon^0(x)\right)^{q/(q-1)} |u_\varepsilon|^q \\
=&\left(\int_{|x|\le r} + \int_{|x|>r}\right) P_\varepsilon^0(x)^{q/(q-1)} |u_\varepsilon|^q \\
\le& \int_{|x|\le r} \left(P_\varepsilon^0(x)\right)^{q/(q-1)} |u_\varepsilon|^q + m^{q/(q-1)} \int_{|x|>r} |u_\varepsilon|^q \\
=& \, o(1)
\end{aligned}$$

as $\varepsilon \to 0$. Thus by (8.51) one has $\|v_\varepsilon\|^2 \to 0$, that is, $h_\varepsilon(t_\varepsilon e) \to \mathscr{J}_m(t_0 e)$. Consequently,

$$\int P_\varepsilon^0(x) G(|w_\varepsilon|) \to 0$$

as $\varepsilon \to 0$. This, jointly with (8.49), shows

$$\widetilde{\Phi}_\varepsilon(w_\varepsilon) = \mathscr{T}_m(w_\varepsilon) + o(1) = \mathscr{T}_m(u_\varepsilon) + o(1),$$

that is,

$$I_\varepsilon(t_\varepsilon e) = J_m(t_0 e) + o(1)$$

as $\varepsilon \to 0$. Then, since

$$J_m(t_0 e) \le \max_{v \in E_e} \mathscr{T}_m(v) = J_m(e) = \gamma_m,$$

we obtain by using (8.48) and (8.50)

$$\gamma_m \le \lim_{\varepsilon \to 0} c_\varepsilon \le \lim_{\varepsilon \to 0} I_\varepsilon(t_\varepsilon e) = J_m(t_0 e) \le \gamma_m,$$

hence, $c_\varepsilon \to \gamma_m$. □

Proof of Lemma 8.28 Given $\varepsilon > 0$, let $\{u_n\} \subset \mathscr{N}_\varepsilon$ be a minimizing sequence: $I_\varepsilon(u_n) \to c_\varepsilon$. By the Ekeland variational principle we can assume that $\{u_n\}$ is in fact a $(PS)_{c_\varepsilon}$-sequence for I_ε on E^+ (see [12, 18]). Then $w_n = u_n + h_\varepsilon(u_n)$ is a $(PS)_{c_\varepsilon}$-sequence for $\widetilde{\Phi}_\varepsilon$ on E. It is clear that $\{w_n\}$ is bounded, hence is a $(C)_{c_\varepsilon}$-sequence. We can assume without loss of generality that $w_n \rightharpoonup w_\varepsilon = w_\varepsilon^+ + w_\varepsilon^- \in \mathscr{K}_\varepsilon$ in E. If $w_\varepsilon \neq 0$ then $\widetilde{\Phi}_\varepsilon(w_\varepsilon) = c_\varepsilon$. So we are going to show that $w_\varepsilon \neq 0$ for all small $\varepsilon > 0$.

For this end, take $\limsup_{|x|\to\infty} P(x) < \kappa < m$ and define

$$P^\kappa(x) = \min\{\kappa, P(x)\}.$$

Consider the functional

$$\widetilde{\Phi}_\varepsilon^\kappa(u) = \frac{1}{2}\left(\|u^+\|^2 - \|u^-\|^2 - \omega\,|u|_2^2\right) - \mathscr{F}_\varepsilon(u) - \int P_\varepsilon^\kappa(x)G(|u|)$$

and as before define correspondingly $h_\varepsilon^\kappa : E^+ \to E^-$, $I_\varepsilon^\kappa : E^+ \to \mathbb{R}$, $\mathscr{N}_\varepsilon^\kappa$, c_ε^κ and so on. Following the proof of Lemma 8.27, one finds

$$\lim_{\varepsilon\to 0} c_\varepsilon^\kappa = \gamma_\kappa. \tag{8.52}$$

Assume by contradiction that there is a sequence $\varepsilon_j \to 0$ with $w_{\varepsilon_j} = 0$. Then $w_n = u_n + h_{\varepsilon_j}(u_n) \rightharpoonup 0$ in E, $u_n \to 0$ in L^q_{loc} for $q \in [1,3)$, and $w_n(x) \to 0$ a.e. in $x \in \mathbb{R}^3$. Let $t_n > 0$ be such that $t_n u_n \in \mathscr{N}_{\varepsilon_j}^\kappa$. Since $u_n \in \mathscr{N}_\varepsilon$, it is not difficult to see that $\{t_n\}$ is bounded and one may assume $t_n \to t_0$ as $n \to \infty$. By (P_0), the set $A_\varepsilon := \{x \in \mathbb{R}^3 : P_\varepsilon(x) > \kappa\}$ is bounded. Remark that $h_{\varepsilon_j}^\kappa(t_n u_n) \rightharpoonup 0$ in E and $h_{\varepsilon_j}^\kappa(t_n u_n) \to 0$ in L^q_{loc} for $q \in [1,3)$ as $n \to \infty$ (see [1]). Moreover, by virtue of Lemma 8.17, $\widetilde{\Phi}_{\varepsilon_j}(t_n u_n + h_{\varepsilon_j}^\kappa(t_n u_n)) \le I_{\varepsilon_j}(u_n)$. We obtain

$$\begin{aligned}
c_{\varepsilon_j}^\kappa &\le I_{\varepsilon_j}^\kappa(t_n u_n) = \widetilde{\Phi}_{\varepsilon_j}^\kappa(t_n u_n + h_{\varepsilon_j}^\kappa(t_n u_n))\\
&= \widetilde{\Phi}_{\varepsilon_j}(t_n u_n + h_{\varepsilon_j}^\kappa(t_n u_n)) + \int \left(P_{\varepsilon_j}(x) - P_{\varepsilon_j}^\kappa(x)\right)G\left(|t_n u_n + h_{\varepsilon_j}^\kappa(t_n u_n)|\right)\\
&\le I_{\varepsilon_j}(u_n) + \int_{A_{\varepsilon_j}} \left(P_{\varepsilon_j}(x) - P_{\varepsilon_j}^\kappa(x)\right)G\left(|t_n u_n + h_{\varepsilon_j}^\kappa(t_n u_n)|\right)\\
&= c_{\varepsilon_j} + o(1)
\end{aligned}$$

as $n \to \infty$. Hence $c_{\varepsilon_j}^\kappa \le c_{\varepsilon_j}$. By (8.52), letting $j \to \infty$ yields

$$\gamma_\kappa \le \gamma_m,$$

which contradiction with $\gamma_m < \gamma_\kappa$. □

Proof of Lemma 8.29 Since $\mathscr{L}_\varepsilon \subset B_\Lambda$ for all small $\varepsilon > 0$, assume by contradiction that, for some $\varepsilon_j \to 0$, $\mathscr{L}_{\varepsilon_j}$ is not compact in E. Then we can choose $u_n^j \in \mathscr{L}_{\varepsilon_j}$ be such that $u_n^j \rightharpoonup 0$ as $n \to \infty$, as done for proving the Lemma 8.28, we gets a contradiction.

Now let $\{u_n\} \subset \mathscr{L}_\varepsilon$ such that $u_n \to u$ in E, and recall $H_0 = i\alpha \cdot \nabla - a\beta$, by

$$H_0 u = \omega u + Q_\varepsilon(x)A_{\varepsilon,u}^0 u - \sum_{k=1}^{3} Q_\varepsilon(x)\alpha_k A_{\varepsilon,u}^k u + P_\varepsilon(x)g(|u|)u$$

one has

$$\begin{aligned}|H_0(u_n - u)|_2 \le \omega\,|u_n - u|_2 &+ \left|Q_\varepsilon(x)\left(A^0_{\varepsilon,u_n}u_n - A^0_{\varepsilon,u}u\right)\right|_2 \\ &+ \sum_{k=1}^{3}\left|Q_\varepsilon(x)\alpha_k\left(A^k_{\varepsilon,u_n}u_n - A^k_{\varepsilon,u}u\right)\right|_2 \\ &+ |P_\varepsilon(x)\,(g(|u_n|)u_n - g(|u|)u)|_2\end{aligned} \tag{8.53}$$

A standard calculus shows that

$$\begin{aligned}\left|Q_\varepsilon\cdot\alpha_k\left(A^k_{\varepsilon,u_n}u_n - A^k_{\varepsilon,u}u\right)\right|_2 &\le |Q|_\infty\,|u_n|_\infty^{1/6}\left|A^k_{\varepsilon,u_n} - A^k_{\varepsilon,u}\right|_6 |u_n|_{5/2}^{5/6} \\ &\quad + |Q|_\infty\,|u_n - u|_\infty^{1/6}\left|A^k_{\varepsilon,u}\right|_6 |u_n - u|_{5/2}^{5/6}\end{aligned}$$

and

$$\begin{aligned}&|P_\varepsilon\cdot(g(|u_n|)u_n - g(|u|)u)|_2 \\ \le\,& |P|_\infty\,|g(|u_n|) - g(|u|)|_\infty^{\frac{1}{2}}\left|(g(|u_n|) - g(|u|))^{1/2}u_n\right|_2 \\ &+ |P|_\infty\,|g(|u|)|_\infty\,|u_n - u|_2\,.\end{aligned}$$

By Lemma 8.15 and the fact that $u_n \to u$ in $L^q(\mathbb{R}^3, \mathbb{C}^4)$ for all $q \in [2, 3]$, one gets $|H_0(u_n - u)|_2 \to 0$, so $u_n \to u$ in $H^1(\mathbb{R}^3, \mathbb{C}^4)$. With Lemma 8.25, $u_n \to u$ in $W^{1,q}$ for all $q \in [2, \infty)$. □

Part 2: Concentration

The proof relies on the following lemma. To prove it, it suffices to show that for any sequence $\varepsilon_j \to 0$ the corresponding sequence of solutions $u_j \in \mathscr{L}_{\varepsilon_j}$ converges, up to a shift of x-variable, to a least energy solution of the limit problem (8.17).

Lemma 8.30 *There is a maximum point x_ε of $|u_\varepsilon|$ such that* $\mathrm{dist}(y_\varepsilon, \mathscr{P}) \to 0$ *where $y_\varepsilon = \varepsilon x_\varepsilon$, and for any such x_ε, $v_\varepsilon(x) := u_\varepsilon(x + x_\varepsilon)$ converges to a least energy solution of (8.17) in $W^{1,q}$ as $\varepsilon \to 0$ for all $q \ge 2$.*

Proof Let $\varepsilon_j \to 0$, $u_j \in \mathscr{L}_j$, where $\mathscr{L}_j = \mathscr{L}_{\varepsilon_j}$. Then $\{u_j\}$ is bounded. A standard concentration argument (see [11]) shows that there exist a sequence $\{x_j\} \subset \mathbb{R}^3$ and constant $R > 0$, $\delta > 0$ such that

$$\liminf_{j\to\infty}\int_{B(x_j,R)}|u_j|^2 \ge \delta.$$

Set

$$v_j = u_j(x + x_j).$$

Then v_j solves, denoting $\hat{Q}_j(x) = Q(\varepsilon_j(x + x_j))$, $\hat{A}^k_{\varepsilon,u_j}(x) = A^k_{\varepsilon,u_j}(x + x_j)$ and $\hat{P}_j(x) = P(\varepsilon_j(x + x_j))$,

$$H_0 v_j - \omega v_j - \hat{Q}_j \hat{A}^0_{\varepsilon,u_j} v_j + \sum_{k=1}^{3} \hat{Q}_j \alpha_k \hat{A}^k_{\varepsilon,u_j} v_j = \hat{P}_j \cdot g(|v_j|) v_j, \tag{8.54}$$

with energy

$$\begin{aligned} S(v_j) &:= \frac{1}{2}\left(\|v_j^+\|^2 - \|v_j^-\|^2 - \omega |v_j|_2^2\right) - \hat{\Gamma}_j(v_j) - \int \hat{P}_j(x) G(|v_j|) \\ &= \widetilde{\Phi}_j(v_j) = \Phi_j(v_j) = \hat{\Gamma}_j(v_j) + \int \hat{P}_j(x) \widehat{G}(|v_j|) \\ &= c_{\varepsilon_j}. \end{aligned}$$

Additionally, $v_j \rightharpoonup v$ in E and $v_j \to v$ in L^q_{loc} for $q \in [1, 3)$.

We now turn to prove that $\{\varepsilon_j x_j\}$ is bounded. Arguing indirectly we assume $\varepsilon_j |x_j| \to \infty$ and get a contradiction.

Without loss of generality assume $P(\varepsilon_j x_j) \to P_\infty$. Clearly, $m > P_\infty$ by (P_0). Since for any $\psi \in C_c^\infty$

$$\begin{aligned} 0 &= \lim_{j\to\infty} \int \left(H_0 v_j - \omega v_j - \hat{Q}_j \hat{A}^0_{\varepsilon,u_j} v_j + \sum_{k=1}^{3} \hat{Q}_j \alpha_k \hat{A}^k_{\varepsilon,u_j} v_j - \hat{P}_j g(|v_j|) v_j \right) \bar{\psi} \\ &= \lim_{j\to\infty} \int (H_0 v - \omega v - P_\infty g(|v|) v)\, \bar{\psi}, \end{aligned}$$

hence v solves

$$i\alpha \cdot \nabla v - a\beta v - \omega v = P_\infty g(|v|) v.$$

Therefore,

$$S_\infty(v) := \frac{1}{2}\left(\|v^+\|^2 - \|v^-\|^2 - \omega |v|_2^2\right) - \int P_\infty G(|v|) \ge \gamma_{P_\infty}.$$

It follows from $m > P_\infty$, by Lemma 8.17, one has $\gamma_m < \gamma_{P_\infty}$. Moreover, by Fatou's lemma,

$$\lim_{j\to\infty}\int \hat{P}_j(x)\widehat{G}(|v_j|) \geq \int P_\infty \widehat{G}(|v|) = S_\infty(v).$$

Consequently, noting that $\hat{\Gamma}_j(v_j) = o(1)$ as $j \to \infty$,

$$\gamma_m < \gamma_{P_\infty} \leq S_\infty(v) \leq \lim_{j\to\infty} c_{\varepsilon_j} = \gamma_m,$$

a contradiction.

Thus $\{\varepsilon_j x_j\}$ is bounded. Hence, we can assume $y_j = \varepsilon_j x_j \to y_0$. Then v solves

$$i\alpha \cdot \nabla v - a\beta v - \omega v = P(y_0)g(|v|)v. \tag{8.55}$$

Since $P(y_0) \leq m$, we obtain

$$S_0(v) := \frac{1}{2}\left(\|v^+\|^2 - \|v^-\|^2 - \omega|v|_2^2\right) - \int P(y_0)G(|v|) \geq \gamma_{P(y_0)} \geq \gamma_m.$$

Again, by Fatou's lemma, we have

$$S_0(v) = \int P(y_0)\widehat{G}(|v|) \leq \lim_{j\to\infty} c_{\varepsilon_j} = \gamma_m.$$

Therefore, $\gamma_{P(y_0)} = \gamma_m$, which implies $y_0 \in \mathscr{P}$ by Lemma 8.17. By virtue of Lemma 8.25 and (8.47) it is clear that one may assume that $x_j \in \mathbb{R}^3$ is a maximum point of $|u_j|$. Moreover, from the above argument we readily see that any sequence of such points satisfies $y_j = \varepsilon_j x_j$, converging to some point in $\mathscr{P}$ as $j \to \infty$.

In order to prove $v_j \to v$ in E, recall that as the argument shows

$$\lim_{j\to\infty}\int \hat{P}_j(x)\widehat{G}(|v_j|) = \int P(y_0)\widehat{G}(|v|).$$

By (g_2) and the exponential decay of v, using the Brezis-Lieb lemma, one obtains $|v_j - v|_\sigma \to 0$, then $|v_j^\pm - v^\pm|_\sigma \to 0$. Denote $z_j = v_j - v$. Remark that $\{z_j\}$ is bounded in E and $z_j \to 0$ in L^σ, therefore $z_j \to 0$ in L^q for all $q \in (2, 3)$. The scalar product of (8.54) with z_j^+ yields

$$\langle v_j^+, z_j^+\rangle = o(1).$$

Similarly, using the exponential decay of v together with the fact that $z_j^{\pm} \to 0$ in L^q_{loc} for $q \in [1, 3)$, it follows from (8.55) that

$$\langle v^+, z_j^+ \rangle = o(1).$$

Thus

$$\|z_j^+\| = o(1),$$

and the same arguments show

$$\|z_j^-\| = o(1),$$

we then get $v_j \to v$ in E, and the arguments in Lemma 8.29 shows that $v_j \to v$ in $W^{1,q}$ for all $q \geq 2$. □

Part 3: Exponential Decay

See the following Lemma 8.32.

For the later use denote $D = i\alpha \cdot \nabla$ and, for $u \in \mathcal{L}_\varepsilon$, write (8.22) as

$$Du = a\beta u + \omega u + Q_\varepsilon(x)A^0_{\varepsilon,u}u - \sum_{k=1}^{3} Q_\varepsilon(x)\alpha_k A^k_{\varepsilon,u}u + P_\varepsilon(x)g(|u|)u.$$

Applying the operator D on both sides and noting that $D^2 = -\Delta$, we get

$$\begin{aligned}\Delta u = a^2 u &- \left(\omega + Q_\varepsilon(x)A^0_{\varepsilon,u}(x) + P_\varepsilon(x)g(|u|)\right)^2 u \\ &- D\left(P_\varepsilon(x)g(|u|)\right)u - D\left(Q_\varepsilon A^0_{\varepsilon,u}\right)u \\ &+ \sum_{k=1}^{3}\left(Q_\varepsilon A^k_{\varepsilon,u}\right)^2 u + \sum_{k=1}^{3} D\left(Q_\varepsilon A^k_{\varepsilon,u}\right)\alpha_k u \\ &+ 2i\sum_{k=1}^{3} Q_\varepsilon A^k_{\varepsilon,u}\partial_k u.\end{aligned} \tag{8.56}$$

With the fact that

$$\Delta |u|^2 = \bar{u}\Delta u + u\Delta\bar{u} + 2|\nabla u|^2$$

and $\overline{\alpha_k u} \cdot u = \alpha_k u \cdot \bar{u}$ one deduces

$$\begin{aligned}\Delta |u|^2 = & 2a^2 |u|^2 - 2\left(\omega + Q_\varepsilon(x) A^0_{\varepsilon,u}(x) + P_\varepsilon(x) g(|u|)\right)^2 |u|^2 \\ & + 2\sum_{k=1}^{3} \left(Q_\varepsilon A^k_{\varepsilon,u}\right)^2 |u|^2 + 2i \sum_{k=1}^{3} \sum_{\substack{1\le j\le 3 \\ j\neq k}} \partial_j \left(Q_\varepsilon A^k_{\varepsilon,u}\right) (\alpha_j \alpha_k u) \cdot \bar{u} \\ & + 4\Im \sum_{k=1}^{3} Q_\varepsilon A^k_{\varepsilon,u} \partial_k u \cdot \bar{u} + 2 |\nabla u|^2 .\end{aligned}$$

In addition, setting

$$f^0_\varepsilon(x) := \max\left\{\left|Q_\varepsilon(x) A^k_{\varepsilon,u}(x)\right| : k = 0, 1, 2, 3\right\},$$

$$f^1_\varepsilon(x) := \max\left\{\left|\nabla\big(Q_\varepsilon(x) A^k_{\varepsilon,u}(x)\big)\right| : k = 0, 1, 2, 3\right\},$$

one has

$$\left|2i \sum_{k=1}^{3} \sum_{\substack{1\le j\le 3 \\ j\neq k}} \partial_j \left(Q_\varepsilon A^k_{\varepsilon,u}\right) (\alpha_j \alpha_k u) \cdot \bar{u}\right| \le c_1 f^1_\varepsilon(x) |u|^2$$

and

$$\left|4 \operatorname{Im} \sum_{k=1}^{3} Q_\varepsilon A^k_{\varepsilon,u} \partial_k u \cdot \bar{u}\right| \le c_2 f^0_\varepsilon(x) \left(|\nabla u|^2 + |u|^2\right).$$

Hence

$$\begin{aligned}\Delta |u|^2 \ge & \left(2a^2 - 2\left(\omega + Q_\varepsilon A^0_{\varepsilon,u} + P_\varepsilon(x) g(|u|)\right)^2 + 2\sum_{k=1}^{3} \left(Q_\varepsilon A^k_{\varepsilon,u}\right)^2\right) |u|^2 \\ & - c_1 f^1_\varepsilon(x) |u|^2 - c_2 f^0_\varepsilon(x) \left(|\nabla u|^2 + |u|^2\right) + 2 |\nabla u|^2 .\end{aligned}$$

Observe that for $\varepsilon > 0$ sufficiently small, by (8.45), we get

$$c_2 \left|f^0_\varepsilon\right| < 2,$$

hence

$$\begin{aligned}\Delta |u|^2 \geq & 2\Big(a^2 - \big(\omega + Q_\varepsilon A^0_{\varepsilon,u} + P_\varepsilon \cdot g(|u|)\big)^2 + \sum_{k=1}^{3}\big(Q_\varepsilon A^k_{\varepsilon,u}\big)^2\Big)|u|^2 \\ & - c_1 f_\varepsilon^1(x)\,|u|^2 - c_2 f_\varepsilon^0(x)|u|^2.\end{aligned} \tag{8.57}$$

This together with the regularity results for u implies there is $M > 0$ satisfying

$$\Delta |u|^2 \geq -M\,|u|^2\,.$$

By the sub-solution estimate [10, 17], one has

$$|u(x)| \leq C_0\Big(\int_{B_1(x)} |u(y)|^2\,dy\Big)^{1/2} \tag{8.58}$$

with C_0 independent of x and $u \in \mathscr{L}_\varepsilon$, $\varepsilon > 0$ small.

Lemma 8.31 *Let v_ε and $\hat{A}^k_{\varepsilon,u_\varepsilon}$ for $k = 0, 1, 2, 3$ be given in the proof of Lemma 8.30. Then $|v_\varepsilon(x)| \to 0$ and $\left|\hat{A}^k_{\varepsilon,u_\varepsilon}(x)\right| \to 0$ as $|x| \to \infty$ uniformly in $\varepsilon > 0$ small.*

Proof Arguing indirectly, if the conclusion of the lemma is not held, then by (8.58), there exist $\delta > 0$ and $x_j \in \mathbb{R}^3$ with $|x_j| \to \infty$ such that

$$\delta \leq |v_j(x_j)| \leq C_0\Big(\int_{B_1(x_j)} |v_j|^2\Big)^{1/2},$$

where $\varepsilon_j \to 0$ and $v_j = v_{\varepsilon_j}$. Since $v_j \to v$ in E, we obtain

$$\begin{aligned}\delta &\leq C_0\Big(\int_{B_1(x_j)} |v_j|^2\Big)^{1/2} \\ &\leq C_0\Big(\int |v_j - v|^2\Big)^{1/2} + C_0\Big(\int_{B_1(x_j)} |v|^2\Big)^{1/2} \to 0,\end{aligned}$$

a contradiction. Now, jointly with (8.45), one sees also $\left|\hat{A}^k_{\varepsilon,u_\varepsilon}(x)\right| \to 0$ as $|x| \to \infty$ uniformly in $\varepsilon > 0$ small. □

Lemma 8.32 *There exist $C > 0$ such that for all $\varepsilon > 0$ small*

$$|u_\varepsilon(x)| \leq Ce^{-\frac{c_0}{2}|x - x_\varepsilon|}$$

where $c_0 = \sqrt{(a^2 - \omega^2)}$.

Proof The conclusions of Lemma 8.31 with (8.57) allow us to take $R > 0$ sufficiently large such that

$$\Delta |v_\varepsilon|^2 \geq \left(a^2 - \omega^2\right) |v_\varepsilon|^2$$

for all $|x| \geq R$ and $\varepsilon > 0$ small. Let $\Gamma(y) = \Gamma(y, 0)$ be a fundamental solution to $-\Delta + \left(a^2 - \omega^2\right)$. Using the uniform boundedness, we may choose that $|v_\varepsilon(y)|^2 \leq \left(a^2 - \omega^2\right) \Gamma(y)$ holds on $|y| = R$ for all $\varepsilon > 0$ small. Let $z_\varepsilon = |v_\varepsilon|^2 - \left(a^2 - \omega^2\right) \Gamma$. Then

$$\begin{aligned} \Delta z_\varepsilon &= \Delta |v_\varepsilon|^2 - \left(a^2 - \omega^2\right) \Delta\Gamma \\ &\geq \left(a^2 - \omega^2\right) \left(|v_\varepsilon|^2 - \left(a^2 - \omega^2\right) \Gamma\right) = \left(a^2 - \omega^2\right) z_\varepsilon. \end{aligned}$$

By the maximum principle we can conclude that $z_\varepsilon(y) \leq 0$ on $|y| \geq R$. It is well known that there is $C' > 0$ such that $\Gamma(y) \leq C' \exp(-c_0 |y|)$ on $|y| \geq 1$, we see that

$$|v_\varepsilon(y)|^2 \leq C'' e^{-c_0 |y|}$$

for all $y \in \mathbb{R}^3$ and all $\varepsilon > 0$ small, that is

$$|u_\varepsilon(x)| \leq C e^{-\frac{c_0}{2} |x - x_\varepsilon|}$$

as claimed. □

Now, with the above arguments, we are ready to prove Theorem 8.8.

Proof of Theorem 8.8 Going back to system (8.16) with the variable substitution: $x \mapsto x/\varepsilon$, Lemma 8.28 jointly with Lemma 8.25, shows that, for all $\varepsilon > 0$ small, Eq.(8.16) has at least one least energy solution $w_\varepsilon \in W^{1,q}$ for all $q \geq 2$. In addition, if $P,\ Q \in C^{1,1}(\mathbb{R}^3)$, with (8.56) and the elliptic regularity (see [10]) one obtains a classical solution, that is, the conclusion (i) of Theorem 8.8. And Lemma 8.29 is nothing but the conclusion (ii). Finally, the conclusion (iii) and (iv) follow from Lemma 8.30 and Lemma 8.32 respectively. □

References

1. Ackermann, N.: A nonlinear superposition principle and multibump solution of periodic Schrödinger equations. J. Funct. Anal. **234**, 423–443 (2006)
2. Busca, J., Sirakov, B.: Symmetry results for semilinear elliptic systems in the whole space. J. Differ. Equ. **163**, 41–56 (2000)
3. Chavel, I.: Eigenvalues in Riemannian Geometry. Academic Press, Cambridge (1984)

4. de Figueiredo, D.G., Yang, J.F.: Decay, symmetry and existence of solutions of semilinear elliptic systems. Nonlinear Anal. **33**(3), 211–234 (1998)
5. Ding, Y.H.: Variational Methods for Strongly Indefinite Problems. Interdisciplinary Mathematical Sciences, vol. 7. World Scientific, Singapore (2007)
6. Ding, Y.H., Wei, J.C.: Stationary states of nonlinear Dirac equations with general potentials. Rev. Math. Phys. **20**, 1007–1032 (2008)
7. Ding, Y.H., Xu, T.: On semi-classical limits of ground states of a nonlinear Maxwell-Dirac system. Calc. Var. Partial Differ. Equ. **51**(1), 17–44 (2014)
8. Esteban, M.J., Séré, E.: Stationary states of the nonlinear Dirac equation: a variational approach. Commun. Math. Phys. **171**(2), 323–350 (1995)
9. Esteban, M.J., Georgiev, V., Séré, E.: Stationary solutions of the Maxwell-Dirac and the Klein-Gordon-Dirac equations. Calc. Var. Partial Differ. Equ. **4**(3), 265–281 (1996)
10. Gilbarg, D., Trudinger, N.S.: Elliptic Partial Differential Equations of Second Order. Springer, Berlin (1998)
11. Lions, P.L.: The concentration-compactness principle in the calculus of variations: the locally compact case, Part II. AIP Anal. Non Linéaire **1**, 223–283 (1984)
12. Pankov, A.: Periodic nonlinear Schrödinger equations with applications to photonic crystals. Milan J. Math. **3**, 259–287 (2005)
13. Pistoia, A., Ramos, M.: Locating the peaks of the least energy solutions to an elliptic system with Neumann boundary conditions. J. Differ. Equ. **201**, 160–176 (2004)
14. Ramos, M., Tavares, H.: Solutions with multiple spike patterns for an elliptic system. Calc. Var. Partial Differ. Equ. **31**, 1–25 (2008)
15. Ramos, M., Yang, J.: Spike-layered solutions for an elliptic system with Neumann boundary conditions. Trans. Am. Math. Soc. **357**, 3265–3284 (2005)
16. Reed, M., Simon, B.: Methods of Mathematical Physics, vols. I–IV. Academic Press, Cambridge (1978)
17. Simon, B.: Schrödinger semigroups. Bull. Am. Math. Soc. **7**, 447–526 (1982)
18. Willem, M.: Minimax Theorems. Birkhäuser, Basel (1996)
19. Xu, T.: A local energy estimate for a class of strongly indefinite functionals. Minimax Theory Appl. **7**(2), 381–408 (2022). Special Issue for The Nehari Manifold: Theory, Applications, and Related Topics

The manufacturer's authorised representative in the EU is Springer Nature Customer Service Centre GmbH, Europaplatz 3, 69115 Heidelberg, Germany. If you have any concerns regarding our products, please contact ProductSafety@springernature.com

Printed and bound by CPI Group (UK) Ltd, Croydon, CR0 4YY
30/09/2026
02225297-0014